HARMONY, PERSPECTIVE AND TRIADIC COGNITION

The big question in the science of psychology is: Why are human cognition and behavior so different from the capabilities of every other animal species on Earth – including our close genetic relations, the chimpanzees? This book provides a coherent answer by examining the aspects of the human brain that have made triadic forms of perception and cognition possible. Mechanisms of dyadic association sufficiently explain animal perception, cognition and behavior, but a three-way associational mechanism is required to explain the human talents for language, toolmaking, harmony perception, pictorial depth perception and the joint attention that underlies all forms of social cooperation.

Norman D. Cook has authored three books on human psychology: *Stability and Flexibility* (1980), *The Brain Code* (1986) and *Tone of Voice and Mind* (2002). He has also published articles in numerous journals, including *Nature, Perception, Journal of Experimental Psychology, Brain, American Scientist, Behavioral Science, Empirical Aesthetics, Music Perception, Spatial Vision, Cognitive Science, Brain and Language, Brain and Cognition, Consciousness and Cognition* and *Neuroscience*. He is currently a professor of cognitive psychology at Kansai University (Osaka, Japan).

Harmony, Perspective and Triadic Cognition

Norman D. Cook
Kansai University

CAMBRIDGE UNIVERSITY PRESS
Cambridge, New York, Melbourne, Madrid, Cape Town,
Singapore, São Paulo, Delhi, Tokyo, Mexico City

Cambridge University Press
32 Avenue of the Americas, New York, NY 10013-2473, USA

www.cambridge.org
Information on this title: www.cambridge.org/9780521192132

© Norman D. Cook 2012

This publication is in copyright. Subject to statutory exception
and to the provisions of relevant collective licensing agreements,
no reproduction of any part may take place without the written
permission of Cambridge University Press.

First published 2012

Printed in the United States of America

A catalog record for this publication is available from the British Library.

Library of Congress Cataloging in Publication data
Cook, Norman D.
  Harmony, perspective and triadic cognition / Norman D. Cook.
    p. cm.
  Includes bibliographical references and index.
  ISBN 978-0-521-19213-2 (hardback)
  1. Cognitive psychology. 2. Consciousness. I. Title.
  BF201.C68 2012
  153–dc23   2011039143

ISBN 978-0-521-19213-2 Hardback

Cambridge University Press has no responsibility for the persistence or accuracy of URLs for
external or third-party Internet Web sites referred to in this publication and does not guarantee
that any content on such Web sites is, or will remain, accurate or appropriate.

CONTENTS

Preface — *page* ix

1. Introduction — 1
    1.1. The Basic Question — 5
    1.2. Triadic Perception, Triadic Cognition and Triadic Social Interaction — 7
    1.3. Triads versus Dyads — 12
    1.4. Musical Harmony — 15
    1.5. Pictorial Depth Perception — 16
    1.6. Tool Use — 17
    1.7. Language — 19
    1.8. Consciousness — 21
    1.9. Other Issues — 22

2. Human Hearing: Harmony — 26
    2.1. Tonality and Dissonance — 28
    2.2. Tension and Instability — 38
    2.3. The Modality of Triads — 64
    2.4. The Affective Valence of Major and Minor — 80
    2.5. Traditional Harmony Theory — 94
    2.6. This Is Your Brain on Harmony — 108
    2.7. Why Not Before? — 115
    2.8. Conclusions — 117

3. Human Seeing: Perspective — 120
    3.1. Stereoscopic Vision: Two Static Points of View — 122
    3.2. Motion Parallax: Two Sequential Points of View — 123
    3.3. Pictorial Depth Perception — 125

3.4. Linear Perspective	133
3.5. Shadows and Shading	147
3.6. Historical Perspective on Shadows	155
3.7. A Reclassification of Depth Cues	157
3.8. "Perspective as Symbolic Form"	160
3.9. Variations on the Illusion of Depth	164
3.10. This Is Your Brain on Reverse Perspective	171
3.11. Conclusions	173
4. Human Work: Tools and Handedness	**175**
4.1. Stones as Tools	176
4.2. Toolmaking and Handedness	177
4.3. The Division of Labor Between the Cerebral Hemispheres	184
4.4. Brain Size	187
4.5. Trimodal Cortical Regions	192
4.6. Hafted Tools	204
4.7. The Behavioral Neurology of Tool Use	206
4.8. Conditional Associations	208
4.9. Causality	211
4.10. Conclusions	213
5. Human Communication: Language	**216**
5.1. The Tripartite Architecture of Language	217
5.2. Behavioral Neurology	222
5.3. The Evolution of Language	224
5.4. Subjects, Objects, Verbs	235
5.5. Universal Grammar	241
5.6. Conclusions	250
6. Consciousness	**255**
6.1. The Main Question	258
6.2. Three Levels of Discussion	262
6.3. Five Approaches to Subjectivity	264
6.4. The Neurophysiological Solution	274
6.5. Implications	295
6.6. Consciousness Is Understood, Self-Consciousness Is Not	301
6.7. Conclusions	303
7. Other Human Talents	**307**
7.1. Rhythm Perception	307

7.2.	Face Perception	309
7.3.	Joint Attention	311
7.4.	Moral Minds	313
7.5.	Intelligent Neural Networks	317
7.6.	Color Perception	319
7.7.	Mental Rotation	321
7.8.	Subitizing	323
7.9.	Four-Body Cognition?	325
7.10.	Trigonometry	327

8. Conclusion   329

References   337
Index   351

PREFACE

Most of the ideas presented here are based on experimental results obtained in the twenty-first century, but genuine progress in cognitive psychology began in the 1950s, and the cognitive revolution has deep roots going back to Renaissance Europe. Moreover, a still-controversial aspect of human psychology – known since antiquity, but "rediscovered" in the 1960s – concerns cerebral laterality. The importance of hemisphere differences for specifically *human* cognition was brought to my attention when I heard Julian Jaynes lecture on the evolution of consciousness in 1969 – fascinating, erudite and persuasive, but far from the experimental science that cognitive psychology was gradually becoming. That stimulus eventually led me to rather inconclusive attempts at exploring laterality issues in Sendai, Japan, and Oxford, England, but I was later fortunate to join a group of neuropsychologists in Zurich, Switzerland, where hemisphere differences were an essential aspect of the study of abnormal human behavior. There, in the good company of Thedi Landis, Marianne Regard and Peter Brugger, I found research topics that were both experimentally tractable and conceptually fundamental and, over the past two decades in Zurich, Philadelphia and Osaka, I have been able to address what I believe to be the "big question" in psychology: What is unusual about the human mind?

It turns out that brain laterality is only one part of the story, but the most important topics in human psychology involve the "left hemisphere talents" of language and tool use and the "right hemisphere talents" of music and art. Unrelated to questions of brain localization, what they have in common is a "triadic" cognitive foundation, which is described in detail in the following chapters. Others have previously commented on the significance of triadic processes for higher cognition, but I introduce several new aspects and try to show how the step from simple (dyadic) associations to three-way associations is the huge cognitive leap of *Homo sapiens*. Explicating the nature of triadic cognition is the primary motivation for writing this book.

My arguments concerning high-level cognition will probably not settle easily with two fringe groups. The first are those who attempt to explain the special status of human beings on Earth by postulating the existence of supernatural forces, and then documenting that assumption by reference to sacred texts. I acknowledge the good intentions of religious thinkers in their emphasis on the "higher" human sentiments, but I deny that their arguments have any explanatory power. On the contrary, logical coherency is lost as soon as tradition, cultural norms or metaphysical speculations are treated as indisputable facts. At best, the preachers end up entertaining their flocks – stating and restating the assumptions they started with – but they do not further our understanding of what it means to be a human being.

The other fringe group – and the one I try to engage in the pages that follow – is the disorganized cabal of scientists who have come to believe that there are no qualitative differences between man and beast. I acknowledge their sincere efforts at the scientific explication of the phenomena of humanity in a material world, but the behaviorist research program comes to an abrupt and totally unsatisfactory halt with clever oxymorons: "the talking ape," "the toolmaking ape," "the singing ape." Well, yes, that is who we are, but what makes those rather un-ape-like capabilities possible? Reminding us that our unusual behaviors evolved in a biological world does not explain the underlying psychology.

Compromise between opposing lunacies is not necessarily the road toward truth, but I show that some progress can be made if we do not begin with the presumed answers of either religion or behaviorism.

1

Introduction

> Science! Thou fair effusive ray
> From the great source of mental day,
> Free, generous and refined!
> Descend with all thy treasures fraught
> Illumine each bewilder'd thought
> And bless my labouring mind!
>   Mark Akenside, *Hymn to Science* (1744)

Since Darwin, the emphasis in scientific research on the nature of human beings has been on the *similarities* between us and other animal species. The importance of such research in revealing who we are, as biological organisms, can hardly be overstated, and no modern intellectual can afford to ignore the academic edifice known as evolutionary theory. But, as real as biological evolution is, it still needs to be said that an eight-year-old walking around a zoo shows more psychological insight than an academic in his lab coat who refuses to acknowledge the gargantuan gap that separates *Homo sapiens* from all other animal species. Despite profound, diverse and numerous similarities with monkeys, chimps and gorillas, human beings are special in an objective, verifiable sense. It is not simply the case that "our tribe is different." We inhabit a cognitive world that is utterly beyond what other animals experience; we have certain (easily identified) behavioral capabilities (speech and tool use), certain other (less obvious, but measurable) perceptual capacities (music and pictorial art) and complex social lives that are fundamental to what it means for us to be human beings, but that are entirely absent or barely recognizable in other species. This book is about those unusual, characteristically human, psychological talents.

    The motivation for elaborating on the theme of human "uniqueness" is not some strange need to justify our superiority over animals nor is it a call to return to prescientific, religious thinking in light of our "higher-level" capabilities. The aim is, more simply, to understand our cognitive strengths, and to do so requires examination of the talents that most clearly differentiate us from animal species – even from our closest cousins, the chimpanzees and bonobos. Insofar as we are active participants in our own biological, social and intellectual evolution, any progress in understanding how it is that we have come this far will help us to develop those strengths further and, indeed, allow us to make our best traits truly universal among all human beings.

There is of course no virtue in gloating about the splendid uniqueness of *Homo sapiens* or in enumerating the psychological deficiencies of the laboratory rat, but a focus on the unusual aspects of human behavior is essential for explaining the predominance of human beings on Earth. Surprisingly, just asking why we are so different quickly reveals how little we understand about the specifically *human* aspects of cognition. On the one hand, on the basis of just whispers of physical evidence, paleontologists have reconstructed the steps that our ancestors took in becoming human, and have provided what is a plausible "best guess" about human evolution. The path from squirrel monkey living in the trees; to bipedal hunter/gatherer; to primitive *Homo sapiens* taming fire, constructing huts, growing crops and so on is a fascinating story in its own right. But underlying those gradual anatomical, behavioral and social changes, there must have been changes in mentality that were equally revolutionary and equally decisive in separating those of our ancestors who dared to walk upright, fashion objects with their hands, play with fire and communicate with words – and those who did not. The question that remains largely unanswered is: Just what is the nature of the cognitive revolution that led to this demarcation between thoughtful, creative, ambitious, adventuresome human beings and playful, pleasant, placid, but – in terms of tangible, real-world achievements – rather uninteresting apes?

The biologist will remind us that we are genetically more than 98 percent the same as the chimpanzees, and we can be certain that all of our human talents have biological origins – traces of which can often be found in species that have had remarkably little impact on the world at large. But, even though we also are essentially primates with countless humble cousins among the mammalian line, we are nonetheless unusual – and it is this stark unusualness that is the source of the nagging skepticism about the modern, scientific worldview. Religious dogma simply declares that "Man is special!" and, regardless of the truth or fiction of the process of evolution, there is no further need for believers to glance back: Unseen powers have chosen us, it is said, for a special role in life on Earth. In contrast, evolutionary theory explicitly links us with the material world and ties us to all other biological species. The implication is that, in essence, we are not more unusual than, say, anteaters or jellyfish – remarkable maybe, different for sure, but just another ephemeral leaf on the evolutionary tree of life that has somehow emerged on this lukewarm planet in this typical solar system.

As a consequence, in acknowledging the truth of evolution, we inadvertently accept a quite counterintuitive view of humankind. Despite mountains of behavioral evidence, the entire history of human civilization and our undeniable feeling that we have an internal cognitive life that is starkly different from anything else in the animal world, the modern scientific worldview classifies us squarely with our furry friends, and does not acknowledge any material basis for our fond fantasies of "higher-level" existence. Details aside, the broad strokes of evolutionary theory have thereby left a curious void at the heart of the scientific explanation of human existence. All of our higher-level aspirations – however we may individually formulate them in terms of political activism, spiritual faith, intellectual rigor, creative enterprise or ethically upright behavior – are in the "big picture" of evolutionary theory essentially irrelevant. Beguiled by the scientific rigor of molecular biology, we have modestly accepted the view that we are psychologically a slight variation on the primate theme – another nimble ape adapted to a particular biological niche. "Despite what you may think," the biologist teases us, "the high-minded ideas you say motivate your life are nothing more than delusions – stories constructed after the fact to justify your biological urges to yourself and others!"

To be sure, when seen through the lens of cellular biology, we are quite explicable as just another example of fanatically replicating "selfish genes" – bundles of biological instincts that, by chance and good fortune, have become rather dominant on Earth over the past 100,000 years or so. And when seen through the lens of physics or chemistry, we are typical of all organic life-forms in consisting of mostly carbon compounds and water. But when addressed at the psychological level, our unusualness is striking: The behavior that flows from our unusual minds has surprisingly few antecedents among animal species. Those remarkable talents are deserving of as much careful study by psychologists as scientists have already devoted to clarification of the physical, chemical and biological worlds.

As judged from our complex behavior, we are cognitively unique. We communicate with each other using symbolic languages that make absolutely no sense to other species. We make tools that we then use to change the external world around us but that are as meaningless as tree stumps to other species. And we create and enjoy art and music for reasons that are sometimes hard to explain but nonetheless have deep meaning for us, but no meaning for other species. What precisely is it that makes us behaviorally so unlike our biological cousins? It is not pride that motivates this inquiry, but curiosity – and that simple curiosity concerning how other species think is something that has not been reciprocated! At the psychological level – thinking about thinking – we are alone.

But if we are unusual in many identifiable ways, it is nonetheless a curious fact that we still have difficulties explaining in concrete terms how and why that might be true. Modern science can catalog the differences among animal species – from gross behavior to DNA sequences – but questions about whether or not we actually think differently from monkeys will be answered with blank expressions. "If you want to get into the problem of why human beings are special, you really don't belong in the Faculty of Science, but should move to the Faculty of Arts and Letters, or Theology – or just give up and study Linguistics!" is a sentiment that lies just below the surface in the world of cognitive neuroscience.

Despite the merits of such career advice, however, it is precisely our cognitive unusualness that needs to be clarified to consolidate our evolutionary successes. In order to transcend the fractious, short-sighted debate about which form of human culture should be considered the transient "winner" in modern times, we need to identify the cognitive roots of our behavioral dominance – not by retreating to the philosophical discussions of previous eras, and certainly not by declaring with the linguists (or is it the theologians?) that we alone miraculously have The Word, and all else follows from there. On the contrary, language is one of the enigmas that need to be solved on the basis of the principles of verifiable cognitive neuroscience. What we, as a species, still need is a deeper understanding of what *Homo sapiens* is and how we got here.

It may well be that the philosophers and poets are on target when they discuss the nature of the human soul using the psychological categories and terminology that Shakespeare, Dante and scholars from the Renaissance would have understood, but those insights need to be translated into the lingua franca of the modern world: empirical science. Just what is language (and music and art, fun and work, awe and wonder, justice and responsibility) in a material world? Despite the deep pessimism among some scholars on the possibility of answering such "big questions," genuine progress has been made in cognitive psychology in modern times and some important insights have already been established. The story is still far from complete, but it is today possible to explain – in broad outline and in some detail – how it is that we are simultaneously standard-issue biological entities entrenched in the messy business of biological survival, and yet remarkably "mindful" – perceiving, cogitating and behaving in a material world in ways that are undreamt of among lower species.

In order to clarify the psychological roots of our higher-level behavior, we must address these issues in the somewhat unusual terminology of cognitive psychology and neuroscience. While "thinking about thinking" and wading through the details of some low-level mechanisms, however, it is important that we stay focused on our ultimate goals at the higher end of human psychology – and not get sidetracked by topics that can be dealt with adequately by the sociologist or the biochemist. In other words, cognitive psychology is a scientific field in its own right and can provide answers to questions about the essential operations that allow for human cleverness – not by abandoning cognitive psychology and attributing our mental strengths and weaknesses to sociopolitical or molecular factors, but by addressing them in terms of the underlying cognitive mechanisms themselves. Above all else, we must ask: What are the mental gymnastics that we undertake in using language and tools, in understanding art and music, and in transcending our individual egocentricity to participate in social organizations? The answers are not as simple as pointing out a special gene or an unusual neurotransmitter that explains all, but the good news is that the answers are understandable in the modern scientific terminology that has been developed for discussing cognitive mechanisms.

1.1. THE BASIC QUESTION

The basic question that will be addressed in the pages that follow is: What distinguishes human beings from similarly big-brained animals, such as chimpanzees, bonobos and gorillas, and from our early ancestors of more than 100,000 years ago? Anatomically, physiologically and even genetically, we are all extremely similar, but only *Homo sapiens* embarked on a rampage of cognitive development that has allowed for the unusual behaviors underlying human civilization. We are justifiably proud of those mental talents and, with only infrequent interludes, we exercise and indulge them throughout the waking day – most notably, in behaviors that involve language, tools, music, 3D visualization and social cooperation – often exploring complex combinations of those talents simultaneously. Other species show hints of such abilities, but only we have turned them into lifelong obsessions, full-time careers and the focus of incessant, large-scale social activity. Where did the human talents come from? And how can they be explained on a scientific basis?

Harmony, Perspective and Triadic Cognition is an attempt to answer these "big questions" in human psychology – not by pontificating on the importance of language or symbolic thought, morality or the exalted state of the human spirit but by focusing on the small-scale psychological miracles that we perform daily. The main themes to be discussed here concern the cognitive functions that are universally acknowledged to be well developed in human beings but are totally absent (or present only as faint traces) in other species. In concrete terms, what precisely do we perceive, think and do when we undertake our characteristically human behaviors? What does it mean to use a language? To handle a tool? To understand a photograph? To enjoy music? To lend a hand? In order to answer these questions, a comparison with infrahuman cognition is sometimes relevant, but the main topic must be our understanding of how the individual human mind works. Once we understand the basic cognitive processes occurring in one human brain, then we can address the next level of inquiry: how the human mind allows one individual to cooperate with other human beings in achieving goals that none of us alone could possibly achieve.

As will become clear, the theme of this book is *human* cognition (not generalized primate intelligence, much less rat neurophysiology), and the focus will remain relentlessly on how our minds can use language, manipulate tools, create art and make music. The topic of interest is how we *think* – not the issues of the shape of the pelvis, the larynx or the thumb that allow particularly efficient walking, talking and grasping. Without doubt, we have bodies that are well adapted to our unusual minds, but it is in the realm of the mind where we are special, and where the academic field known as cognitive psychology has the potential to contribute profoundly to our understanding of the human condition.

One of the main lessons of nineteenth-century *biology* was that there are deep *similarities* among all animal species – similarities that can be explained in terms of evolutionary theory. But – given that evolutionary foundation – the *dissimilarities* with even our closest biological cousins present twenty-first-century *psychology* with its greatest challenge. Yes, we resemble vertebrates, mammals and particularly the apes in many ways, but the unusualness of human behavior is the central issue that the science of *psychology* must clarify. If successful, elucidation of the cognitive mechanisms underlying our unusual minds should bring some benefits in the form of practical applications in education, clinical psychiatry and social engineering, but more importantly an understanding of the human mind would provide insight into the ways in which we are more than beasts in the struggle to survive. We already have some reason to be proud of human civilization and our unusual status on Earth, but our propensity for intraspecies murder; for putting each other into locked cages; for placing more importance on rituals than on common decency; for falling into the ridiculous conceits of "papal infallibility," "enlightened perfection" and "sacred words"; and the pervasive tendency to regard a small gang of our compatriots as "decent folk" and the rest of humanity as "infidels" are human foibles not to be overlooked. Just maybe, there's still room for improvement.

1.2. TRIADIC PERCEPTION, TRIADIC COGNITION AND TRIADIC SOCIAL INTERACTION

Most of the topics mentioned in textbook discussions on human cognition will be examined in the following pages. Individual chapters are devoted to language and symbolic thought, tool use (toolmaking and handedness), music (harmony) perception, and pictorial depth perception (seeing 3D depth and mentally manipulating the objects represented in 2D pictures), and brief discussion is made of social cooperation ("joint attention") and the "moral mind." There are in fact technical arguments for maintaining that related behaviors, cognitions and perceptions do occur to some extent in nonhuman species – and such arguments constitute important support for the common evolution of all animal species. But, despite that thread of biological continuity, what is nonetheless striking is that many of our normal ways of thinking are either impossible or notably difficult for all other animal species.

On the one hand, apes, dolphins, parrots, cats and dogs that are raised in essentially human-controlled environments and that receive intensive training can achieve certain behaviors that reveal a potential cleverness not seen in the wild. And even in the wild, modern animal research continues to reveal social hierarchies, vocal communications and tool-usage talents suggestive of higher cognition that, only a few decades ago, was thought impossible for infrahuman species. For driving home the continuity argument known as evolutionary theory, such research is of interest, but, for understanding the underlying cognition, it is far more important to examine what precisely the human forms of those talents are.

For all of us who have received a normal human upbringing, the "intensive training" that was our childhood releases a huge repertoire of sophisticated "higher cognitions" – cognitions that eventually become easy, natural and the focus of our daily lives. "Of course, I can speak a language!

And can understand the 3D scene in that painting! And can hear the sadness in that melody! And can light a match! And can imagine the shape of Florida! And can do simple arithmetic in my head! And can turn my attention to what you point out!" and a thousand other things that tax the upper limits of a chimpanzee trained for months on simple versions of such tasks. Those behaviors are so easy for us that we don't often ponder their significance, but they lie at the heart of all forms of human civilization and are – in the biological world – far from typical.

In fact, many books written during the past century have documented aspects of the evolution of the human talents that are discussed here, but *Harmony, Perspective and Triadic Cognition* is unusual in proposing a cognitive mechanism that underlies them all. That mechanism is the raison d'être for this book and fills in the blank space where typically even the most demanding, hard-headed scientist hopefully pencils in "random mutation" when trying to explain the unusualness of *Homo sapiens*. "Random mutation" is, of course, the accepted evolutionary jargon for "Hey, something remarkable that we haven't a clue about must have happened!" and indeed the genetic argument might boil down to something as simple as a few well-placed errors in DNA replication. But, whatever the genetics, the psychological argument for how we, but apparently not other species, *think* must be explained at the level of cognitive mechanisms. For that reason, the following chapters address the question of what is happening in our minds when we use language, hear harmony, handle a tool, see depth in a flat picture or attend to a conversational topic that someone has just raised. In other words, just as an understanding of biological life requires a close examination of cellular processes (and cannot rely solely on lower-level arguments from physics or chemistry), an understanding of human psychology requires an examination of cognitive processes (and not a reliance solely on the anatomical, cellular or molecular topics favored by nonpsychologists).

The theme that unifies the diverse, characteristically human behaviors discussed here is that, in each case, the individual must deal with *three* distinct streams of information at once – must have "three things in mind" simultaneously. Whether words, musical tones, ideas or baseballs, "juggling" with three objects is simply beyond the capabilities of other species, but it can be mastered by all normal children in a variety of domains – usually through mere exposure to the behavior of adults or early basic training. Juggling three sensory cues, physical objects or thought processes is what this book is about.

To clarify how different we are and, indeed, to justify the notion of "higher" and "lower" cognition, many familiar topics in human psychology will be covered here, but what distinguishes this book is an insistence on framing familiar questions about the unusual character of the human mind in the simplified terminology of cognitive psychology. The coherence of the answers will become apparent individually in consideration of specific human behaviors, but, in every case, the cognitive mechanism that brings these diverse behaviors together is that our uniquely human talents are (merely!) one step more complex than simple "two-component, dyadic associations." I maintain that what we do so easily, so naturally and so well is "triadic cognition."

The forms and contents of triadic cognition are diverse, but its "higher" character can be clearly explained in comparison with *dyadic cognition* – thought processes that involve only two elements. Associations between this and that, pairing between conditioned and unconditioned stimuli, and one-to-one correlations between stimulus and response are sufficient for operant conditioning – and *all* animal species, including *Homo sapiens*, can "think" at that level (Macphail, 1987). In other words, the basic learning capabilities that even a primitive nervous system provides are enough to handle most forms of "two-body" associations. Depending on the perceptual processing that precedes the dyadic associational mechanism, the content of such thinking will differ markedly from species to species, but the simple linkage between, for example, a sensory stimulus and a favorable or unfavorable outcome will be enough for most brains to link the perceptual event with appropriate motor behavior. In terms of neuronal mechanisms, stimulus followed by response, correlation between X and Y, and all other two-element connections are inherently easy and do not tax the learning capacities of the cockroach, much less cats and dogs. But, for all the strengths of associationism (and the associational, Hebbian learning mechanisms that underlie the psychological theory known as behaviorism), "the association between X and Y dependent on the state of Z" is a logic of a higher form – a psychological operation that suddenly separates the men from the boys, the adults from the children and, indeed, us from all other animal species. I will show that the gateway to so-called higher cognition is the ability to handle just this type of three-body "conditional association" – seemingly not such a complex undertaking, but the start of the cognitively unprecedented human revolution nonetheless.

The abstract idea of three-body cognition is inevitably a controversial theme – and most researchers interested in the evolution of the human mind will be anxious to return to discussion of mirror neurons, the acoustics of vowel production, the verbal exchanges in social interactions, and other specific examples of cognitive talents that don't have the aroma of a metaphysical "trinity!" While those concrete issues are all in fact part of the story of human uniqueness, I will show that triadic cognition is the underlying mechanism – and that it provides a conceptual framework within which those specialist topics are best understood ... with absolutely no metaphysics implied! On the contrary, it is the concreteness of the triadic argument that is its strength. Depending on the specific topic, it can be said that various triadic talents are (i) already empirically well understood, (ii) currently the focus of experimental research or (iii) a familiar, if still rather vague, idea. Details and examples are provided in the chapters that follow, but it is the intrinsic "threeness" of the underlying cognition that is so unusual.

Discussion of various three-body computations necessarily leads into the academic fields of linguistics, brain physiology, art history, music theory, developmental psychology, neural networks and other disciplines that are normally of concern to specialists only. The scope is inevitably broad, but I maintain that the core arguments in these diverse fields are not of such extreme complexity that they must remain out of reach to the non specialist. On the contrary, the underlying thesis about triadic cognition is so easily stated that it is spelled out here in this introductory chapter. As a result, the reader should be able to put the more technical and less familiar arguments in the following chapters into their proper contexts – and to understand why it can be said that human cognition is at once endlessly complex, and yet the basic processes are thoroughly easy.

Let it be noted that many scholars have previously advocated "triadic," "ternary," "tripartite," "trichotomous" or "tertiary" cognitive models to explain specific aspects of the human mind. The fact that others have advocated three-way explanations of human cognition does not place them above our normal skeptical evaluation, but it does indicate that the general argument concerning triadic cognition falls more or less within orthodox thinking in several branches of human psychology. Moreover, as demonstrated later, the triadic argument turns out to be surprisingly concrete. Although various abstract models will help guide our search, ultimately the three-way mental gymnastics that we are so good at are comprehensible *because* they are so familiar. Countless examples from everyday life are available and explicit support for the idea that the human psyche has a triadic core comes from experimental results in cognitive psychology and cognitive neuroscience. In other words, the argument is scientific, not philosophical, and provides many opportunities for empirical testing.

In Chapters 2 and 3, the three-body argument is explained in the *perceptual* realm – specifically, in terms of musical harmony and pictorial depth perception, where the threeness of the sensory stimulus can actually be heard or seen. There are also *perceptual* triads in the realms of tool use (Chapter 4) and language (Chapter 5), but both tool use and language also have well-known triadic *cognitive* operations. Indeed, the three-component models of tool use and language have been the focus of much research for many decades and are well established in clinical neurology. By their very nature, the cognitive processes are "once removed" from the sensory stimuli, but because we spend so much of our waking day engaged in these triadic pursuits, the "tripartite architecture" of language and the "three-module" theory of tool use will undoubtedly strike most readers as reasonable, if not downright obvious.

In both the perceptual and cognitive cases, "having three things in mind" is the foundation on which the capacity for characteristically human thought and behavior stands. This is not to say that every trichotomy that we might imagine is a valid example of triadic cognition, and, equally, it is not true that cognition involving two or four or eleven components is unimportant. But the basic idea underlying the triadic hypothesis is that every example of typically human-level cognition – the kinds of ideas that grab our attention, fascinate and entertain us, and sometimes make us concentrate to wrap our minds around – is essentially a three-element process. And it is the three-element process that puts this kind of perception/cognition just out of reach for animal species.

Unlike some of the alluring, but ultimately empty sophistry that can be found in the so-called popular science literature, triadic relations can be *understood* – and insight into triadic processes normally provides that "aha!" moment when we finally grasp the essence of the triadic mechanism. Notably, an understanding of triadic cognition does *not* rely on an appeal to the infallibility of long-dead authorities (those classics that can be understood only in the original Greek!), does *not* rely on the righteousness of sacred texts (the one and only book that contains all knowledge!), and does *not* rely on obscure insights in specialist fields that are distant from human psychology and that most people don't even pretend to understand (that slippery tome on quantum mechanics that remains unopened on the bookshelf!).

Triadic cognition is what we all do – all day, everyday – and if it can't be explained in language we all understand, it simply isn't real. Some mental effort will be required to understand specific examples of triadic processing, but a little concentration will suffice to rearrange concepts and view familiar topics as cognitive triads. These are the minor insights that make learning fun – and precisely the kinds of ideas that, no matter how simplified and no matter how long the training proceeds, our animal friends just "don't get." Triadic processes alone do not, of course, exhaust the talents of our species. Indeed, most concrete examples of "higher-cognition" are typically examples of multiple triads – and often referred to as "hierarchical recursion," but the core process underlying recursion is demonstrably triadic.

The trichotomies that are *not* discussed here are the many religious ideas that, for one reason or another, are organized around triadic descriptions. Ancient metaphysical trinities are of course familiar – Father, Son and Holy Ghost or Brahma, Vishnu and Shiva. Real or otherwise, there is little chance of scientific demonstration of such abstract forces. Even within the field of psychology, there are many tripartite models that cannot be rigorously evaluated. Freud's ideas about Id, Ego and Superego are certainly interesting, but – after nearly a century of proselytizing – empirical testing, much less proof, of the reality of that mental architecture remains an impossibility. The traditional "trilogy of the mind" (cognition, emotion and motivation) (Hilgard, 1980) is a plausible subdivision of the mind, but not (yet) defined with sufficient precision to become a testable theory. Eysenck's three dimensions of personality, MacLean's triune division of the nervous system and Sternberg's three-component model of intelligence are all useful as conceptual models and have helped to drive experimental work, but it is questionable whether there is a material basis for those models that could not as easily be described in a four-component or a seven-component model. To be sure, it is an interesting fact that various psychological, philosophical and religious trinities have proven to be seductive for so many people worldwide over so much of historical time. Apparently we like to think "in threes!" But the following chapters will discuss only those triadic processes where the empirical threeness is unambiguous, where scientific research has indicated a three-component process and where there is a consensus among specialists in the field that a tripartite description is meaningful, useful and has a neuronal basis.

1.3. TRIADS VERSUS DYADS

Many forms of cognition ("lower cognition") are simply two-body associations: X in relation to Y. In contrast, what I refer to as "higher cognition" demands three-way associations: X in relation to Y in light of Z. Such cognition takes many different forms depending on the content, sensory modality and relationships among X, Y and Z, but it is always of a complexity beyond two-body associations.

Moreover, it is crucial for an appreciation of triadic cognition to understand that triadic associations are *not* reducible to "multiple or serial two-body associations." The idea that complex systems, in general, are not simply the sum of their components is of course familiar from many fields of inquiry, but the classic discussion of the "three-body problem" comes from the venerable field of Renaissance astronomy. If we have a sun, a planet and a moon revolving around each other, it is known that their orbits cannot be calculated by considering each of the two-body gravitational attractions one at a time. The system is more than the sum of its parts *because* the relationship between any two parts is affected by the third part. This is not to say that it is all a "mystical oneness" that defies understanding, but we must resist the lure of extreme reductionism and not oversimplify to the point where, in the name of analytic purity, the phenomenon we want to understand is entirely squeezed out of the analysis! In cognitive psychology, the three-body interaction itself is the unitary phenomenon and whatever might be said about the individual pieces or lower-level two-body interactions, there is a higher-level phenomenon that must be discussed in terms of the interactions of three independent components.

In order to clarify the meaning of three-body processes, a comparison with two-body processes is essential, so that the psychology of X-Y associations will be discussed before moving on to more interesting triadic phenomena. Whereas all animal brains are capable of X-Y pairing, the unusual capabilities of the human brain are due to the fact that, whenever we "give thought" to anything, we naturally, habitually and spontaneously undertake a more complex process in which (a minimum of) three separate components are considered. In contrast, two-body calculations involving stimulus–response pairs or conditioned–unconditioned stimulus pairs are often obvious. To be sure, cognition based on dyadic pairs has a certain intricacy of its own (X implies Y, Y implies X, X is negatively correlated with Y, etc.) – that is, the full panoply of X-Y associations (and serial two-body associations) that the experts in animal learning theory study. Moreover, it must be said that such one-to-one associations and correlations are essential building blocks used for the construction of more subtle forms of cognition. But there is an explosion of complexity that comes with a third element – when cognition becomes triadic.

Further increases in complexity of course follow with the addition of a fourth, a fifth and further components, but the cognitively interesting step is from basic (but generally trivial) dyadic associations (hardly worth thinking about!) to the intricate, fascinating and more profound triadic processes that we consciously ponder. This blossoming of mental complexity is apparently unrealized in the animal brain, and is the main reason why most normal people – with nothing but the warmest feelings for their pet cats, dogs, horses, goldfish and parakeets and a genuine fascination with the diversity of animal species – are unwilling to believe that nonhuman species have "minds" of any kind! Nice, cute, loving and clever in a thousand ways, our pets somehow lack the internal cogitation that is typical of our species – the propensity to weigh various factors, try out different perspectives and consider alternative outcomes. This is not malevolent "speciesism," but, more plainly, recognition that animal behavior seems to be explicable at the simpler level of one-to-one associations, whereas human behavior is more complex.

Dyadic associations are easy: "Stop at the red light, go on green!" – but, once learned, such associations hardly require ponderous thought of any kind. The complexity of real-world situations with multiple factors means that judging the implications and weighing the alternatives before taking action is the wiser course. That is where "three-body calculations" come into play and where we, as a species, are generally competent, while other species appear to be permanently stuck in a dyadic mode – and forced into serial attempts of trial and error, relying on simple associations without seeing the relevance of other factors ... the larger context.

Of course, the idea that "context" might be important is hardly a new insight in the annals of psychology, but the hypothesis explored here is that a large part of what is often referred to as "contextual processing" can be explained quite precisely at the level of three-body interactions. In other words, before invoking unspecified, contextual influences that make quantitative analysis impossible, and before pulling the sophisticate's trick of declaring the problem beyond the powers of reductionist analysis because of the presence of delicate factors that scruffy schoolboys are unable to comprehend, some clear progress can be made by focusing on triadic interactions. Three-way effects are difficult enough that very young children and animals cannot handle them, but the complexities of three-way combinations are nonetheless eventually understood by the normal child well before seven years of age and developed, polished and mastered on the road to adulthood. This slightly more nuanced way of thinking is our strength and allows us all to play some small role in civilized society – where "go on green" and "stop on red" are not rules of sufficient subtlety for navigating the traffic jams of the modern world.

Glib demonstration that animals are not so smart is not the motivation for explicating the nature of triadic cognition, but a comparison with other species is an inevitable part of the clarification of what is meant by "higher cognition." Similarly, discussion of the gradual transformation of human thinking over the past 100,000 years – as deduced from archeological relics – is not intended to bolster our boastful modern self-satisfaction, but rather to demonstrate the developmental path that humankind has tread. Early indications of triadic mental processes are clearly recorded in the short history of human civilization – from stone tools, bead necklaces, cave drawings and primitive flutes remaining from Paleolithic times – and many aspects of so-called modern culture are clearly little more than elaborations on those ancient discoveries. So, to understand who we are today, the historical context is as essential as the evolutionary context.

1.4. MUSICAL HARMONY

In *every* musical tradition, combinations of three tones underlie the phenomena of musical harmony and allow all normal people to hear the major and minor modes – and, indeed, the amodal harmonies of "chromatic" music. Whether played as isolated chords, short melodic phrases, or buried in complex and lengthy symphonic compositions, it is *harmony* that glues the musical piece into a coherent whole and gives us a feeling for what musicians call "key" – the musical focus that is totally absent in birdsong. So, if you enjoy music of any kind, you are interested in harmony.

Historically, the use of *simultaneous chords* was an invention of European musicians in the early Renaissance, but it is crucial to understand one simple fact about all forms of music: Any musical tradition that utilizes pitch necessarily exploits the harmonic effects of three-tone combinations. Even when not played as three-tone chords (the Renaissance specialty), sequential tones (melodies) are accumulated and held in short-term memory and imply the harmoniousness that we recognize as the major and minor modes. Since the most common descriptions of music are based on traditional harmony theory from Renaissance Europe, that terminology will be used here, but it is important to note that all musical cultures exploit harmony in creating music.

The emotions elicited by musical harmony are of interest in cognitive psychology primarily because normal listeners *without* musical training and from diverse cultural backgrounds can "hear" the emotional significance of music, in general, and harmony, in particular. Quite aside from scholarly explanations, virtually *everyone* understands music intuitively. This is not to say that musical training and the learning of the common patterns of various musical genres are irrelevant or that the differences among various musical cultures are uninteresting. But, given a rudimentary exposure to almost any form of music, the inherent affect of three-tone combinations becomes obvious to everyone – young children, normal adults, and peoples raised in various musical traditions – while they remain a silent mystery to all animal species that have been experimentally tested.

Discussions of the psychological meaning of music date back at least to Pythagoras and became a full-fledged industry during the Renaissance, but insights into the underlying cognitive mechanisms have been achieved only recently. As discussed in Chapter 2, the explanation of the triadic nature of harmony is built on the foundations of *physical* acoustics, but includes two crucial *psychological* pieces of the puzzle to account for the fact that the seemingly simple acoustical phenomena of "vibrations in the air" can often elicit an emotional response. The psychological arguments concern the three-tone (melodic or chordal) harmonies. Four-tone chords and harmonic cadences are also briefly discussed, but the main point is that pitch triads are the *beginning* of true musicality. Beyond three-tone combinations, real music quickly becomes complex – and explication of the "hierarchical recursion" in music becomes necessary. But the use of simple harmonies is already a huge step above the warbling of birds, whistling in the dark and the rather monotonous Gregorian chants of the Medieval Age. The fact that the major and minor keys of the vast majority of classical and popular music are built from triads means that, before all else, the mystery of triadic harmony must be unraveled in order to understand why music has emotional impact. Even in musical traditions that do not use three-tone chords, three-tone melodic segments often imply major or minor "modality" in a Western, diatonic sense. For this reason, the cognitive psychology of harmony, as explained in Chapter 2, is quite general, and applies equally to music that employs 5-, 7-, 12- and even 22-tone scales.

1.5. PICTORIAL DEPTH PERCEPTION

The visual perception of spatial depth has been studied within three subfields in visual psychology: (i) binocular depth perception (stereopsis), (ii) depth perception through movement (motion parallax) and (iii) so-called pictorial depth perception, that is, the perception of the *illusory* 3D depth in 2D paintings, drawings and photographs. In binocular vision, the left and right retinal images contribute two views of the same object in the visual scene, and the calculated "disparity" between the two images allows for an estimation of depth. In the case of motion parallax, changes in the visual image due to self-motion again provide the raw data for two sequential views of the visual scene, from which 3D structure and the relative depths of objects can be calculated. In both cases, it is the difference between two images that tells us about the 3D shape of the world. Whereas our capabilities for depth perception through binocular stereopsis and motion parallax are shared with all primates, the perception of the 3D structure in static 2D images is an unusual human skill. Pictorial depth perception is thus a classic example of the human ability for higher cognition that has only weak analogs in the animal world.

Chapter 3 briefly discusses the development of pictorial depth perception over the past 50,000 years of human culture. As seen already in prehistoric cave drawings, the overlap, relative size, position, and orientation of objects in a painted visual scene provide the illusion of 3D depth in 2D images. It is noteworthy that those artistic tricks are all essentially two-body perceptions: suggestions of 3D depth that are obtained by contrasting two visual objects – pairwise comparisons of relative size, height and undisturbed wholeness. As important as such 2-body comparisons are, further insights concerning the depiction of realistic 3D scenes were formalized in the Renaissance, when the laws of optics and linear perspective were formulated and exploited in a cultural revolution in fine art. Similar to the Renaissance discovery of harmony, the techniques of linear perspective led to a blossoming of art forms that we still enjoy today.

Although the history and technical details of linear perspective have been discussed by many scholars, the relative importance of two versus three visual cues in producing the illusion of depth has been explicated only recently in human visual psychology. As with musical harmony, the topic of linear perspective leads inevitably into the complexities of experimental design, brain science, and art history, but there is again a profound simplicity to pictorial depth perception that can be explained in terms of the processing of small numbers of visual cues. Many readers will undoubtedly be familiar with the main themes of the psychology of visual perception, and other readers will know the rough outlines of art history. But far fewer readers will be familiar with the significance of two-object vs. three-object visual patterns, and how quite simple configurations of visual cues are built up into the familiar global illusion of 3D depth on a 2D surface. Chapter 3 explains what we perceive when we see depth in a flat picture – and, conversely, what most animal species do not "see" when viewing the same visual image.

1.6. TOOL USE

For explaining the origins of triadic cognition, it is essential to explore the evolution of *Homo sapiens*. Although the triadic nature of tool use can be delineated in the terms of modern cognitive psychology, the fossil record also provides rather clear evidence concerning both toolmaking and tool use at a time when language was either absent or extremely primitive. Those simple tools and indications of how they were first used show the importance of cognitive triads.

The idea that specifically *three* mental processes are essential for using tools has been discussed in the neurology literature for many decades (dating back at least to Hugo Liepmann in 1908) and is strongly supported by modern brain-imaging studies. In brief, in order to use a tool, the individual must continually engage three distinct mental processes. One must (i) imagine a desired end-product (a "goal-state" concerning the intended configuration of physical objects in the external world) and keep this goal "in mind," (ii) perceive the current state of the raw materials in one's immediate surroundings in order to calculate the difference between the desired goal-state and the actual current-state (an "error detector" function) and finally (iii) utilize motor routines to control the sequential movements of the musculature of the hands for the purpose of transforming the current-state into the desired-state.

When engaged in skilled motor activity, conscious attention normally moves back and forth among the three main processes depending upon the current-state of the external reality, but all three must remain "on-line" to achieve the tool-use goal. For example, having imagined the need to drive a nail into a piece of wood, we hit the nail on the head with a hammer, evaluate its depth and angle in the wood and adjust the trajectory of the hammer for the next blow accordingly. In the midst of a carpentry chore, it would be unusual to contemplate which of the three functions is currently in need of our full attention to proceed, but it is clearly the cognitive tension provided by an understanding of (i) what one wishes to perceive and (ii) what one actually perceives that is used (iii) to drive motor routines.

When any of these three functions moves "off-line," we lose the train of thought and cannot continue with the tool-use task until all three functions are again simultaneously in mind. In most normal circumstances, we would pause, take a metaphorical step back and mutter to ourselves, "Now just what am I up to here?" in hope of reestablishing the triadic mind-set. With hammer in hand, it is unlikely that we would not immediately recall what we are engaged in, and we would easily return to the tool-use task. But there are other situations where the coordination of the three processes can be temporarily lost. With muscular exhaustion, the normally automatic triggering of motor routines eventually becomes difficult, and the process of competently swinging the hammer (or stroking the brush over the canvas or plucking strings with the fingers, etc.) becomes a chore in itself, and completion of the higher-level task becomes impossible. Similarly, with a slight "change of heart" regarding the desirability of this project, the intended goal may become unclear. And, in the midst of a tool-use chore with a clear-cut goal and an energized motor system, sweat in our eyes or distracting noise in our ears can prevent the necessary "error-detection" calculation – and again disrupt what should be a routine task. In each of these cases, the mental obstruction is not what we would consider serious pathology, but it is precisely the loss of psychological unity among triadic processes that prevents us from proceeding.

The tripartite model of tool-use by human beings has a long pedigree in behavioral neurology. The kinds of pathology that arise with disturbances of any of the three main neural systems are known clinically, and indeed the behavioral syndromes have led to a neurological model of tool use where three brain systems are involved. Following discussion of the evolution of tool use and toolmaking, this cognitive triad is the main focus of Chapter 4.

1.7. LANGUAGE

Among the three-body cognitions that have been studied explicitly as such, some are "supramodal" and therefore not tied to specific sensory systems, but it is the threeness of the cognition that is most unusual. Above all else, language is the crucial human talent that distinguishes us from all other animal species, and it has two "triadic" aspects that are well-known in linguistics. The first triad is the three quasi-independent processes of language—syntax, semantics and phonetics – that make communication possible. The "tripartite architecture of language" has been advocated most forcefully by Ray Jackendoff (1997), but explicitly supported by Derek Bickerton (1990) and Willem Levelt (1999), among many other linguists. The basic idea is that all forms of natural language are made possible by the simultaneous coordination of the mental processes underlying (i) the conventional rules of word-ordering, known as syntax, (ii) the conventional linkage of meanings to specific words, known as semantics and (iii) the rapid encoding and decoding of speech sounds, known as phonetics. Depending on which linguist one consults, any one of these three primary functions may be advocated as the "key" to human language capabilities, but there is a striking consensus among linguists that language must be described under those three headings: language competence is a triadic cognitive process. When all three processes are in working order, human beings can think and communicate using language, and, when any one of these processes is dysfunctional because of the unfamiliarity of a foreign language or breaks down through brain damage, characteristic cognitive and communicative disorders arise.

Another important linguistic triad concerns the triadic structure of syntax. Basic language competence requires that a speaker knows how to use the correct sequence of, to begin with, subjects (S), objects (O) and verbs (V) in order to communicate specific chains of causality through words. Syntactic rules differ by the language, but every human language has a set of SOV (or comparable) conventions through which the roles of specific entities and their dynamics can be specified. Through those conventions, meaning can be conveyed to any other human being who understands the same conventions. The innate human capability to learn and use SOV syntax is the linguistic manipulation that most clearly distinguishes human language from the dyadic signaling of animal species. Stated in the slightly less comfortable jargon of "phrase structure grammar," I argue in Chapter 5 that this is the essence of Universal Grammar, as advocated by Noam Chomsky since the 1950s. Imbedding triads within other triads leads to the recursive complexity of language, but the crucial point is that recursion is triadically hierarchical, not simply dyadic "chaining" or the serial addition of word components into meaningful sentences.

Chimpanzees and bonobos, for all their notable cleverness, lack the oral dexterity to produce the auditory phonemes of a complex language system; as a consequence, they have not developed the mental machinery for the rapid-fire sequencing of sounds that underlies both the production and understanding of speech. On the other hand, the parrot, for all its notable phonetic talents, seems to lack the capability for syntactic organization and thus the ability to string words together in the proper sequence to explicate causality. Those birds show remarkable talents in phonetics and even semantic associations, but they struggle with even the simplest of three-component grammatical transformations that are a normal part of the grammatical manipulations of human languages. And in the world of artificial intelligence, machines with engineered phonetic capabilities and rather sophisticated rules of syntax can exhibit seemingly-intelligent, human-like language performance in restricted domains. Although silicon brains are impressive in many ways, they still have overwhelming difficulties in handling "meaning" – the semantics of everyday concepts. So, although the engineering problem of associating specific word-sounds with specific meanings has been solved, the hierarchical structure of meaning and the complex semantics of human "common sense" that normally gives coherency to our everyday conversations remain unsolved problems. The ape, the parrot and the robot each have their undeniable strengths, but a higher-level coordination of the three essential processes is yet lacking, while it is an innate part of the mental apparatus of *Homo sapiens*.

Delineation of the neuronal coordination of these processes and their relationships remains a difficult research problem in cognitive neuroscience, but the importance of language for normal human existence has meant that the disturbances of language have been studied in clinical neurology for well over 100 years. In this regard, it is noteworthy that the behavioral syndromes associated with localized brain-damage have resulted in a classification system based explicitly upon the independence of syntax, semantics, and phonetics. In other words, the known categories of clinical abnormalities have provided an empirical base that is also essentially triadic in character. Quite aside from any notion of human uniqueness or the triadic nature of human cognition, neurologists have found that linguistic competency involves specifically *three* language processes.

1.8. CONSCIOUSNESS

Chapter 6 is a brief interlude away from issues of "how we think" in order to consider the topic of subjective consciousness. A chapter on consciousness in a book on cognition is necessary as a means of curtailing the mistaken idea that the "specialness" of the human mind is due to a nonmaterial dimension that is otherwise unknown in the biological world. The subjective feeling of whatever we perceive, think, or do is of course for each of us individually the essence of who we are – but as much could be said for any animal organism, from frog to ape to our next-door neighbor. We are all "special" in the sense of feeling the importance of our own mental processes. But the specific, species-level "specialness" of human beings lies more concretely in our new forms of *cognition* and only secondarily (if at all) in the subjective feeling of the content of that cognition. In other words, we are cognitively special, *not* because of how that cognitive activity feels but because it is "computationally" unusual in bringing together the contents from three (or more) informational streams.

The perceptual and cognitive arguments of Chapters 2–5 emphasize the *differences* between human and infrahuman mental processes, but the basic point about subjective consciousness in Chapter 6 concerns our deep neuronal *similarities* with other animal species. We simply have no scientific grounds for believing that the subjective experience of our own neuronal activity (consciousness, awareness, subjectivity, qualia, soulfulness – whatever we want to call it) is qualitatively different from that which occurs in the brains of any other sentient being with a nervous system. Perhaps we utilize more neurons or synchronize them more efficiently than monkeys or dolphins – and experience "richer qualia" as a consequence. And we certainly attach far more verbal labels to the thoughts running through our heads than do animals without language capabilities. Nonetheless, despite the importance of our triadic cognitive processes for human behavior, the subjective *feeling* of cognition – whatever the number or complexity of contributing neurons – cannot differ in a qualitative sense from the neuronal operations in animal brains insofar as all forms of mental activity are simply the consequence of the firing of neurons. Those neuronal events will have simple or complex behavioral effects dependent on details concerning the wiring circuitry, the temporal synchronization of the neurons involved and the anatomical structure of the body which those neurons control, but there is no known basis for maintaining that the subjective *feeling* of using neurons differs from individual to individual or from animal species to animal species. So, as important as the topic of subjectivity is for a proper understanding of brain functions (the main topic of Chapter 6), the crucial distinction between man and monkey will not be found there. In essence, it is not how we *feel*, but what we *think*, that makes us so unusual.

1.9. OTHER ISSUES

There are many other triadic phenomena that deserve mention, if not chapter-length discussion. Among these, the most important is the triadic cognition of "joint attention" that is at the heart of human child development, in particular. The basic idea is that the cognition underlying human social development is *not* "the child's focus on X," but "the child's focus on X through another's eyes." That is to say, characteristically human learning is not memorization of dyadic pairs (external event and self), but understanding of dyadic pairs in relation to a social context. From the infant's point of view, the minimal context is, to begin with, "mom" – and is eventually generalized to all other people. That psychological state is what Michael Tomasello (1999, 2003) explicitly calls "triadic." He and co-workers have shown that triadic communication is crucial for the normal learning experiences of human infants, but – surprisingly – is *not* the normal mode of learning for other species, even the apes.

Triadic interactions are also a normal part of adult social interactions – where it is not simply you-versus-me (who is dominant here?) but you and me (with our inevitable, dyadic, dominance issues) *and* our topic of mutual concern: our joint attention on a third object, theme or phenomenon. Noteworthy is the fact that joint attention plays little role in the social behavior of other primates, whose social interactions are essentially dyadic – almost exclusively: (i) one-on-one, competitive, dominance struggles or (ii) one-on-one sexual encounters. It is a surprising fact that, with the exception of certain chimpanzee "geniuses" raised in human homes, apes do not even "point!" Forget about complex behaviors such as language, tool use and recursive logic, even the simple act of raising a finger to indicate the location of an object of interest to another ape is not a normal part of the chimp behavioral repertoire. In contrast, literally or metaphorically pointing out an object or event in order to share it as the topic of our common concern is a spontaneous behavior of very young human infants and later becomes an important part of all forms of human social cooperation. The triadic nature of social life and the triadic nature of what has been called the "moral mind" are discussed in Chapter 7.

Chapter 7 also contains brief discussion of the educational implications of three-body cognition. That is, in all fields of characteristically human endeavor – from language to music to sports to mathematics – there are fundamental, preliminary issues involving the training of two body associations. Those fundamentals must come first and, in their respective realms, "practice makes perfect." Such dyadic training is the essential groundwork underlying all higher-level functions, but two-body associations themselves are conceptually simple – and best learned at a very young age. Once those associations have been mastered, however, all of the important issues of *meaning* involve putting the two-body associations into a wider cognitive context – and there the more demanding, but more interesting, "three-body calculation" is essential. The educational argument is therefore that the content of three-body cognition can and should be made explicit and, moreover, conveyed as the centerpiece of "higher" education. Learning the discipline of triadic processes, not rote memory or motor performance, should be the focus of educational systems hoping to foster the development of intelligent human beings. "Two-body" training comes first, but training without context is mindless memorization, at best, and animal tricks, at worst. It is only when the basic dyadic associations are imbedded into larger contexts that we find the meaning of higher cognition.

Finally, the significance of the quantum leap from two-body associations to three-body cognitions is discussed in relation to the prospect of yet another jump into the world of four-body effects. That evolutionary argument is necessarily speculative – thoughts on the future of human cognition – but it is a logical implication of the prior discussion of two- and three-body processes.

In discussing the cognition that is traditionally thought to be characteristic of human beings, *Harmony, Perspective and Triadic Cognition* covers a wide range of topics. In fact, some animal species show hints of related cognitive capabilities, but there is no new lesson to be learned from getting embroiled in debates concerning the relative magnitude of the differences among animal species. The argument concerning evolutionary continuity has been established beyond a reasonable doubt, regardless of whether or not hidden triadic talents in animal behavior remain to be discovered. Whatever cross-species comparisons may yet reveal about animal intelligence, human behaviors involving language, tools, music, pictorial depth perception and social cooperation indicate the reality of a sizable gap between the normal mentality of most people and the normal mentality of all animal species. Most psychologists today would undoubtedly acknowledge that gap, but many would be tempted to "explain it away" as a set of successful adaptations that are as inexplicable as any of the many evolutionary steps that have led from primitive cellular life forms to the complexities of dinosaurs, birds, monkeys and human beings: "We are nothing less, but nothing more, than a Swiss Army Knife of semi-miraculous talents all bundled into the human form!" For some, that type of egalitarian agnosticism may be a sufficient explanation of human nature, but I find that concrete explication of triadic cognition provides deeper insights.

To conclude this introduction concerning the overall aims of this book, an evolutionary argument emphasized by John R. Anderson (1983), Merlin Donald (1991) and Michael Tomasello (2003) is worth mentioning. That is, the historical record indicates that bipedal hominids with large brains were thriving in Africa more than three million years ago, but the first clear indications of symbolic thought did not emerge until as recently as 50,000 years ago. In other words, although our ancestors went through many gradual changes in body form during those millions of years, concrete evidence of modern human intelligence is essentially absent until about 2,000 generations ago. The scientific paradox is that, from an evolutionary perspective, a mere 50,000 years is far too brief to produce *multiple* fortuitous mutations resulting in *multiple* rewirings of the human brain that could account separately for the emergence of language, toolmaking, music, art, social cognition and so on. On the contrary, in terms of conventional evolutionary theory, we would expect that *one* crucial revolution in brain functioning led to the various cognitive talents of modern human beings. For this reason, the idea that the emergence of triadic associational mechanisms is the one source that has made possible various human capabilities is much closer to evolutionary orthodoxy than the "many hopeful monsters" alternative.

By their very nature, evolutionary speculations remain rather inconclusive, but setting any argument about the emergence of mind into the historical, biological context is nonetheless essential. Many questions concerning human evolution remain unanswered. Some are perhaps unanswerable, but the present book provides an explicit hypothesis concerning the unusual nature of the human mind – invoking one set of three-element neuronal mechanisms that, in turn, underlie a variety of our most significant cognitive talents. An attraction of the triadic hypothesis is that we are not left with the unsatisfactory situation of needing to assume unspecified mutations for each and every one of the unusual talents of *Homo sapiens* – mutations that produce higher-cognition essentially out of the blue, but with no indication of where those cognitive talents came from or how they might be related to one another. Although the evolutionary argument is yet incomplete, the coherence of the triadic hypothesis in explaining a variety of human talents suggests that the transition from monkey behavior based on dyadic associations to human psychology employing revolutionarily new triadic processes was the important "single-step" that led eventually to human civilization. Moreover, that one step has a plausible neuronal foundation in terms of the unprecedented convergence of auditory, visual and touch information at the cortical level (Chapter 4) – an explicable neurological change that does not require us to put faith in either biblical or Darwinian miracles.

2

Human Hearing: Harmony

Music is created in all known human cultures and is an integral part of the daily lives of many people worldwide. Historically, one of the most remarkable findings concerning music has been the discovery of primitive flute-like musical instruments made from animal bones, dating from 10,000 to 40,000 years ago (Figure 2.1). Although the actual musicality of such instruments is uncertain, the number of holes in the shaft of the bone indicates that three or four distinct tones were produced by our early caveman ancestors.

It therefore seems certain that melodies have been played in human communities for many millennia. The simultaneous use of two or more such instruments would allow for the production of consonant and dissonant intervals, but concrete indications of the development of polyphonic musical sounds are not found until ancient Greek discussions of the psychological character of pitch intervals. And it is not until the development of written musical notation around 1000 AD (Figure 2.2) that two different pitches were an explicit (written) part of musical culture. From that time onward, the gradually increasing complexity of pitch combinations is well documented in the historical record, but, from a modern perspective, what remains surprising is that the use of three simultaneous pitches emerged only quite recently (~1300 AD).

Even if three-tone simultaneous harmonies were not used prior to the European Renaissance, that does not mean that harmony was not perceived in more ancient music. Although we can only speculate on what aspects of musical melodies were most highly valued in the past, today it is a verifiable fact that major, minor and chromatic harmonies can be heard from pitches played not only as simultaneous chords, but also as sequential melodies. In this regard, the story of music is essentially the story of harmony.

The perception of harmony is perhaps the easiest and the most familiar of human perceptual talents that has little or no analog in the animal world.

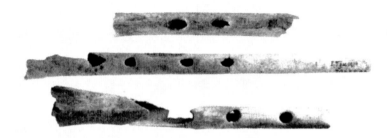

FIGURE 2.1. Flutes made from bird bones dating from more than 20,000 years ago (Photograph by F. d'Errico, Institute of Prehistory, University of Bordeaux, used with permission). The spacing and similar sizes of the holes arranged linearly along the bone surface are suggestive of man-made artifacts.

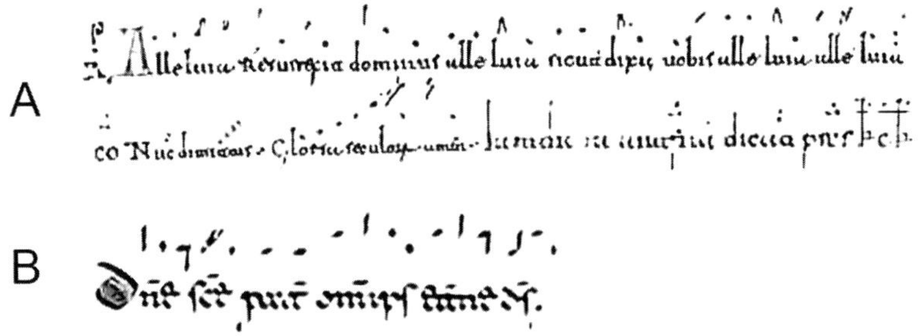

FIGURE 2.2. Two examples of early musical notation, prior to the establishment of the modern notational system by Arezzo in the eleventh century. Rises and falls in pitch over time are evident. None include pitch intervals, much less chords. (A) Musical notation in an eleventh-century Mozarabic antiphoner from Spain (© The British Library Board, add ms 30850, f.105v). (B) Musical notation from a twelfth-century missal (© The British Library Board, Egerton 3511, f.173v, used with permission).

A discussion of "real music" – from folk songs to symphonies – would be a massive undertaking well beyond the scope of this book; however, the basic phenomena of harmony in music can be explained rigorously by consideration of isolated pitches, pitch intervals, pitch triads and simple harmonic cadences, in turn. As will become clear, each level of auditory perception has its own fascination, but it is the step from two-tone intervals to three-tone harmonies that constitutes the huge leap into "musicality." That is, we first experience the musical phenomenon of "modality" (or "mode") – the distinct sonority of major and minor chords and melodies – at the level of pitch triads. Whereas many animal species have "sharper hearing" than human beings – and can perceive fainter sounds or higher frequencies or smaller changes in auditory pitch – no animal that has been tested experimentally has demonstrated an ability to perceive the major or minor modality of three-tone harmonies. That is surprising because, for us, hearing major and minor (or the absence of major or minor modality) is fundamental to the appreciation of virtually all forms of music. So, let us examine just what it is that we hear when we hear harmony.

2.1. TONALITY AND DISSONANCE

Musically, isolated tones are not normally very interesting, but the entire scientific story of music begins with their wave structure. As simple as any isolated tone may seem to be perceptually, it is a physical phenomenon of some complexity due to the presence of so-called upper partials (overtones or higher harmonics). It is this *one* acoustical fact concerning upper partials that is unfamiliar to many music listeners, but an understanding of the complex upper partial structure of tones adds a dimension that ultimately leads to a deeper appreciation of the phenomena of harmony.

The crucial point is simply that what we hear as a "single tone" actually consists of several related components (partials or harmonics). The basic pitch of an isolated tone can be described in terms of its fundamental frequency (F_0, defined in terms of cycles per second, or Hertz). The F_0 itself is a single frequency that can be illustrated as a sine wave (Figure 2.3). Associated with the F_0, however, are several upper partials – F_1, F_2, F_3 and so on – that are multiples of the fundamental frequency. So, if the F_0 is middle-C (261 Hz), then F_1 is 522 Hz, F_2 is 783 Hz and so on, with the upper partials usually having gradually weaker amplitudes. (Eventually, the upper partials become so weak that they can be ignored, but most discussions of musical tones include the effects of at least the first five or six partials.) Any musical sound in the real world will necessarily be composed of a set of these so-called partials. As a consequence, what enters our ears and what we hear as a single pitch is in fact a rather complex set of vibrations consisting of multiple frequencies, as illustrated in Figure 2.3.

The entire "upper partial story" would be extremely simple if all of the partials were separated by octaves. Given an F_0 of some frequency, F_1 at twice the frequency would be one octave higher, F_2 would be two octaves higher, and there would be no complications. In fact, however, pitch is scaled logarithmically, so that upper partials sometimes fall on pitches identical to the F_0 (e.g., C and C′) and sometimes not (e.g., C and G′). Unfortunately, this fact makes the acoustical description of tones a little bit

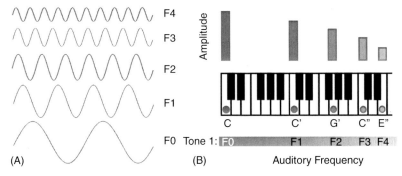

FIGURE 2.3. (A) Any oscillating tone has a fundamental frequency (F0), and a series of upper partials (F1, F2, F3, F4, etc.). (B) A tone at middle-C (F0 = 261 Hz) has upper partials at C′ (F1 = 522 Hz), G′ (F2 = 783 Hz), C″ (F3 = 1044 Hz), E″ (F4 = 1305 Hz) and so on, with ever weaker amplitudes. What we hear as "the pitch" is normally the F0.

complex, but it also makes the phenomena of pitch combinations musically interesting.

Given that any isolated tone played on a musical instrument is actually a set of partials, what can be said about the intrinsic psychological "meaning" of a sustained tone? As a physical phenomenon, the energy of a wave is proportional to its frequency. Higher frequency tones therefore have higher energy – and this is a fact given by the physical universe. Similarly, any tone of fixed frequency has greater or lesser power depending on its amplitude. Loud noises mean more to biological beings than soft noises because they indicate that the source of the vibrations is likely to be closer than low amplitude sounds. Together, the auditory frequencies and amplitudes of the partials contained within any tone contribute to its overall "tonality" ("tonalness" or "tonal coherency"). Unlike the beeps and hums of electronic equipment, isolated tones can be rich, luxurious and complex – largely as a consequence of their upper partial structure (Figure 2.4), but, without a larger musical context, it is difficult to maintain that an isolated tone has intrinsic "meaning."

Whatever significance isolated tones may have in a purely physical sense, when they are combined with other tones, their psychological meaning will change. The simplest way to state this truism is to note that tones are "context-sensitive." A reverberating C# with the upper partials of an oboe may be a beautiful tone on its own, but if it is played within the context of a musical piece in the key of C, the beautiful resonance of C# will strike us as an ugly violation of the coherency of the music. C# just doesn't belong there! To allow and to justify that intrusion, and to repair the damage done

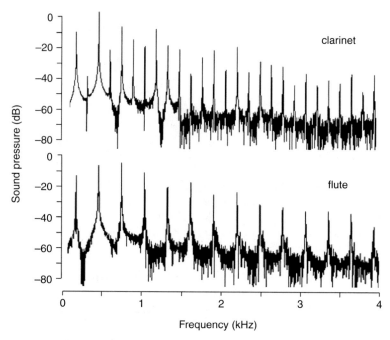

FIGURE 2.4. The upper partial structure of the clarinet and flute, when playing the same pitch at 220 Hz. Note the gradually descending strength of the upper partials in both instruments, as well as a gradual increase in "noise" between the partial peaks (P. Dickens et al., 2007, image used with permission).

to the coherency of the music in the key of C, the composer will need to labor and the listener will need to be patient. Given sufficient compositional skill and a willing listener, the music may still work as music, but the general principle is that isolated tones do not have fixed meanings; they differ with differing musical contexts. And it is for this reason that we must gradually build an understanding of the factors that contribute to a musical context.

As was the case with isolated tones, a two-tone interval is normally described in terms of its two fundamental frequencies, Fo_1 and Fo_2, but in fact both of the tones have upper partials associated with them. (The number and amplitude of the upper partials will differ depending on the musical instrument, but pure, isolated sine wave Fos are rare.) This means that, although we may say that the two tones together produce an "interval of a fourth" – meaning two fundamental frequencies that are five semitones apart – there are normally many more, weaker tones entering our ears and affecting our perceptions. In Figure 2.5, we see that, although only two tones

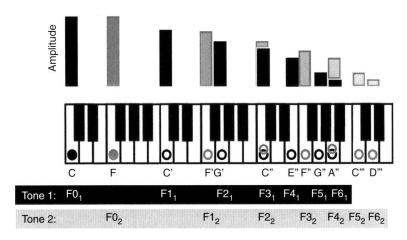

FIGURE 2.5. Illustration of the locations of the upper partials of an interval of a fourth (C-G). Only the tones indicated by the two solid circles are played on the keyboard, but many upper partials (identified as open circles) are also produced and are detected at the cochlear membrane in the ear. Note that the scalar pitch of the first upper partial (F1) is the same as the F0 (F0_1 = F1_1 = C and F0_2 = F1_2 = F), but further upper partials often land on different pitches (F0_1 = C, but F2_1 = G″, and F0_2 = F, but F4_2 = A″).

have been struck on the keyboard, many upper partials are also part of the auditory stimulation.

As a consequence of the complexity of the upper partial structure of even a two-tone interval, what enters the ear and causes vibration of the cochlear membrane is not simply the F0_1 and F0_2 tones themselves but rather a smorgasbord of partials. Depending on the configuration of that smorgasbord along the dimension of pitch, we hear some degree of overall consonance or dissonance that makes even simple intervals musically more interesting than isolated tones.

So, although intervals are normally described simply as two-tone combinations, already at this level there is a new perceptual phenomenon that is more than the sum of the two isolated tones. Particularly when any two tones are played simultaneously, we inevitably hear the consonant/dissonant relationship between them. In most cases, we can still discern that in fact two distinct tones are being played, but the consonance or dissonance of the two tones now strikes us as musically more important than the tonality of either of the isolated tones alone. There is a higher-level interaction between the two tones – and the interaction itself dominates our perception.

The perceptual phenomena of consonance and dissonance have been investigated by several generations of psychologists. Experimentally, what is consistently found is that normal listeners hear a mildly "unpleasant," "unsettled," "grating" dissonance whenever two tones are one or two semitones apart (Figure 2.6). Two tones separated by 10 or 11 semitones are also notably dissonant, despite the fact that they do not lie close to one another. Moreover, relatively mild dissonance is also reported for an interval of six semitones (the tritone).

Given this general pattern of dissonance perception, the psychologist's task is to explain such effects. From experimental and theoretical work in the 1960s, it was found that, using a theoretical curve with the approximate shape of that shown in Figure 2.7A, the "total dissonance" for any combination of tones could be predicted by adding up the effects of the dissonance among all pairs of partials. The truly remarkable result of such modeling is that, simply by including the first few upper partials for each of the two fundamental frequencies, the theoretical dissonance curve (Figure 2.7B) immediately comes to resemble the experimental curve shown in Figure 2.6.

The successful modeling of interval perception was a triumph of reductionist science that is still heralded in the textbooks today. Using an extremely simple model that describes the perceptual fact that small intervals of one or two semitones are dissonant (Figure 2.7A), the total dissonance of an interval of any size can be calculated simply by adding the effects of upper partials. The resultant theoretical curve (Figure 2.7B) indicates that relatively strong dissonance will be heard at 6, 10, 11 and 13 semitones, as experimentally known (Figure 2.6).

The effects of the summation of dissonance when 1~6 partials are included in the theoretical calculations are shown in Figure 2.7B for all intervals between 0 and 13 semitones. Clearly, the curves gradually become more complex as more upper partials are included in the calculations. When only F0 is considered, the basic model predicts (or, more precisely, simply restates the experimental fact) that small intervals (1~2 semitones) are perceived as dissonant, but the inclusion of upper partials produces small peaks of dissonance in the three regions of dissonance reported by nearly all normal listeners.

In fact, the theoretical curves show several interesting properties. Most importantly, regardless of how many upper partials are included, minima of *decreased* dissonance are consistently found at 7 and 12 semitones. These are the two intervals – a fifth and an octave – that are used in virtually every musical tradition worldwide and are heard by all listeners as consonant, "pleasant" and "beautiful" intervals. In other words, the very simple model

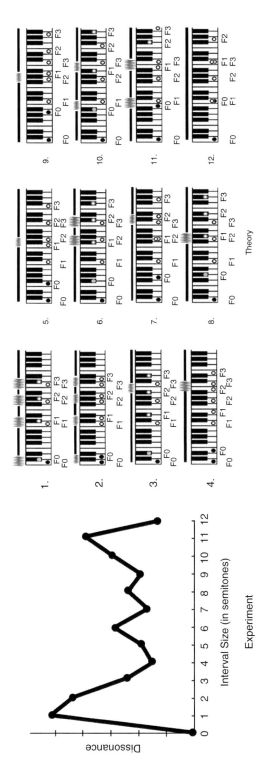

FIGURE 2.6. Experimental and theoretical aspects of interval perception. On the left are shown typical results on the perception of two-tone intervals by normal listeners. The dissonance at 1–2 semitones and that at 10–11 semitones is striking, whereas that at 6 semitones is weaker but consistently found. On the right are shown the first 12 intervals with their upper partials (F0~F3) and the location (indicated as vertical zig-zags) of "dissonances" among the partials. Note that there are various "sharp dissonances" due to semitone combinations (the tall zig-zags), and "mild dissonances" due to whole-tone combinations (the short zig-zags). The greater the number of dissonant pairs, the less consonant is the interval.

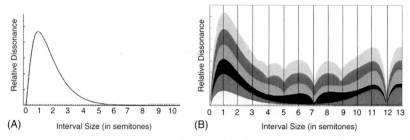

FIGURE 2.7. (A) The "dissonance model" obtained by averaging the evaluations of the dissonance of various small intervals (<6 semitones), as reported by a large number of normal listeners (Plomp & Levelt, 1965). The model summarizes the fact that small intervals of 0.5–2.5 semitones are heard as dissonant, whereas very small intervals (<0.5 semitones) and intervals of ~3–5 semitones are heard as more or less consonant. (B) The dissonance curves obtained from calculations using the dissonance curve in (A), and including 1~6 partials. The white curve in the lower left-hand corner is the theoretical model [the same curve as in (A)], whereas the other curves are the theoretical results when the consonance/dissonance among upper partials (F1–F5) are also included. The amplitude of the partials is assumed to decrease as $1/n$.

of *dissonance* perception (Figure 2.7A) predicts that intervals of 7 and 12 semitones are the most *consonant*. Somewhat less clearly (and somewhat dependent on the exact shape of the dissonance curve), the model also predicts small decreases in dissonance at (or near to) many of the intervals of the most familiar five-tone pentatonic and seven-tone diatonic scales (at intervals of 3, 4, 5, 7, 9 and 12 semitones) (Figure 2.7B).

Clearly, the number of the upper partials included in the calculation has a significant influence on the shape of the curves, and it is also found that the relative strength (amplitude) of the upper partials has a small influence (Figures 2.7 and 2.8). Unless quite unusual upper partial structures are assumed, however, the dissonance curves in these figures are all rather similar – with the main peaks and troughs arising at similar locations.

We can therefore conclude that, simply as a consequence of the dissonance of small intervals (Figure 2.6) and the reality of upper partials (Figures 2.3–2.5), various tone combinations will have varying degrees of consonance and dissonance. That alone is hardly surprising, but what is surprising is that the very simple model curve in Figure 2.7A implies the rather complex structure of the upper curve in Figure 2.7B – which shows a remarkable correspondence with the locations of the (consonant) intervals in the common musical scales. In other words, the human auditory system hears a complex pattern of consonance and dissonance as a direct consequence of its sensitivity to the nearness of the two tones in small intervals.

The match between the minima of dissonance and the tones of musical scales means that the spacing of the tones in musical scales is *not* an arbitrary invention – but reflects an acoustical reality that the human auditory system can perceive: some tone combinations have less dissonance, and music that is constructed with those intervals is consequently more pleasing to the human ear. The creation of "pleasant music" turns out to be a good deal more complex than simply avoiding dissonance – and indeed there are a variety of different scales that use different intervals – but it is clear that the amount of consonance/dissonance employed to make music will be an important factor in determining its popularity among typical listeners.

Much theoretical and experimental work on interval perception has been reported since the 1960s, and there are ongoing skirmishes among academics concerning the exact shape of the dissonance curve and the optimal tuning of diatonic scales (Duffin, 2007; Frosch, 2002; Isacoff, 2001). The main lesson concerning upper partials, however, was already known to Helmholtz in the nineteenth century: The musicality of scales is due largely to the relative consonance of their intervals, and the relative consonance of the intervals is due to our (generally, unwitting) perception of the upper partial structure of tones.

The successful prediction of the locations of the most *consonant* intervals (Figures 2.7 and 2.8) on the basis of a very simple *dissonance* model is widely interpreted as indicating that the perception of musical consonance/dissonance was probably the main factor that led to the development of the most common scales used in various musical cultures around the world. Not only are intervals of an octave and a fifth used in nearly all musical traditions, but other musical scales (e.g., the pentatonic scale) can be viewed as subsets of the seven-tone diatonic scale – with certain scales steps omitted.

Calculations of the relative consonance of the intervals of the diatonic scales were done by Plomp and Levelt in 1965 and Kameoka and Kuriyagawa in 1969. Those researchers also pointed out the close match with the peaks of consonance produced by interval sizes calculated as the ratios of small integers, attributed to Pythagoras (Figure 2.8). The match is in fact not precise, but, given the typical human tolerance for small deviations from "perfection," the dissonance models and the Pythagorean calculations can be viewed as rather good fits – and suggest that, regardless of whether musicians play to strike the "perfection" of integer ratios or play to avoid the broader humps of relative dissonance, the acoustics of vibrating strings provides the ear with five relatively "pleasant" intervals (in addition to unison and the octave).

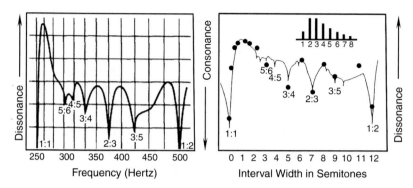

FIGURE 2.8. The peaks of maximal dissonance obtained using the Plomp and Levelt model (1965) (left) and the Kameoka and Kuriyagawa model (1969) (right) (images used with permission). The small circles are experimental data points – clearly indicating a strong confirmation of the theoretical model.

Insofar as the minima of consonance have an acoustical basis, it is no surprise that, when given a choice, even mice and Rhesus monkeys will select environments in which consonant rather than dissonant intervals are sounded. Very little of the high-level art of music can in fact be explained at this rather crude level, but, at the same time, the low-level physiology of dissonance perception cannot be ignored. We and many of our mammalian cousins have a slight preference for consonance over dissonance – and this preference is due principally to our sensitivity to the higher harmonics in sustained tones.

It should be noted, however, that the locations of maximal consonance in Figures 2.7 and 2.8 do not correspond to any known musical scale. To obtain a scale that is actually used, tone intervals need to be added or subtracted – and that is precisely what inventive musicians have done throughout history. The fact that physical acoustics does *not* uniquely produce real scales, much less the Western diatonic scales or the full 12-tones of the keyboard octave, indicates that some form of human manipulation must be included to establish a musical tradition. In this regard, Leonard Bernstein's famous declaration that the "physical universe" has provided us with diatonic music is clearly stating the case too strongly. The physical universe provides some nonarbitrary raw materials, but those raw materials can in fact be assembled in various ways.

So, what is the inherent "meaning" of an interval? Similar to the phenomena of isolated tones, the auditory frequencies of two-tone intervals determine the energy level of the interval and their amplitudes determine its total power. Strictly as a physical phenomenon, a musical interval has greater meaning for us if it is played at a higher rather than a lower volume within the range of auditory frequencies that we can hear. There is, however, a further psychological property of intervals that is a direct consequence of the interaction of the two tones – and is not deducible directly from the sum of the qualities of the isolated tones. That is the perceptual quality of consonance/dissonance. As shown previously, consonance and dissonance are a complex function of the spacing between the fundamental frequencies due to the presence of upper partials.

If tones did not have upper partials (or the upper partials arose solely at octave intervals), the dissonance of intervals would be a simple function of the distance between the fundamental frequencies. As a consequence of the gradually decreasing distance (number of semitones) between upper partials, however, the relationship between interval size and total dissonance is not that simple. On the contrary, with increases in interval size there are "irregular" ups and downs in the dissonance curve – indicating a complex relationship between interval size and our perception of the overall pleasantness of the interval.

Although the quality of any interval can be described solely in terms of its overall consonance, there are in fact many different consonant intervals that have noticeably different implications within any given piece of music. So, do the different intervals have different inherent psychological "meanings" that are more specific than their relative overall "pleasantness?" This is an old question – answered variously by the ancient Greeks, Renaissance musicians and modern-day psychologists, but with little agreement over the centuries. Even such hard-headed scientists as Helmholtz (1877/1954) offered his psychological impressions of the meaning of intervals – suggesting that, for example, the interval of a second is "rousing or hopeful," a major third is "steady or calm," a fourth is "desolate or awe-inspiring," a fifth is "grand," while that of a seventh is "piercing or sensitive" and so on. More recently, David Huron (2006) has canvassed professional musicians and found that intervals of a fourth are sometimes described as "awkward," a fifth "towering" and a major sixth as "airy and open."

It is of genuine interest that there is some consensus among musicians on the usage of such adjectives when talking about music, but the problem with all descriptions of this kind is that the perceptual qualities of any interval are highly context-dependent. Sit down at a piano and invent a little tune of 10 or 12 notes in the key of C-minor – and then play an interval of a major third (C-E). It sounds out of place, grating and unexpected – anything but "steady or calm!" The music theorist will of course respond that, "The tune was in a *minor* key, so that the *major* third must be avoided" – but that is precisely the point. If the context is so all-important, what can be said that is objectively true and unambiguous about the "intrinsic meaning" of any isolated interval? The answer, apparently, is very little.

In this respect, all of the psychological descriptions of intervals in the music perception literature must be considered suspect. Except for the subjective evaluation of the relative consonance or dissonance of intervals when played in isolation, completely outside of any musical context, it is doubtful that intervals have fixed "meanings." The evaluation of pitch tonality and judgments of interval consonance are already topics of human *perception* – not physical acoustics – and our perceptions are dependent on the musical context within which the tone or interval is played. For this reason, we need to go one step further in order to fix those perceptions within a minimal tonal context.

2.2. TENSION AND INSTABILITY

The basic terminology used to describe isolated pitches, pitch intervals and their perception was introduced earlier. The only concept not already familiar to any high school student is that of the so-called upper partials. In brief, any tone sounded in the real world will have not only a dominant pitch – normally corresponding to its fundamental frequency – but also various upper partials associated with it. The number of upper partials and their relative amplitudes can vary and will have influence on our perception of the pitch – ultimately telling us whether the sound came from a clarinet, a flute or a piano, from the vocal tract of Winwood or Sting, or is a computer-generated sine wave. Similarly, when we hear two pitches sounded simultaneously, the perceived consonance or dissonance of the musical interval will be determined not only by the difference between the two fundamental frequencies, but also by the number, strength and spacing of their upper partials.

It should therefore not be surprising that the perception of chords – whether three-tone triads, four-tone tetrads or more complex harmonic cadences – is also influenced by upper partials. Their effects will be considered in detail in this chapter, but the first topic to discuss is the spacing among the fundamental frequencies themselves. Similar to the phenomena of musical intervals, when we play a three-tone chord, the frequencies with the greatest amplitude are usually those of the three distinct notes that we choose with our fingers and that are noted in the musical score. The higher harmonics tag along for free and give the chord a "richness" that we might call its overall "sonority."

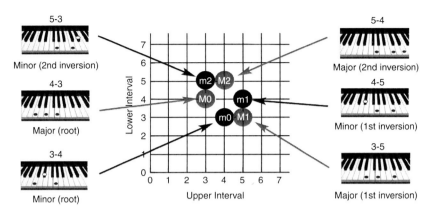

FIGURE 2.9. A small triadic grid. The locations of the most familiar major (M0, M1, M2) and minor (m0, m1, m2) triads in their three inversions are shown. They are all located at the intersection of lines indicating intervals of 3, 4 or 5 semitones. Examples of these chords in the key of C and their interval structures are shown on the left and right.

So, let us begin the discussion of harmonic sonority by considering the set of all possible combinations of three tones, as represented on the "triadic grid" (Cook & Hayashi, 2008; Cook, 2009). Figure 2.9 shows a small triadic grid on which major and minor chords are defined by the size of their two intervals. The vertical axis of the grid represents the lower interval and the horizontal axis represents the upper interval. For example, the major chord in root position has a lower interval of four semitones (e.g., C to E) and an upper interval of three semitones (E to G) (grid position 4–3). Of course, the C-major chord can also be played in inverted positions with intervals of 3–5 and 5–4 semitones, and still maintain its major sonority. The root position of the minor chord is found at grid position 3–4, and its first and second inversions at 4–5 and 5–3.

Larger grids of the same kind are shown in Figure 2.10, where it can be seen that the major and minor chords can be played in various configurations with intervals of 3 to 21 semitones. Of course, the starting note of such triads (middle-C or elsewhere within the range of human auditory sensitivity) is unimportant here: The relevant factor for understanding the basic patterns of triadic harmonies is solely the number of semitone steps between the tones. Note that, although these same chord structures could also be depicted on a piano keyboard or guitar (Figure 2.11), the peculiarities of the construction of musical instruments mean that the interval structure of triads is more easily understood when shown on the triadic grids.

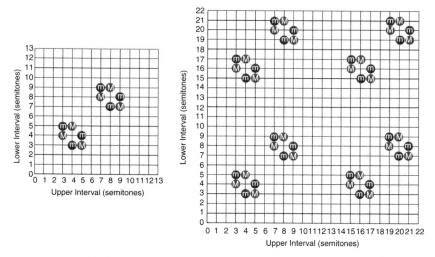

FIGURE 2.10. Medium-sized and large triadic grids. A variety of major (M) and minor (m) chords can be played over two or three octaves and still retain their characteristic major and minor qualities. Note that the relative positions of the major and minor triads remain the same, regardless of location on the grid.

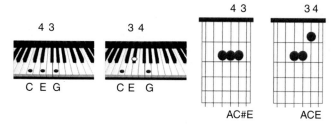

FIGURE 2.11. Examples of major and minor chords on the piano and guitar. Interval structure (in semitones) is indicated by the numbers.

Although still recognizably major or minor, the sonority of these triads in their different inversions and played over different ranges differs somewhat and their musical usages also differ, so we must take note of their detailed interval substructure. Traditional harmony theory does precisely that using a specialized vocabulary that relates tones and intervals to the musical scales. The scalar pitches themselves are referred to in terms of their harmonic character – tonic, dominant, subdominant and so on, which are, in turn, defined in terms of the given musical key. It is, to be sure, a complex and fascinating system, but in the present account such notions from traditional music theory will be avoided (insofar as possible), and all harmonic structures will be described in terms of their semitone substructure.

In principle, any of the major and minor triads can be played over the span of several octaves and retain their major or minor sonority. For the present purposes, however, the basics of harmony will be illustrated using only those triads that are played over one or two octaves (as illustrated in medium-sized grids with the axes running from 0 to 13 semitones, Figure 2.12). Any recognizable triad – and a good many unrecognizable ones as well – can be specified on the grid as the point of intersection of the vertical and horizontal lines that indicate the semitone steps. The complete set of major triads are then seen to be the chords with intervals of 4–3, 3–5, 5–4, 8–7, 7–9 and 9–8 semitones, while the minor chords have intervals of 3–4, 4–5, 5–3, 7–8, 8–9 and 9–7 semitones.

In addition to the major and minor chords, there are of course many other triads with established names in harmony theory. Labels for the most common triads are shown in Figure 2.12, and indicate that there are two clusters where all of the common chords lie. Unlike more advanced discussions concerning musical usage, there is absolutely no controversy concerning the *structure* of the chords shown in Figures 2.9 through 2.12, their locations on the triadic grid and their common labels.

The major and minor triads are the chords that provide the harmonic framework for most "heavy" classical and nearly all "light" popular music; they are used again and again either as triads or as triads with repetition of the triadic pitches an octave higher or lower. The other locations on the triadic grid include chords of varying utility and beauty – and quite a few chords that are simply avoided in most types of music. Whatever our subjective evaluation of the sonority of any of these triads, however, they have certain structural properties that can be described in an objective way.

Note that the triadic grids are illustrated with precisely equal semitone steps (the 12-tone equitempered tuning system). Other tuning systems would imply small shifts (<5 percent) of the grid lines, but, as will become apparent later, the arguments concerning harmony developed here are more fundamental than the subtleties of the different tuning systems. The questions that need to be answered about human hearing are basic: Why do we hear musically resolved, major and minor chords as being harmonies of unusual clarity, purity and beauty – and not simply polyphonic cacophony? And why are the augmented, diminished and suspended chords all inherently "unresolved" – interesting and beautiful in their own ways, but lacking the composure and finality of the major and minor chords? The specialist issues of tuning will be left for those in search of optimal purity

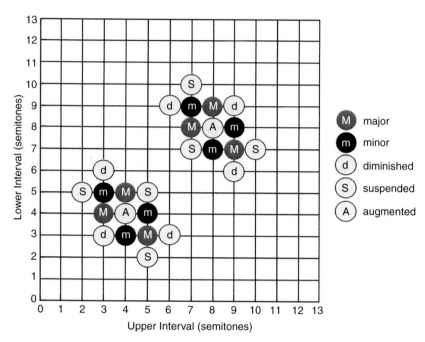

FIGURE 2.12. The triadic grid with the labels of the 13 most common triads when played within one octave (intervals of 2–6 semitones) or within two octaves (intervals of 6–10 semitones). Note the following points about terminology. The so-called augmented chord is essentially the major chord in root position (intervals of 4 and 3 semitones) with the interval of a fifth (7 semitones) "augmented" by a semitone; this produces a chord with two intervals of a major third (4 semitones each). Inversions of the augmented chord lead to chords with identical structures (always two intervals of 4 semitones each). The diminished chord is essentially a minor chord in root position (intervals of 3 and 4 semitones) with the fifth "diminished" by a semitone; this produces a chord with two intervals of a minor third (3 semitones each). Inversions of the diminished chord lead to chords with 3 and 6 semitones and 6 and 3 semitones, both of which are also referred to as diminished chords. The suspended fourth chord is, by definition, a major chord with the major third omitted, and the interval of a fourth included (a chord with intervals of 5 and 2 semitones). Inversions of the suspended fourth chord lead to chords with intervals of 2 and 5 semitones and 5 and 5 semitones). In the music literature, there is some confusion on the naming of this chord – and which chord is the root and which are inversions, but these three chords will be referred to as suspended chords here.

when playing isolated tones or pitch intervals on particular instruments and will not be addressed here. Instead, for the present discussion of harmony, the equitempered grid will be used, with the understanding that the triads with the greatest sonority would sometimes be located just off of the grid lines if different tunings were used.

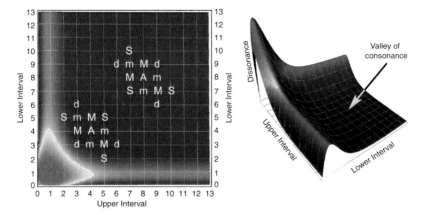

FIGURE 2.13. A triadic grid showing the total dissonance inherent to any triad containing intervals of 0 to 13 semitones. Only the dissonance effects among the fundamental frequencies (Fos) are included in the calculations. The darker regions have low dissonance, whereas the lighter regions indicate higher dissonance. When either interval is less than 2 semitones in magnitude, dissonance is present, but this falls off rapidly, so that all of the common triads – major (M), minor (m), diminished (d), augmented (A), and suspended (S) chords – are located in the dark "valley of consonance."

So, given the interval structure of the triads as illustrated on the triadic grid, what can be said about their relative "sonority" or "harmoniousness"? How can the stability of the triads – resolved or unresolved – be explained and how can the commonly perceived positive and negative emotional tone of the major and minor chords be accounted for? These are the important questions for anyone interested in the psychology of harmony.

Let us begin by examining the topic of the dissonance contained in various chords. Since triads can be viewed as three intervals among the three tones, the obvious first step in trying to explain their sonority is to add up the dissonance of the intervals to obtain the total dissonance. Will that explain their overall harmoniousness? Of course, the inclusion of the consonance/dissonance among all pairs of upper partials will make the calculation of "total dissonance" a bit more complex than examining only the relationships among the fundamental frequencies, but the basic question is essentially the same: Does the summation of interval effects explain polyphonic harmony?

Figure 2.13 illustrates the summed dissonance of the intervals when only the fundamental frequencies of the three tones are considered. We see that there are two strips of relatively strong dissonance when either interval is one or two semitones in size. As shown on the right-hand side of

Figure 2.13, the relative dissonance of the triads can be seen more clearly by viewing the map obliquely. We then see an extremely steep peak of dissonance when both intervals are one semitone in size and two high ridges when either interval is one or two semitones. The remainder of the triadic grid is found to be a region of dark gray, low dissonance – where most of the common triads lie.

Note that the calculation of the total dissonance of the three-tone chords is made here for *every* combination of tones between 0.0 and 13.0 semitones at intervals of one-tenth of a semitone ("10 cents" in the lingo of musical acoustics). So doing, the dissonance maps have a fairly smooth topology – with peaks of dissonance, valleys of consonance and gentle transitions between the heights and depths.

What does the topography of the dissonance map tell us? Most clearly, it says that an explanation of chord perception in general *cannot* rely solely on the dissonance among the fundamental frequencies. Such a view would imply that *all* of the common triads – lying in the deep valley of consonance (at grid intersections indicating intervals of 3.0 or more semitones) – are more or less equally consonant. Perceptually, that is simply not true: The major and minor triads are normally heard as relatively stable, final and resolved, but other triads (that do not have small intervals of one or two semitones) are typically heard as tense and unresolved (e.g., the augmented chord at position 4–4 on the triadic grid, the suspended chord at 5–5, and the diminished chords at 3–6 and 6–3). These are not necessarily "unmusical" chords, but they are not often perceived as being as "stable," "sonorous" or "beautiful" as the major and minor chords. In other words, making pleasant harmonies is *not* simply a matter of avoiding the dissonance of small intervals; other factors are also involved.

Since we already know (Section 2.1) that the upper partials have a strong influence on the perception of intervals themselves, the next step in exploring triad perception is to bring the upper partials into the picture. Figures 2.14–2.16 are precisely that: dissonance maps that include gradually more and more of the upper partials in the calculations.

The basic effects of adding upper partials on the topology of the dissonance maps are already evident in Figure 2.14 with the addition of F_1. By adding further upper partials, the fine structure of the maps gradually gets more complicated, but the general pattern remains more or less intact. That is, there are regions of strong dissonance (when either of the intervals is small) and expanses of relatively strong consonance (where all of the common triads lie). The upshot is that while we have good reason to think that small intervals make chords less stable because of dissonance effects, the total dissonance of the triads – regardless of how many upper partials

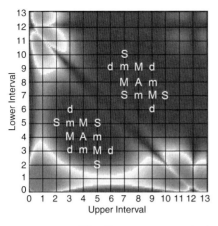
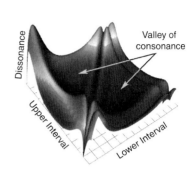

FIGURE 2.14. The dissonance map obtained when both the fundamental frequencies (F0) and the first set of upper partials (F1) are included in the calculations. Note that the two ridges of dissonance along the x- and y-axes have somewhat more complex structure than when upper partials are not considered, and the "valley of consonance" in Figure 2.13 is now divided into two regions where all of the common triads are located.

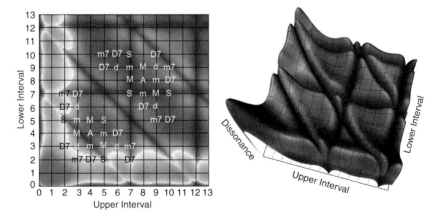

FIGURE 2.15. The dissonance map when the effects of F0~F2 are included in the calculations. The two valleys of consonance now show more complex structure but remain as regions of relatively low dissonance. On the left are shown the locations of the most common triads, including the dominant seventh (D7) and minor seventh (m7) chords when played as triads.

are included in the theoretical calculations – does not give clear indication why we so easily distinguish between the major and minor chords or between the resolved (major and minor) chords and the unresolved (diminished, augmented and suspended) chords. Judging solely from the

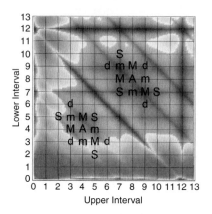 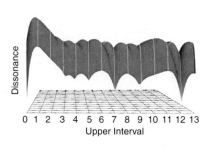

FIGURE 2.16. The dissonance map when F0~F4 are considered. Because the total dissonance increases with the addition of further upper partials, the shading suggests stronger dissonance over the entire triadic grid, but the common triads still lie in valleys of relatively strong consonance. Note that, when the dissonance map is viewed from the side, the total dissonance of two-tone intervals (with the lower interval set to zero) is identical to the dissonance of the dissonance curve in Figure 2.7. Again, small decreases in dissonance are found at most of the intervals of the diatonic scales.

"total dissonance" among the partials, *all of the common triads* are rather consonant, but, perceptually, that just isn't so!

It should be noted that some music theorists have arrived precisely at this point, and have proceeded to adjust the tuning of the intervals (Pythagorean tuning, just tuning, mean-tone tuning, etc.) and the relative amplitudes of the upper partials in the hope of finding an ideal combination of factors that would show why only the major and minor chords are stable, resolved and "beautiful" and the others are unstable, unresolved and inherently "less beautiful." In fact, no such manipulation gives the empirically known sequence of sonority. The well-known dissonance models in the music perception literature inevitably predict (Table 2.1) that either the augmented or the suspended chords contain less sensory dissonance than even the major or minor chords and should therefore be perceived as more consonant. That is perceptually *not* the case. Although rarely commented on in the music perception literature, this failure of the interval dissonance models is clear indication that, whatever merits those models may have in explaining the structure of scales and the perception of intervals themselves, they cannot explain harmony. As Philip Ball (2010, p. 167) has commented, "The fact that the harmoniousness of a chord is not simply the sum of the consonance or dissonance of the intervals it contains is widely recognized

TABLE 2.1. *Empirical findings and theoretical predictions concerning the harmonic sonority of the common triads*

	Empirical sonority			Theoretical sonority			
	Incidence in classical music	Evaluation in laboratory experiments		Predicted by various interval models			
	Eberlein (1994)	Roberts (1986)	Cook et al. (2007)	P&L (1965)	K&K (1969)	Parncutt (1989)	Sethares (1999)
Major	1 (51%)	1	1	2	2	2	2
Minor	2 (37%)	2	2	2	2	3	2
Diminishled	3 (9%)	3	4	5	4	4	4
Suspended 4th	4 (2%)	–	3	1	1	–	1
Augmented	5(<1%)	4	5	4	5	1	5

Empirical data and theoretical calculations are available for all of the inversions of these chords (Cook & Fujisawa, 2006). P&L denotes the model of Plomp and Levelt (1965), and K&K denotes the model of Kameoka and Kuriyagawa (1969). Note that Roberts had experimental subjects evaluate the stability of isolated chords (played one at a time in random order), but she did not include the suspended chords in the experiment. Similarly, Parncutt (1989) did not calculate their relative sonority. The experimental data from Cook et al. (2007) include the suspended chords.

but poorly understood." Precisely, but let us not shirk from discussion of this crucial issue concerning music perception.

To be fair to the theorists who have worked on the problems of interval perception, it should be said that these (historically, important) theoretical models were developed explicitly to explain the relative consonance/dissonance of tone intervals themselves, rather than triadic sonority – and, as models of *interval* perception, they have proven to be useful. It is nonetheless instructive to see that, if these highly acclaimed explanations of interval perception are applied directly to the problems of harmony perception under the assumption that chords are nothing more than the summation of their interval effects, the theoretical results are quite wrong. Nobody hears the augmented chord or the suspended chords to be more stable than the major and minor chords! The sonority of particularly the suspended chords is musically interesting, and they have many uses in jazz and folk music, but they are perceived as requiring resolution and are typically followed by a major or minor chord. It must, therefore, be concluded that computational models that predict the augmented or suspended chords to be more harmonious than the major and minor chords are calculating something other than the harmonic sonority that most people perceive and highly value in music.

The predictive anomalies of these models are most apparent if we compare the actual incidence of these chords, as used in a large sample of classical music, with the theoretical predictions of the interval models (Table 2.1). We find that, when composers sit down to write music, they choose major and minor triads over the other triads at a rate of nearly 9-to-1. To be sure, the dominance of major and minor chords in the music of the early Renaissance has slowly given way to a bit more diversity and acceptance of other triads, but most popular music worldwide still utilizes the familiar sonority of major and minor chords far more than the spicier unresolved chords.

To make a long story short, it turns out that the fine-tuning of models that consider only the summation of interval consonance/dissonance to explain harmony does *not* allow for a reproduction of the perceptual facts: no "ideal" tuning system, no unique combination of upper partials and no set of weightings for the upper partials will produce strong consonance uniquely at the major and minor chords, and greater dissonance elsewhere (Cook & Fujisawa, 2006). This was already apparent from the total dissonance plots shown in Figures 2.13–2.16. In brief, the interval dissonance models have not succeeded in explaining even the rudimentary facts about musical *harmony*.

This "negative result" – the failure to explain triad perception solely on the basis of dyad dissonance – has had far-reaching consequences for discussions of the psychology of music. Because the relative sonority of the triads (known for a historical certainty and verifiable by anyone through five minutes of experimentation at a piano keyboard) cannot be explained on the basis of dissonance alone, many music psychologists have, in effect, abandoned the attempt to explain music in acoustical terms. In dismissing acoustics, however, theorists are then forced to conclude that the overwhelming use of certain harmonies is simply a learned habit – adherence to an arbitrary cultural "idiom" without acoustical foundations. In what has become the conventional wisdom concerning pitch perception, acoustics is said to play a role only up to two-tone combinations, while more complex pitch phenomena are essentially "learned." Such a view has often been expressed, as in recent editorials in *Nature*: "It may even be that acclimatization to a convention can completely override [the] acoustic facts" (Ball, 2008). "Our emotional response to particular scales or chords seems likely to be acquired from exposure to a particular culture" (McDermott, 2008). "The objective organization of sounds is only loosely related to how minds interpret those sounds" (Huron, 2008). And "scale and harmonic structures depend on learning" (Trainor, 2008).

The radical (and, as shown later, incorrect) conclusion from academia that the perception of harmony is primarily due to culture is a direct consequence of the inability of the interval models to explain harmony (Table 2.1). Culture trumps acoustics, it is said, and "harmoniousness" is *not* a matter of the perception of auditory vibrations, but rather one of memory – recalling associations between such vibrations and the human activities in societies producing those sounds. In that view, both musicians and nonmusicians are acculturated to hear, for example, minor chords as harmonious simply because they are so often used in the kinds of music that we listen to. Similarly, we hear the augmented chord as "unpleasant" – quite regardless of its acoustical consonance – because it is rarely used in popular music and virtually never played on the radio! The universal preference for major and minor chords is, according to the cultural hypothesis, simply a matter of preferring the familiar.

But acoustics fails to explain harmony only if we assume that interval effects are the only relevant factors. On the one hand, the dissonance of small intervals (Figures 2.6–2.8) coherently explains why there are diatonic scales in the first place, and why certain tunings of the diatonic intervals sound particularly consonant, and others less so. Sensory dissonance also explains why most chords containing small intervals are not harmonious, and why nearly all popular music relies heavily on the low-dissonance "common triads" to produce pleasant musical effects. But, on the other hand, the interval models do not explain the relative sonority of even the common triads, where "sensory dissonance" is essentially absent. In that view, acoustics holds firm up to and including two-tone intervals, but then falls apart. Although rarely stated so baldly, this view of pitch perception is popular for one and only one reason: A "cultural" explanation of harmony perception is thought necessary simply because the dissonance models (Table 2.1) are clearly wrong! If "bottom-up" acoustics can't explain harmony, what else is left except culture?

The results shown on the dissonance maps (Figures 2.14–2.16) clearly indicated that dissonance does indeed play a role in determining which of the many possible triads sound harmonious. The major and minor chords are composed of intervals of at least three semitones, and, as a consequence, they are all rather consonant. At the same time, however, it must be noted that the unsettled, unresolved nature of other chords consisting of similarly consonant intervals cannot be explained solely in terms of "sensory dissonance." This is the essence of the old problem of the interval dissonance models (Table 2.1): they are useful but do not explain much about music!

Theoretically, it is perhaps conceivable that culture simply overpowers lower-level biological and psychological effects. But, in the realm of harmony perception, the notion that "cognition trumps perception" is hard to sustain. The antidote to the idea that our sense of harmonic sonority is culturally "learned" can be experienced by playing, for example, an augmented chord (C-E-G#) followed by a major (C-E-G) or minor (C-E-A) chord ... and listening. Try it! And try it in reverse – moving from a major or minor chord to the augmented chord. There is unmistakably something intrinsic to the augmented chord – something in its acoustical structure – that nearly all normal listeners hear and react to. Most people – musicians and nonmusicians – will say something like: "I hear some 'dissonance' that fades away with transition to a major or minor chord," but, as shown later, the unsettled nature of the augmented chord turns out to be something slightly different from "dissonance."

However one subjectively feels about the beauty or otherwise of the augmented chord, it is a mistake to follow the cultural theorists here! To pretend that what we perceive is a "learned bias" and not something ringing in our ears is to deny the sensations of our own auditory apparatus – truly a senseless argument. If the perceived lack of sonority of the augmented chord is a "social construct" and, having grown up in cultures that don't often employ the augmented chord we have inadvertently come to hear a reasonably consonant chord as unpleasant because it is culturally "unacceptable," how could we draw any conclusions about interval dissonance/consonance or, indeed, any other sensory phenomenon? If auditory psychophysics does not rely on the perceptual judgments of human beings, what is there left to evaluate except the sublime beauty of theory?! Clearly, this line of argumentation is a slippery slope – a theoretical world where human perceptions are no longer of interest. Let us not go there.

Instead, let us ask what other acoustic factors might be involved. Clearly, the first issue to examine after two-tone intervals is three-tone configurations. Whereas the structural question concerning *interval* perception was simply: "How close are two tones (and their upper partials) to one another?" a different structural feature of the triads might also contribute to harmonic stability. What could that be?

A possible answer is apparent if we examine the interval structure of three well-known chords that have sonorities often described as tense, ambiguous and unsettled: (i) the diminished chord in root position, (ii) the augmented chord and (iii) the so-called suspended chord in second inversion (Figure 2.17). All of the intervals in these chords are themselves consonant – 3, 4 or 5 semitones – but they somehow add up to something less sonorous,

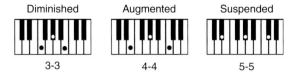

FIGURE 2.17. Tension chords, each containing two *equal* intervals (intervallic equidistance).

less resolved and less settled than their consonant intervals would suggest. If asked to evaluate these chords, most people find that they are clearly less sonorous than the major and minor chords (Table 2.1). Despite their relatively low "sensory dissonance," they evoke a feeling of unresolved tension that has been repeatedly found in laboratory experiments with people from various cultures. Perceptually, this is an unambiguous fact, rarely discussed in the harmony textbooks, but an explanation can be found in a classic book from 1956 on the psychology of music by Leonard Meyer. There, he suggested that the tension in these and similar chords (played as chords or as short melodies) is due to the *equivalence of two neighboring intervals* (Figure 2.17). Why would equivalent intervals produce this effect?

According to Meyer, tension is perceived because it is unclear how to "group" the three equally spaced tones. It is the uniformity of the steps between the tones that makes it difficult to hear a "tonal center." In contrast, when a three-tone combination has a larger interval and a smaller interval, we naturally group the two tones of the smaller interval together, and hear the more distant tone as being alone. Generally speaking, we do not think about this grouping of tones and the issue of the relative size of the intervals in a chord simply does not rise to consciousness. On the contrary, similar to the perception of visual illusions, we become aware of only the overall impression – not the "how" or the "why." In the case of three tones equally spaced along the dimension of pitch, all we hear is the inherent tension, but Meyer's argument is that the *symmetry* of the three tone configuration gives the harmony its ambiguous, unsettled character.

The deep question from Gestalt psychology is "why" such symmetry would cause us to hear tension? Easy answers in both the visual and auditory realms are not available, but the phenomena themselves are clear. In visual psychology, the Necker cube is a good example of inherent ambiguity in the visual domain. When all of the lines that form a cube are drawn with equal thickness (Figure 2.18A), we are given no information concerning the relative nearness of the different faces of the object. Typically, viewers see the cube as oriented one way or the other and then, after a short time,

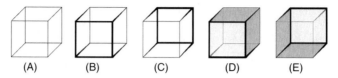

FIGURE 2.18. The depth ambiguity of the Necker cube. (A) In the symmetrical case, with the thickness of all lines the same, it is uncertain which face is nearer to the observer and which is further away. In (B) and (C), the thicker lines appear closer so that the orientation of the cube does not as readily oscillate between the two interpretations. When the transparency of the sides of the cube is also manipulated (D and E), it becomes progressively more difficult to alternate between the two equivalent views of the cube – and the balance shifts to asymmetrical.

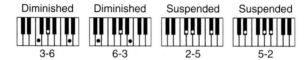

FIGURE 2.19. Tension chords with two *unequal* intervals. Do these counter examples disprove the idea of tension due to equivalent intervals? (Note that the 4–4 interval structure of the augmented chord remains unchanged when "inverted" by shifting one or two tones an octave higher.)

they see it switch to a different configuration. When, however, the nearness of the edges is indicated with lines of different widths or sides of different transparency, the ambivalent symmetry is broken and one interpretation is more strongly favored over the other (Figure 2.18B–18E).

The tension chords exhibit a similar instability when played either melodically or harmonically: Three equally spaced tones produce a sense of unresolved, up-in-the-air ambiguity, but if one of the tones is altered to give the asymmetry of unequal intervals, the ambiguity is resolved and the harmony takes on a relatively stable character.

Meyer's idea that symmetry leads to tension seems to apply nicely to the three chords shown in Figure 2.17, but what about the various inversions of those chords? As shown in Figure 2.19, inversions of diminished and suspended chords – all of which have a notably unsettled, "tense" character – have *unequal* intervals, so does Meyer's argument break down already with these simple counterexamples?

The answer is that, if we bring the upper partials into consideration, then we find that there is an abundance of equal intervals in *all* of the unresolved triads (diminished, augmented and suspended in all of their inversions), while equal intervals are *not* found in *any* of the major and minor

(A) major (B) minor (C) diminished (D) augmented (E) suspended

FIGURE 2.20. The interval substructure of the triads of Western harmony, with the size of the intervals noted as small integers. Paired intervals showing "intervallic equivalence" (gray boxes) are seen for the diminished chord in root position, the augmented chord and the suspended chord in second inversion.

(A) major (B) minor (C) diminished (D) augmented (E) suspended

FIGURE 2.21. The interval substructure of the triads with the first set of upper partials shown as half notes. Interval sizes in semitones are shown as small integers. *None* of the major and minor chords, but *all* of the diminished, augmented and suspended chords show repeating intervals of the same size (gray boxes). This intervallic equidistance is apparently the origin of the unsettled "tension" of these latter chords.

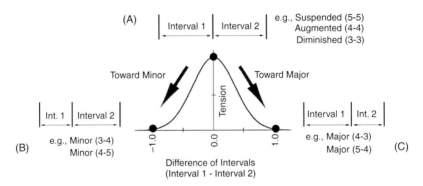

FIGURE 2.22. The tension curve and examples of resolved and unresolved chords. (A) Triads with two equal intervals, and therefore high tension. (B) and (C) Triads with unequal intervals, and therefore low tension.

triads. This acoustical fact of diatonic harmony is illustrated in Figures 2.20 and 2.21 in musical notation.

Depicting the interval structure of triads in musical notation is fine for the musician, but the symmetry/asymmetry of chords is more easily understood if we return to an acoustical description. In fact, Meyer's idea of intervallic equivalence can be expressed as a psychophysical model and used to calculate the tension inherent to any given combination of tones. Specifically, using the Gaussian curve shown in Figure 2.22, the harmonic

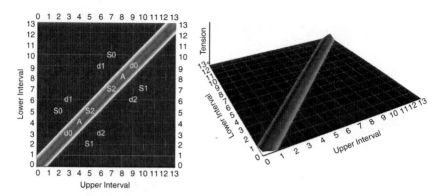

FIGURE 2.23. The theoretical tension of triads when only the fundamental frequency is considered. The diagonal strip of high tension is due to the presence of two intervals of the same size.

"tension" of any triad can be calculated (Cook, 2002). When the difference in the size of the two intervals is zero, there is maximal tension; when the perfect symmetry is broken, we slide down to some lower level of tension (i.e., toward resolution).

Since we already know that the upper partials have an influence on our perception of isolated pitches and pitch intervals, we can anticipate that the perception of three-tone chords will also be affected by the upper partials. Structurally, we know that there are equally spaced tones among the upper partials of certain chords (Figure 2.21) – suggesting that tension, ambiguity and "unresolvedness" will be heard in any chord containing such intervallic equivalence.

So, if the perceptual feature of harmonic tension is due to equal-sized intervals, it should be possible to determine the "total tension" in any chord using the model curve by calculating the "difference of intervals" among every threesome of partials in the triad. Depending upon how many of the upper partials are included, the results will differ somewhat, so a variety of "tension plots" are shown in Figures 2.23–2.25. Again, the triadic grid is used, and the different levels of tension are color-coded.

As was the case with the dissonance maps, the tension maps gradually become more complex as the effects of additional upper partials are included. When only Fo is considered (Figure 2.23), there is a diagonal strip of high tension – above and below which tension is absent. This strip indicates the unsettled character of the diminished chord in root position (intervals 3–3 and 9–9, d_o), the augmented chord (intervals 4–4 and 8–8, A) and the suspended chord in second inversion (intervals 5–5 and 7–7, S_2) but leaves the

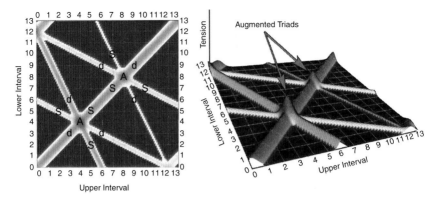

FIGURE 2.24. The theoretical tension of triads when both F0 and F1 effects are included. The same diagonal strip as in Figure 2.23 shows the regions of most pronounced tension, but there are now indications of increased tension along other oblique lines due to intervallic equivalence among the upper partials. Note that these strips of increased tension extend to all inversions of the tension chords (d, A, S).

perceptual tension of the other unresolved chords (6–3 and 9–6, d_1; 3–6 and 6–9, d_2; 5–2 and 10–7, S_0; 2–5 and 7–10, S_1) unexplained.

It is remarkable, however, that, with the addition of only the first set of upper partials, various oblique lines on the tension maps indicate higher tension at *all* interval combinations that correspond to the augmented, diminished and suspended triads in *all* of their inversions (Figures 2.24, 2.25 and 2.29). The two adjacent maps shown in Figure 2.25 show the locations of the unresolved tension chords and the resolved major and minor chords. Clearly, the major and minor triads lie at sites just off of the high-tension strips, while the tension chords lie directly on the tension ridges.

In addition to Meyer's comments on intervallic equidistance, a few others have noted the unusual harmonic character of the symmetrical triads. Jones (1974, p. 49), for example, has stated that "Composers from Debussy to Hindemith have frequently made use of chords built of perfect fourths or fifths.... These chords, like all other chords that are built of equal intervals ... have a certain sameness of sonority that quickly leads to monotony unless they are used along with other, less symmetrical chord types." Despite a general neglect of the notion of the symmetry/asymmetry of the triads, Figures 2.23–2.25 clearly show that symmetry accounts quite nicely for the perceptual quality of harmonic tension. Indeed, the success of the calculations in indicating that the diminished, augmented and suspended triads have high tension – in all of their inversions and when played

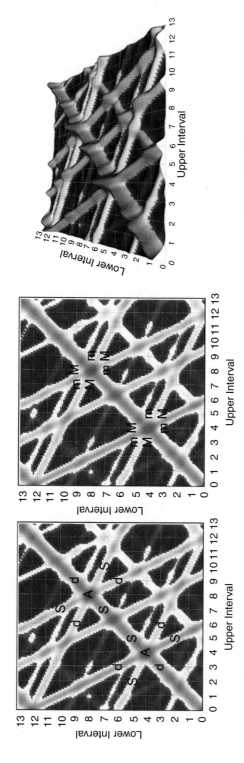

FIGURE 2.25. The harmonic tension of triads when F0–F2 effects are considered. The augmented chords (A) at both locations show the highest tension. Note that there is also a peak of tension at 6.5–6.5 semitones – a triad that cannot be played using the 12-tone equitempered scales but that can be exploited in music using quarter-tone scales.

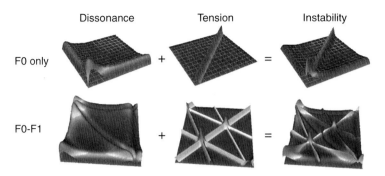

FIGURE 2.26. The addition of dissonance and tension effects to obtain total "instability" scores for the triads. In the upper row are shown the dissonance, tension and instability maps when only F_0 is considered. In the lower row are shown these maps when both F_0 and F_1 are included in the calculations.

over one or two octaves – suggests that the total harmonic "stability" or "instability" of triads may be a consequence of *two independent acoustical factors*. The first is interval dissonance and has been acknowledged to be an important part of music perception at least since Helmholtz's work in the nineteenth century. The second factor is triadic tension – and is explicitly a three-tone effect.

The next step in explaining the overall sonority of the triads is therefore to bring together dissonance and tension to show both two-tone and three-tone effects on the same maps. The addition of the dissonance and tension factors to give an estimation of the total harmonic "*instability*" of any chord is shown in Figure 2.26 and the instability results for three-tone chords with one, two and three partials are shown in Figure 2.27. As was the case with both dissonance and tension, the absolute values of "instability" in the instability maps change slightly as upper partials are added (and therefore the color-coding also changes), but it is a remarkable result that the major and minor chords, in all of their inversions and when played over one or two octaves, remain in regions of relatively high stability (low instability), while the tension chords do not.

The conclusion that can be drawn from these calculations of the total instability of the triads is that the major and minor chords are particularly stable for straightforward *acoustical* reasons (Figure 2.28). Most of the other commonly used triads are rather consonant, but they simultaneously show relatively high tension values ... and are therefore perceived as somewhat unstable – ultimately demanding resolution to the major and minor chords that, without exception, exhibit greater stability.

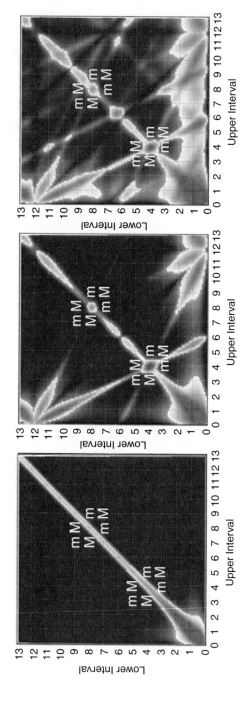

FIGURE 2.27. The effects of the upper partials on the total instability of triads. On the left are shown the results when only F0 is considered. In the middle are the results when F0 and F1 are included in the calculations, and on the right are the results for F0–F2. All 12 of the major and minor chords lie in regions of low instability.

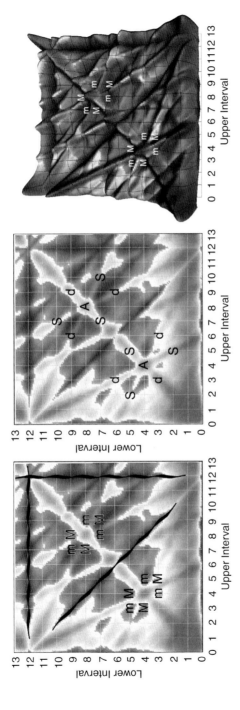

FIGURE 2.28. Total harmonic instability when the effects of interval dissonance and triadic tension are added together (F0–F4). The major (M) and minor (m) chords fall on islands of stability, whereas the unresolved tension chords (d, A, S) lie around those islands. Note that most of the regions of extreme (dark gray) stability involve octave intervals (where either interval is 12 semitones or the sum of the two intervals is 12 semitones). On the right is shown a 3D screenshot from the Seeing Harmony software (Cook & Hayashi, 2008).

The instability calculations imply that the musicians of the fourteenth century who first employed simultaneous three-tone chords were not simply the lucky inventors of a musical idiom that has proven popular. On the contrary, they were discoverers – musicians who somehow became sensitive to the symmetry/asymmetry in the acoustical patterns of *three-tone configurations*, while their Medieval predecessors remained enthralled by lower-level *interval* effects. This distinction between dyadic and triadic effects remains an issue in discussions of harmony today, but it is a misunderstanding to maintain that either effect alone explains harmony or that one factor is more important than the other. When music employs pitch intervals, it is essential that their tuning is "perfect" and that the sweetest, most consonant combination of the two tones is found. But when music includes triads, the interval effects become secondary, and the tuning of the chord *as a chord* becomes the primary perceptual event. Of course, if there is a starkly dissonant interval in the triad, its dissonance will be the most salient feature of the chord. But when the intervals in a chord are not overwhelmingly dissonant, then the relative spacing of neighboring intervals, not the location of the tones relative to the tonic, becomes of central concern. In essence, harmoniousness is a *triadic phenomenon* – not simply the summation of isolated pitch "tonality" and not simply the summation of interval "consonance."

What the Renaissance discoverers of harmony achieved, therefore, was a shift in focus – away from the relative "perfection" or "imperfection" of the consonance of intervals. As important as interval perception continues to be for music, in general, Renaissance musicians successfully drew attention to the higher-level issue of triadic structures – the symmetry or asymmetry of three-tone configurations (or, equivalently, the relative size of neighboring intervals). What is of historical interest is the fact that, while Renaissance *musicians* were busy discovering new kinds of polyphonic music that employed primarily asymmetrical three-tone configurations, Renaissance *theorists* remained obsessed with intervals – and devoted their theoretical energies to justifying why three-tone chords are "consonant" or "dissonant" due to the use of certain intervals. Rather than address triadic harmony on its own terms as a "three-tone phenomenon," the theorists stuck with the theoretical framework that had sufficed for discussion of intervals (i.e., the relative "perfection" of tone dyads). Although aware of the existence of the upper partials, music theorists such as Gioseffo Zarlino (1517–1590) and Jean-Philippe Rameau (1683–1764) were vociferous advocates of the "explain-everything-in-terms-of-intervals" school of thought. Rameau, for example, explicitly noted that "the power of the major and minor chords is obtained by the use of the major or minor third.... Thus, we can attribute the power of harmony to these intervals" (1722/1971, p. 123).

Despite the continuing influence of traditional ideas on modern harmony theory, Rameau's explanation of harmony solely in terms of interval effects is incorrect and easily shown to be incorrect if the various inversions of the major and minor chords are considered: A major chord in root position (e.g., C-E-G) contains an interval of a "major third" (C-E), but the "power" of that interval in producing the sonority of the major mode comes to nothing if the third tone is A instead of G (a "major sixth" above the tonic instead of a "fifth"). The triad (consisting of a major third and a major sixth) is then transformed into a minor chord! A similar effect is found with the minor triad in root position, which has the sonority of a *major* chord when a minor sixth instead of a fifth is used, despite the power of the minor third!

The complex theoretical system known as traditional Western diatonic harmony theory goes through various Ptolemaic strategies to explain away these "anomalies," but the powerful thirds are demonstrably influenced by contextual effects and many of the basic phenomena of harmony simply cannot be explained in terms of the effects of isolated intervals. In fact, the deficiencies of traditional harmony theory were well known already in the late sixteenth century. One of the early debates that ended up in print was between the great Italian music theorist, Zarlino, and Vincenzo Galilei (the father of Galileo Galilei). Zarlino defended the interval model of harmonic beauty, but in rebuttal Galilei promulgated the "Laws of Galilei" – one of which was that "two successive consonances of the same size do not produce a consonance" (Heilbron, 2010, p. 10). Easily demonstrated, perceptually obvious and neatly stated, but Galilei's Law from 1588 remained dormant until independent discovery by Leonard Meyer in 1956.

The Renaissance theorist's obsession with interval effects continues to have an unwarranted influence on modern discussions of harmony. Still today, many music textbooks go no further than general arguments to the effect that major and minor chords are acoustically rather "consonant," while their specifically major or minor character remains a "puzzle." Authors who are forthright about the dilemma of trying to equate harmoniousness with the summation of interval consonance are then faced with the problem of either retreating to the notion that acoustics is irrelevant or declaring that learning, training and the influence of musical traditions are so strong that the otherwise "valid" acoustical arguments of Pythagoras and Helmholtz are nonetheless fully negated by these cultural factors. In either case, the entire "explanation" of harmony perception thereby collapses into statements about what kinds of harmony are in fact used in music – "what the tradition demands!" – with nothing remaining of the "why." Having abandoned the acoustical argument, there is nothing more to say except that "This is harmony as we know it in the Western idiom!" But erudite knowledge is not the same as understanding, and to declare that what we hear with our ears is actually recollection of centuries of common usage is to confuse the issue of auditory perception with that of the familiarity of cultural norms.

In any case, it is worth repeating that nearly all music in diverse musical traditions around the world employs multiple (>2) pitch combinations – played either as harmonies or, more frequently, as melodies. Depending on the sequence and duration of those pitches, there will inevitably arise moments of strong harmonic implication. Regardless of whether a musical tradition makes use of the major/minor/chromatic labels of so-called Western harmony or the more complex description of the Ragas in Indian music, three-tone combinations will frequently and repeatedly result in precisely the harmonic effects discussed in traditional diatonic harmony theory. It is for this reason that both Leonard Meyer (1956) and Leonard Bernstein (1976) have argued that nearly all forms of polyphonic music are essentially diatonic. It is of course not the case that all folk traditions use major or minor chords, but whenever three nondissonant tones are played in succession, the implied harmony will be, in the terminology of harmony theory, either major or minor or chromatic.

Before discussing the important issue of the major/minor modality of triads, the dissonance of the so-called seventh chords should also be mentioned. The most frequently employed four-chords are the dominant seventh (e.g., C-E-G-A#) and the minor seventh (e.g., C-D#-G-A#). Both have a recognizable major or minor character (since they are built on the major and minor root triads), but the presence of the seventh adds some unsettled jazziness to the harmony. Note that seventh chords are, by definition, tetrads, but they can be played (and perceived as seventh chords) using only three notes (e.g., C-E-A# and C-D#-A#). That is, a three-note seventh chord consists of the tonic and the seventh itself, plus either the third or the fifth (with the omitted fifth or third being "understood," but not played). The abbreviated forms of the seventh triads are not often discussed in textbooks on harmony, but they are frequently employed in real music, and their names are displayed on the LCD monitor of most electronic keyboards for certain combinations of just three tones, as shown in Figure 2.29. After the major and minor chords themselves (in their various inversions), the seventh chords are in fact among the most frequently used triads in folk and pop music.

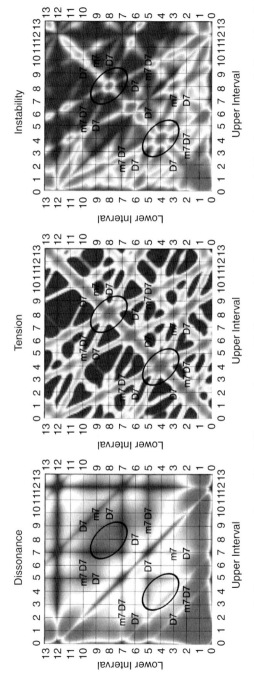

FIGURE 2.29. In dissonance, tension and instability maps drawn with four partials, the seventh chords (dominant seventh, D7, with a distinct major sonority, and the minor seventh, m7, with a distinct minor sonority) are found to lie in regions of slightly greater dissonance surrounding regions where the major and minor triads lie (the oval regions).

Harmonically, the seventh chords are good examples of slightly dissonant chords that are perceived as rather sonorous, especially when played over a range of 14–18 semitones, and therefore do not demand instant resolution. Their positions on the triadic grids suggest that they produce only mild dissonance despite the fact that the seventh note of the diatonic scale lies only a whole-tone below the first upper partial of the tonic.

2.3. THE MODALITY OF TRIADS

The preceding sections concerning dissonance and tension – and their summation to give a total instability score – indicate that there are *acoustical* grounds for considering the major and minor triads to be musically more sonorous than the other triads. Undoubtedly, the learning of musical styles and habituation to the different kinds of harmonies, scales and tuning systems in various musical traditions also influence how we perceive pitch combinations. But whatever additional effects are produced by culture, there are clearly three-tone features of chords that have little to do with culture and everything to do with acoustics. In effect, the perception of triadic structures leads to a sense of greater or lesser harmonic instability that is a consequence of both two-tone dissonance and three-tone tension.

So far, however, this discussion of the acoustical regularities underlying triadic sonority has not addressed the most important question concerning harmony: Why do major and minor harmonies feel so different from one another? All of the major and minor triads are rather stable chords because they lie in broad valleys of low dissonance and, moreover, in smaller pockets of low tension. But, if interval dissonance and triad tension were the only factors determining harmoniousness, we should expect that all the major and minor chords would sound rather similar – all small variations on the theme of "triadic stability." On the contrary, however, the labels "major" and "minor" were invented *because* there was something perceptually different about these two classes of sonorous chords. In English, French and Italian, the major/minor distinction suggests differences in size, and in German *Dur* and *Moll* literally mean "hardness" or "strength" (the origin of *durability* in English) and "softness" or "weakness" (*mollify* in English).

Despite the fact that all of the major and minor triads contain intervals of three, four or five semitones, they feel different and are often described in terms of a slightly positive emotional valence for the major chords and a negative valence for the minor chords. There is something "bigger," "stronger," "harder" and "brighter" about major chords, and something "smaller," "weaker," "softer" and "darker" about minor chords. (Ask anyone!

FIGURE 2.30. When asked to indicate which face more accurately depicts the feeling of isolated chords, children choose the happy face for major chords and the sad face for minor chords (see Terwogt & van Grinsven, 1991, and Trainor & Trehub, 1992).

Ask a child! But, if you are expecting an answer this week, you better not ask a music theorist.) Whatever adjectives we might prefer for describing these triads in specific musical settings, it is clear that major and minor harmonies are, affectively, *not* variations of one another.

Experiments with children as young as 4 or 5 years who have had no formal training in music have also revealed the remarkable "bias" for associating the minor chords with negative emotions and the major chords with positive emotions. When asked to choose between the two faces shown in Figure 2.30 in response to isolated chords, young children will pick the smiling face for the major chords and the frowning face for the minor chords. Clearly, even those naïve subjects have been raised in a social milieu where this association is common, but equally clearly they demonstrate that "learning the Western idiom" does not depend on years of practice at the keyboard or familiarization with formal harmony theory. Apparently, it is quickly and effortlessly absorbed through casual listening to typical diatonic music.

In the psychology laboratory, the affective response to such chords has been measured many times (e.g., Hevner, 1936; Kastner & Crowder, 1990; reviewed by Scherer, 1995; Gabrielsson & Juslin, 2003) and clearly demonstrates that both musicians and nonmusicians hear the difference between the major and minor modes. Typical results of three such experiments in our laboratory are shown in Figure 2.31.

Why do three tones resonating in our ears suggest something about size, texture or strength? More specifically, what is the structural feature of chords that allows people to so unambiguously identify the three major chords as different from the three minor chords? We have already seen that the perceptual stability/instability of the triads of traditional diatonic music

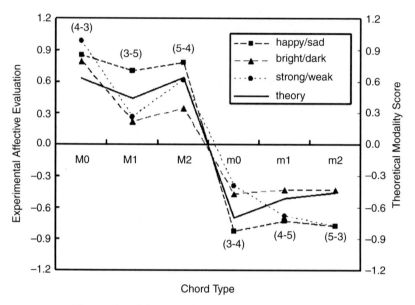

FIGURE 2.31. The results of three experiments in which 20 (18 or 66) undergraduate nonmusicians evaluated the bright/dark (happy/sad or strong/weak) quality of 72 (24 or 12) isolated major (M0, M1, M2) and minor (m0, m1, m2) chords presented in random (pseudorandom or fixed) order in various keys and at various pitch heights. Indications of differences among the inversions of these chords are also of interest, but, in any case, the affective distinction between major and minor is clear. (The thick solid line shows predictions of our model, Cook & Fujisawa, 2006).

can be explained in acoustical terms – provided that both interval dissonance and triad tension are considered. So, for many musicians and nonmusicians alike, the conclusion that "harmony is a result of acoustics" will seem obvious and hardly worth mentioning, but in fact several centuries of music theory have not provided an answer to the question of harmonic modality (mode). Indeed, some commentators contend that the perception of major and minor harmony may be little more than an arbitrary, culture-specific "habit" with no claim to universality – a local custom that has insidiously infiltrated to all parts of the world for political, not musical, reasons! As Philip Ball (2010, p. 273) states it: "It is even possible that music's emotive power is itself in part merely a tacit cultural consensus."

Well, maybe, but if the perceived *stability* of triadic harmonies can be deduced solely from the acoustical structure of three-tone chords (Section 2.2), then it would be incorrect to maintain that the perception of harmoniousness is nothing more than "a tacit cultural consensus" or that listeners have been brainwashed into believing there is something different about resolved and unresolved chords, when acoustically they are all similar. On the contrary, particularly the results shown in the instability maps support the idea that the major and minor chords – those chords that are traditionally considered to be musically the most beautiful and that historically have been most frequently employed in diatonic music – exhibit relatively low levels of intervallic dissonance *and* simultaneously relatively low levels of triadic tension.

So, we are clearly on the right track in considering both two- and three-tone effects when discussing harmony, but the stability/instability argument alone does not explain the perceptual difference between major and minor chords. It is an empirical fact that most people perceive a slightly positive affect in the major mode (triads, scales, keys) and a slightly negative affect in the minor mode (and the same can be said for the Indian Ragas that use scales related to the major and minor diatonic scales, Chordia & Rae, 2008). This perception can of course be suppressed and even reversed through rhythms, timbres and lyrics that tell a different story, but if all else is held constant, a direct comparison of major and minor triads will consistently point in the direction of positive versus negative affect. That difference is one of the longest standing puzzles of Western harmony. It is also one of the most important because the positive and negative affect associated with harmony are crucial focal points where music has meaning for us: We "feel" major and minor chords at an emotional level that is clearly more significant than the pleasant meandering of birdsong, the chortle of splashing water in a mountain stream or the harmless whistling of children trying to keep the ghosts away! Individual tones may be pure and isolated intervals may be consonant, but without harmonic mode such auditory candy does not rise to the level of emotional engagement.

Many musicians object – quite rightly – to a simple happy/sad characterization of the major and minor modalities for the good reason that real music (as distinct from chords played in the psychology laboratory) is often a mixture of major and minor harmonies, as well as moments of dissonance and unresolved tension. Perhaps a few major chords to announce a sports victory or a few minor chords to lament a lost love can be effectively played as a relentless sequence of chords in the same major or minor mode, but music that connoisseurs would describe as interesting, subtle, nuanced, and ultimately effective as music often includes mixtures of major, minor and tension chords. Such music is purposely constructed to elicit an affective atmosphere with many twists-and-turns, intimations of positive or negative affect and the highlighting of internal contradictions before resolving to the affect of unambiguous major or minor harmonies. Music without tonal resolution can still be interesting – from the trance music of Bali to the geometrical intellectualisms of Webern – but most music, including modern classical and jazz, uses major and minor harmonies to resolve tensions and produce moments of emotional release and composure (Narmour, 1990).

So, let us acknowledge that real music is exceedingly complex and not easily summarized with dichotomous adjectives! Still, the pitch components that underlie real music can be examined in rather small packages and analyzed for regularities that contribute to the experience of emotion. Before we declare that music is ineffable, mysterious and beyond scientific analysis or "just an arbitrary local custom," let us look for the acoustical basis for our ability to distinguish between major and minor – and then try to explain why we feel the acoustical factors in the ways that we do.

A simple model for calculating the *tension* inherent to any three-tone combination was shown in Figure 2.23, where the relative size of the two intervals was the essential factor (Cook, 2001, 2002). It was then demonstrated how tension calculations can be used to differentiate between the *stability* of the major and minor chords, on the one hand, and the inherent *instability* of the diminished, suspended and augmented chords, on the other. What was already evident from Figure 2.23 is that, from a state of "intervallic equivalence" (with its inherent perceptual tension), there are two – and only two – directions to move in order to reduce the tension. As soon as the two intervals in a triad differ by one semitone, the tension of the symmetrical chord disappears – and the asymmetrical triad "resolves." Since the only available directions toward resolution from tension correspond to major and minor harmonies (lower interval > upper interval, or vice versa), it is useful to reformulate the tension model such that moving away from tension will result in a quantitative measure of the degree of positive "majorishness" or negative "minorishness" of any three-tone chord (Fujisawa, 2004), as shown in Figure 2.32.

The two types of resolution can be distinguished using a "modality curve," where major modality produces a positive valence score and minor modality a negative one. The basic meaning of the curve in Figure 2.32 can again be described simply in terms of a "difference of intervals" argument. That is, if the two intervals in a triad are the same size, the difference between them is of course zero (along the horizontal axis), and the "valence score" is also zero (along the vertical axis). When either interval is slightly larger than the other, the valence score will rise or fall to a maximum or minimum of 1.0 or −1.0 when the difference is exactly 1.0 or −1.0 semitone – and then return to zero again if the difference is two or more semitones.

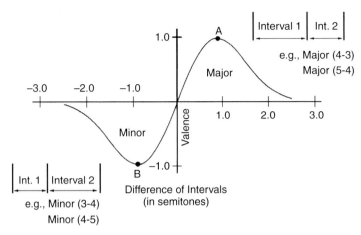

FIGURE 2.32. The modality curve. The difference in the magnitude of the intervals (lower minus upper) of a triad will determine its positive (major) or negative (minor) valence. Point A is the maximum valence score (obtained when the difference in interval size is 1.0). Point B is the minimum valence score (-1.0). When both intervals are the same size, the affective valence is zero (Fujisawa, 2004; Cook & Fujisawa, 2006).

As was true for the tension curve, the modality curve uses the relative size of the two intervals in a triad to calculate the degree to which it has the ring of a major or minor chord. Unlike the tension model, where the question was essentially: "Are the two intervals of the triad the same size?" the modality model asks the question: "Which interval is larger?" Application of this simple idea to real chords, however, requires once again that we bring the effects of the upper partials into the picture.

To begin with, consider the valence scores obtained when only the fundamental frequencies of the three tones are used in the calculation (Figure 2.33). We see that there is an elevated ridge of "major modality" when the lower interval is one semitone larger than the upper interval – and a depressed valley of minor modality parallel to it. The shading in the figure indicates that four of the six resolved major chords (intervals of 4-3, 5-4, 8-7 and 9-8 semitones) and four of the six minor chords (intervals of 3-4, 4-5, 7-8 and 8-9 semitones) have appropriate (positive or negative) valence scores, but the remaining four triads are located in regions that have valence scores near to zero – theoretically, neither major nor minor, which is contrary to what we know from musical experience.

When upper partials are also considered, however, the match between the theoretical modality maps becomes remarkably consistent with our perception of major and minor triads (Figure 2.34). Already with consideration

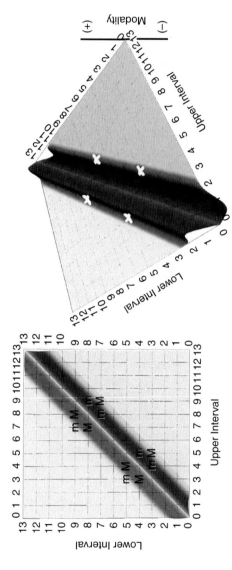

FIGURE 2.33. The modality maps when only F0 is included in the calculations. Eight of the 12 major and minor chords have appropriate "valence scores" (+1 for major chords and −1 for minor chords), but two major chords and two minor chords (x's on the grid to the right) have valence scores near to zero. The inclusion of upper partials in the calculation of modality solves this problem.

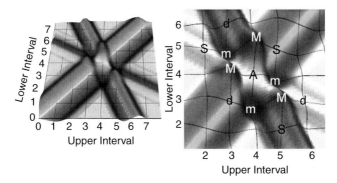

FIGURE 2.34. Three "peninsulas" of elevated major modality and three "troughs" of minor modality are found when both Fo and F1 are included in the calculations. All of the minor triads have negative modality scores and all of the major chords have positive scores. Note that the tension chords (d, A, S) lie at "sea level" with modality scores near to zero.

of only the first set of upper partials, we find that there are regions of positive and negative valence at all 12 locations of the major and minor triads on the large triadic grid (Figure 2.35).

The conclusion that can be drawn from the modality maps is that, among the upper partials of all of the *major* chords, there is a predominance of triadic structures where the lower interval is larger than the upper interval. *Minor* chords show the opposite structural feature. This simple regularity of the major and minor chords can be seen if we examine interval sizes for the Fos and F1s of these chords when written in musical notation (Figure 2.36).

It is noteworthy that most of the dominant seventh (D7) and the minor seventh (m7) chords fall in regions of elevated and depressed modality, respectively – as would be expected of these slightly unstable, major- and minor-related chords. More precisely, of the 18 seventh chords displayed in Figure 2.37, 13 are found in regions of the appropriate modality (negative for minor seventh and positive for dominant seventh chords). The remaining five (one minor seventh and four dominant seventh triads) lie in regions of extremely weak modality (valence scores near to zero). Unlike the major and minor chords themselves (whose positive/negative valence scores remain unchanged regardless of the number or strength of the upper partials, $n > 1$), the seventh chords with valence scores near zero fluctuate between small positive and small negative values depending upon the upper partial structure. Perceptually, the major or minor character of these seventh chords is correspondingly weak.

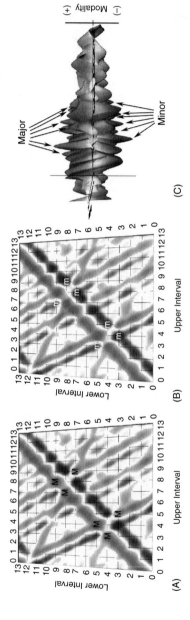

FIGURE 2.35. Calculation of the total "modality" of three-tone combinations produces positive values for all major chords (M) and negative values for all minor chords (m). Unresolved chords (augmented, diminished, etc.) have modality scores near to zero. These computational results clearly demonstrate that the perception of major, minor and chromatic harmonies is not simply a matter of acculturation; there is an acoustical basis for the common perceptions of chords (see the back cover for a color version).

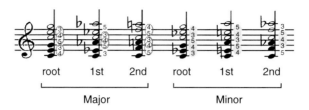

FIGURE 2.36. The interval substructure of the major and minor chords. In all of the major chords there are neighboring intervals where the lower is one semitone larger than the upper (unfilled ovals). In all of the minor chords, the lower of the two neighboring intervals is one semitone smaller than the upper interval (gray ovals).

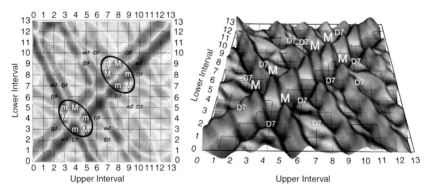

FIGURE 2.37. The modality of the seventh chords. The dominant seventh (D7) chords lie in regions of strong positive modality (3-2, 7-3, 9-5, 10-9, 2-7), or near-zero modality (2-4, 4-6, 6-2, 10-6, 8-10, 6-8, 5-10), making them modally ambiguous. The minor seventh (m7) chords lie in regions of strong negative modality (7-2, 2-3, 9-10) or near-zero modality (10-5, 3-7, 5-9), with similar implications for their perceptual qualities.

As was most evident in Figure 2.35, the modality maps are covered with bands of major and minor modality, so it is natural to ask why there aren't many more versions of major and minor chords. If the modality model of Figure 2.32 is accurate, we might expect that, with the exception of the amodal tension chords (where intervallic equidistance is predominant), almost any combination of three tones would be "modal" in a major or minor direction. Why are only 30 of the 169 sites on the triadic grid traditionally acknowledged to be major or minor?

The answer is that, although dissonance and modality are independent features of all chords, interval dissonance is the more salient, lower-level effect. As a consequence, even when relatively strong modality scores are

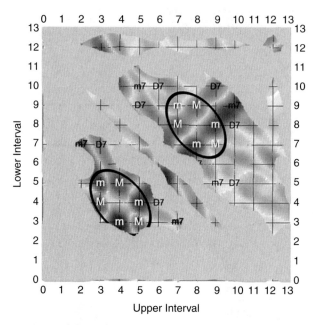

FIGURE 2.38. The modality map when all regions of dissonance are removed. Most of the ungrayed locations are either the locations of the common triads, or they are regions where there are octave repetitions of pitches. It is interesting to note the modality scores of various regions that are not traditionally labeled as major or minor. These include "minor modality" at 12-3, 8-4, 4-12 and 3-9 semitones and "major modality" at 12-10, 12-4, 10-5, 8-12, 6-10 and 4-8 semitones. None of these triads are major or minor in traditional harmony theory, but are predicted to be so from the acoustical model discussed here.

obtained, dissonance dominates the perception of the vast majority of all possible triads. Of course, regions of high tension have weak modality, but, in many other locations, there is strong modality that, theoretically, should have the "ring" of major or minor, but perceptually it does not. For example, some regions with major-chord-like valence (triads with intervals of 2-1, 6-5, 7-6 or 11-10 semitones) or minor-chord-like valence (triads with intervals of 1-2, 5-6, 6-7 or 10-11 semitones) are overpowered by their strong (semitone) dissonance.

The perceptual negation (masking) of modality brought about by the greater salience of dissonance can be illustrated by graying out all regions on the modality maps where high dissonance is found (Figure 2.38). So doing, we are left with strong modality scores only for a relatively small number of chords containing combinations of *consonant* intervals.

In this view of harmony, the tension and modality of triads are perceptual phenomena that "compete" at the same level of complexity – due to our perception of the "difference of intervals." When interval size is identical, tension is heard; when intervals differ by one semitone, modality is heard. As a consequence, any three-tone combination will necessarily fall into one of three general categories: major, minor or tension. Our ability to hear the modality or the amodal tension will, however, be masked whenever two-tone dissonance becomes strong. When the lower-level effects of interval dissonance (particularly, a semitone dissonance) are present, any underlying modality effect is perceived as less salient than the dissonance effects that cry out for immediate resolution.

If dissonance is the factor that prevents us from hearing the hidden modality of certain three-tone chords, it is of interest to note that the dissonance maps indicate relatively broad expanses of consonance when triads are played over a range larger than one octave (see Figures 2.14–2.16). It is perhaps for this reason that post-Renaissance harmonies have gradually extended to chords containing 9th, 11th or 13th intervals. They are normally played in open, as opposed to close, form (lying in the upper right-hand quadrant of the triadic grid) and thereby relatively distant from the regions of strong dissonance where either interval is a semitone or a whole-tone. When triads are played over a pitch range of more than one octave, the inherent dissonance among certain pitch classes is lessened and the underlying modality can be more easily perceived.

Provided that upper partials are included in the modality calculations, there is no doubt that there is a simple structural regularity that determines the major/minor character of triads. In other words, an *acoustical feature* calculated directly from the structure of chords has been identified and shown to underlie the concept of modality in traditional harmony theory. The simplicity of the structural rule will come as a surprise to anyone familiar with the complexities of traditional theory because the traditional view requires the entire theoretical edifice of Western harmony theory – with special consideration of the roles of major- and minor-thirds in relation to the tonic, and therefore an understanding of key and the use of scales to establish key. Although it can be said that traditional theory successfully *describes* modality through such complex arguments, the simplicity of the acoustical explanation suggests that there may be a more direct route to understanding.

What then is the functional relationship between the amodal tension chords and the major and minor chords? Let us consider a concrete example: The changes of mode that arise following pitch rises or falls, starting from an augmented chord (Figure 2.39). As was shown in Figures 2.14–2.16, the augmented chord has rather *low* dissonance, but very *high* tension (Figures 2.24–2.27). Presumably because of this high tension, it

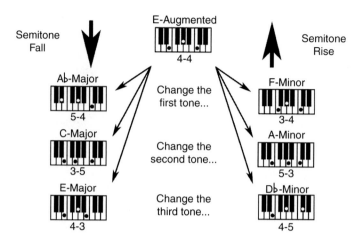

FIGURE 2.39. A semitone *rise* in any tone of an augmented chord results in a *minor* chord, while a semitone *fall* results in a *major* chord (interval structure is noted below each chord).

is not often used in either traditional classical or popular music. In fact, Eberlein (1994) examined the musical scores of music from Bach, Schubert, Mozart and other classical composers and found that, among the thousands of triads noted in the tablature, the augmented triad accounted for less than 0.5 percent. It was seriously unpopular!

The augmented chord is nonetheless of interest because all of its nearest neighbors on the triadic grid are major or minor. In other words, starting with the unresolved tension of the augmented chord, lowering or raising any of its tones by one semitone will transform its unstable tension into the resolved stability of a major or minor chord. As shown in Figure 2.39, there is a noteworthy pattern in what happens when any of its tones are changed: *Raise any tone, and one proceeds to a minor chord; lower any tone, and one ends up with a major chord.*

In other words, given the starting point of the tension of intervallic equidistance, the rising or falling *direction* of pitch movement of any tone determines the mode of resolution: major (downward) or minor (upward). It is a truly remarkable regularity of diatonic harmony in general that pitch changes *in any of the tension chords* (diminished, augmented or suspended in all of their inversions) give similar results (see Table 2.2).

Let us examine the pattern of shading illustrated in Table 2.2. All of the dark gray cells are structurally related to the major mode (major or dominant seventh chords); all of the light gray cells are in the minor mode (minor or minor seventh chords); and all of the off-white cells in the central column

TABLE 2.2. *The major and minor chords in relation to the tension chords*

	Chord after lowering one tone			Tension chord	Chord after raising one tone		
	n-n-d	n-d-n	d-n-n		n-n-u	n-u-n	u-n-n
Interval structure (in semitone units)	3-2	2-4	4-3	3-3	3-4	4-2	2-3
	4-3	3-5	5-4	4-4	4-5	5-3	3-4
	5-4	4-6	6-5	5-5	5-6	6-4	4-5
	6-5	5-7	7-6	6-6	6-7	7-5	5-6
	7-6	6-8	8-7	7-7	7-8	8-6	6-7
	5-1	4-3	6-2	(~5-2)	5-3	6-1	4-2
	2-4	1-6	3-5	(~2-5)	2-6	3-4	1-5
	8-7	7-9	9-8	8-8	8-9	9-7	7-8
	9-8	8-10	10-9	9-9	9-10	10-8	8-9
	3-5	2-7	4-6	(~3-6)	3-7	4-5	2-6
	6-2	5-4	7-3	(~6-3)	6-4	7-2	5-3
	n-n-d	n-d-n	d-n-n		n-n-u	n-u-n	u-n-n
Labels from music theory (alternatives are possible)	dom7	dom7	major	dim	minor	–	min7
	major	major	major	aug	minor	minor	minor
	major	dom7	–	sus4	–	–	minor
	–	–	–	tritones	–	–	–
	–	dom7	major	sus4	minor	–	–
	–	major	dom7	(sus4)	minor	–	–
	dom7	–	major	(sus4)	–	minor	–
	major	major	major	aug	minor	minor	minor
	major	dom7	dom7	dim	min7	–	minor
	major	dom7	dom7	(dim)	min7	minor	–
	dom7	major	dom7	(dim)	–	min7	minor

Key: d, downward semitone shift; n, no change; u, upward semitone shift; (), Inversions of tension chords.

are tension chords. The white cells are the few remaining triad structures that do not have traditional labels related to their modality.

Starting from the inherently ambiguous modality of the tension chords, a rise in pitch pushes the triad into minor modality, a fall results in major modality. This pattern is a direct consequence of the lawfulness of traditional diatonic harmony, but it is a remarkable fact that, because the feature described here as harmonic "tension" is not recognized in traditional theory, the structural relationships among the major, minor and amodal tension chords are not mentioned in any of the traditional harmony theory textbooks.

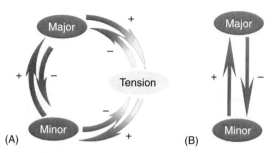

FIGURE 2.40. (a) The Cycle of Modes. The plus and minus symbols indicate rises or falls in semitone steps. If all chords containing interval dissonances are avoided, semitone *rises* lead from tension to minor to major and back to tension harmonies indefinitely, whereas semitone *falls* show the reverse cycle. (b) The traditional view of the major and minor chords is only part of the cycle.

The connections among all of the *consonant* triads (major, minor and chromatic) can also be illustrated as a Cycle of Modes (Figure 2.40a) (Cook, 2002). The traditional view of mode relationships is that the essential difference between the major and minor chords is the semitone shift that can transform major chords into minor chords, and vice versa (Figure 2.40b). That view is of course correct, as far as it goes, but implicitly dismisses all other triads as irrelevant "dissonances" – which, technically, is not correct. By bringing the (unstable, but not dissonant!) diminished, augmented and suspended fourth chords into a broader theory of harmony, it is clear that there is an endless Cycle of Modes that can be traversed by raising or lowering triad tones one at a time. Provided that chords containing semitone dissonances are avoided, the cycling will entail repeated transitions from tension to major to minor and back to tension (with falling tones) or transitions from tension to minor to major to tension (with rising tones).

The relationships among the common triads discussed previously can go unnoticed if we use the common labels from music theory for chords, simply because the "tension chords" are not traditionally viewed as a coherent set. The pattern becomes immediately obvious, however, if we group the diminished, augmented and suspended chords together because of their common structural feature of intervallic equidistance. So doing, it is seen that there is an endless cycling of the three distinct modes that occurs with semitone rises or falls in triad tones (Figure 2.41).

A less succinct way of expressing the cycle relationship among the triads is shown in Figure 2.42, where the major/minor/tension character of the triads is color-coded on the triadic grid. What is evident from the diagonal strips is that, starting almost anywhere on the grid, vertical or horizontal

FIGURE 2.41. Possible transitions among the modes with rises or falls of one semitone in each triad. Many alternatives are possible, and the sequence extends indefinitely.

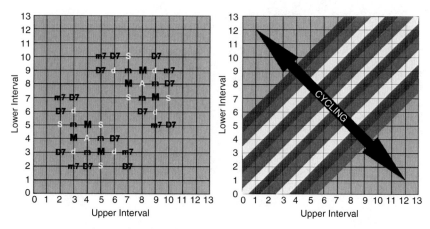

FIGURE 2.42. The Cycle of Modes represented on the triadic grid. Changes of mode correspond to changes of shading. The darkest diagonal stripes indicate the locations of the minor (and minor seventh) chords; the next darker stripes are major (and dominant seventh) chords; and the lightest stripes are where the tension chords lie.

movement (corresponding to semitone rises or falls) results in a color change – representing a change of mode. The only caveat to note here is that many of the grid sites underneath the modal strips represent locations that entail pitch-class repetitions (e.g., 7–5) or dissonant intervals (e.g., 1–5). Some of those sites have labels from harmony theory (major seventh triads, triads with a flatted fifth, etc.), but whatever inherent modality they may have is generally well-hidden by the dissonance of small intervals. In contrast, the major and minor chords, and the dominant seventh and minor seventh chords have notable "modality."

Traditional music theorists may be inclined to dismiss the Cycle of Modes as little more than terminological games, but the important point is that, by focusing on the semitone structure of three-tone combinations, the extreme complexities of traditional theory can be greatly simplified. The issue of harmonic modality can be reduced to three fundamental states: major, minor and chromatic tension – all three of which are defined in terms of relative interval sizes. In light of this simplification of traditional theory, let us now ask the most important questions of all in music psychology, questions concerning emotional implications.

2.4. THE AFFECTIVE VALENCE OF MAJOR AND MINOR

What has not yet been addressed is *why* the two modalities have their characteristic affective valences. Is it nothing more than a musical custom, a habit or an entrenched bias for hearing the major chord as somehow strong, bright and positive and the minor chord as weak, dark and negative – but for no good reason, other than the fact that Western music has always been that way? Could it as easily be reversed? And, empirically, is there any musical genre, culture or subculture on Earth where they are in fact reversed? Just one example where major and minor are perceived inversely would prove the case and demonstrate that the affect of major and minor is utterly arbitrary! But asking people who have never heard polyphonic harmony to evaluate chords, or cherry-picking one song from thousands where the harmony points in one direction and the tempo, rhythm, timbre and lyrics point in the opposite direction is far from conclusive.

To most music listeners, simply to question the emotional valence of major and minor is an absurdity. In the vast majority of cases, the emotional implication of minor chords (keys, melodies) seems to be mildly negative, while major is positive – and neither can be "redefined" as the other any more than the color red can be seen as blue by declaring that we have stupidly become biased to see red as "passionate red" due to cultural associations with lust, war or Coca Cola. As unlikely as the "learned association" model may seem, there is a long-standing academic myth that maintains precisely that argument: Minor chords and melodies were, as a matter of pure chance, used in the early Renaissance in ballads of unrequited love and lonesome sunsets – and the association has stuck. Similarly, victory marches and songs of sexual triumph were, so the story goes, by chance sung in major keys – and the subsequent 30 generations of gullible Europeans have been duped into believing that there is some inherent relationship between positive affect and the major mode and negative affect and the minor mode! Peering down on the absurdities of Western culture, Philip Ball (2010) defends that view: "Many people in the West are convinced that there must be something inherently cheerful in the major mode and inherently sad in the minor" (p. 274). "But there's no reason to believe that people will inevitably experience the minor [...] chords as sad and anguished, if they have not learnt those associations" (p. 275). "The fact is that there is no fundamental reason to suppose that minor-key music is intrinsically sad" (p. 276). "All this is hard for Westerners to intuit" (p. 277).

Who knows how common that view is, but two modern music psychologists well respected for their experimental work have also been seduced by the cultural learning theory. In *Sweet Anticipation*, David Huron (2006, p. 402) asks the important question: "Why [do] major and minor chords tend to evoke the *qualia* of 'happy' and 'sad' for listeners experienced with Western music?" but then answers unconvincingly that "One possible explanation is that these chords have prior associations. In the same way that Pavlov's dogs could associate the arbitrary sound of a bell with food, listeners might simply learn to associate the minor chord with sadness." In *Music, Language and the Brain*, Aniruddh Patel (2008, p. 315) elaborates on the same unlikely theme: "[T]he affective quality of the major/minor distinction in Western music is evident only to those who have grown up listening to this music and who have formed associations between these modes and expressive content (e.g., via hearing songs in which words with affective content are paired with music)."

Hearing the sadness in a minor chord because of the lyrics? Pavlov's salivating dogs? These are clearly not serious arguments, but at least Ball, Huron and Patel have addressed the problem! In contrast, there are countless books with titles such as "the science of music," "the physics of music" and "the acoustics of music" that deftly avoid the major/minor question. The erudite music theorist Fred Lerdahl has developed a comprehensive theory of *Tonal Pitch Space* (2001) that is built explicitly on the major mode; the problem of modal affect is never even raised, other than to note that the minor triad is "somewhat conflicted" or "ambiguous" (pp. 80–1). In *Exploring the Musical Mind*, the cognitive psychologist (musicologist and stalwart antiwar activist) John Sloboda (2005, p. 220) pointed out "some unsurprising correlations between certain gross characteristics [of the major and minor chords] and certain emotions" without any further discussion. Well, they may be as unsurprising as the American lust for military conflict, but even the obvious facts demand explanations.

Despite some modern silliness, serious inquiries concerning the cross-cultural perception, postnatal development and the musical foundations of the major and minor modes have, in fact, been made by countless researchers since the Renaissance. Given the acoustical fact that all of the tones of a major chord are found among the first several upper partials of an isolated tone, the stability and pleasantness of the major chord is generally thought to be a "bottom-up, given fact" of diatonic music. (Specifically, the upper partials of middle-C contain both E and G, giving middle-C itself a very slight major sonority.) Unfortunately, the nearly equivalent sonority of the minor chord and its contrasting affective qualities present theoretical difficulties. If the higher harmonics of each and every isolated tone contain a hint of major-chord harmonicity, why do we not hear minor chords as jumbles of dissonance – where the chord itself is minor, but the individual tones are major? And why is it generally a simple task to rework a song written in a major key into a song in a minor key, with little or no sense that the minor version is less musical than the major version? These are puzzles that musicians, mathematicians and psychologists have frequently pointed out, pondered and commented on, but not solved. Meyer (1956) summarizes the dilemma as follows:

> The theoretical and psychological basis for the affective power of the minor mode in Western music has puzzled and perplexed so many excellent musicologists and psychologists that it may seem rash to propose another answer here. But ... any theory which purports to explain the basis for the affective response to music must take account of and attempt to unravel this mystery. (p. 222)

Meyer himself proposed a music theoretical solution based on the idea that there are more possible transitions to and from minor chords than major chords. Quite aside from the hint of dissonance that a minor chord might embody, he noted that there are less rigid rules for chord cadences proceeding from minor chords, than major chords, because of differences inherent to the major and minor scales. He therefore argued that the increased freedom of minor chord transitions implies that they are inherently more "changeable" and "deviant," and therefore have a slightly negative connotation. "From a harmonic point of view, the minor mode is more ambiguous because the repertory of possible vertical combinations is much greater in minor than in major and, consequently, the probability of any particular progression of harmonies is smaller" (p. 226).

Only a reckless goat-herder would dispute Meyer's musical intuitions, but the difficulty with his explanation of the emotional valence of the minor mode is that it assumes that the typical listener has a rather sophisticated schema of allowed chord transitions "in mind" simply from exposure to Western music. Particularly the results of recent developmental studies – in which very young children have also been shown to perceive the positive and negative affect of major and minor chords – would be difficult to explain on that basis. So, without challenging Meyer's argument on music theoretical grounds, let us examine the interval structure of the major and minor chords to find if there is a lower level (i.e., musically less-sophisticated, acoustical explanation).

Here, the Cycle of Modes can be put to good use. Of the three types of harmonies that do not entail interval dissonance, the tension chords are affectively neutral, "amodal" triads, while the major and minor triads have their characteristic positive and negative sonorities. The tension chords certainly elicit the emotions associated with ambiguity and anxiety, and often leave the listener in a state of anticipation expecting some sort of resolution of the uncertainty. But the tension itself does not point in either a positive or a negative, bright or dark, happy or sad, weak or strong direction. Tension

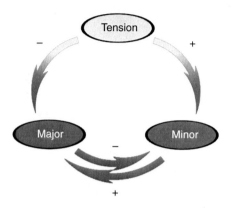

FIGURE 2.43. Steps in either direction around the Cycle of Modes are possible, but the strongest affect arises in the transition from ambiguous (up-in-the-air, still-to-be-decided) tension to major or minor modality. Harmony theory from the early Renaissance was concerned exclusively with the major-to-minor and minor-to-major transitions, while dismissing the amodal tension chords as dissonances.

is an unresolved point of acoustic instability from which we expect progression to the affect of either the major mode or the minor mode through some sort of tonal change. Again, this can be accomplished melodically or harmonically.

As discussed in the previous section, it is an empirical fact of the triadic harmonies made possible by the 12-tone scale that semitone pitch rises from such affective ambiguity imply the negative affect of the minor mode, whereas pitch falls imply the positive affect of the major mode (Table 2.2). Of course, multiple semitone pitch rises and falls can move any triad from any mode to any other mode, but the nearest "local" phenomenon in triadic pitch space from a stance of unresolved, tense uncertainty to one of emotionality involves single steps around the Cycle of Modes. The simplest formulation of the old puzzle of harmonic modality is therefore to ask why the human ear attaches emotional significance to such trivial changes in auditory frequency? *From the affective ambiguity of tension, why do pitch rises lead to negative emotional valence and pitch falls lead to positive valence* (Figure 2.43)?

Surprisingly, the answer to this question is already known in fields unrelated to music and referred to as the "frequency code" (or "sound symbolism"). From the study of animal vocalizations, it is known that there is a strong tendency for animal species to signal their strength, aggression and territorial dominance using vocalizations with a low and/or falling pitch and, conversely, to signal weakness, defeat and submission using a high and/or rising pitch (Morton, 1977). Concrete examples of the frequency code are familiar to most people from the low-pitched growling of aggressive dogs and the high-pitched yelp of defeated, retreating dogs, but it is known to be true for species as diverse as gorillas, elephants, frogs, horses, monkeys and birds – virtually any species that uses changes in vocal frequency for the purposes of social communication.

Ohala (1983, 1984, 1994) has been one of the leading advocates of the idea concerning the inherent sound symbolism of rising or falling pitch. He has noted that

> Animals in competition for some resource attempt to intimidate their opponent by, among other things, trying to appear as large as possible (because the larger individuals would have an advantage if, as a last resort, the matter had to be settled by actual combat). Size (or apparent size) is primarily conveyed by visual means, e.g. erecting the hair or feathers and other appendages (ears, tail feathers, wings), so that the signaler subtends a larger angle in the receiver's visual field. There are many familiar examples of this: threatening dogs erect the hair on their backs and raise their ears and tails, cats arch their backs, birds extend their wings and fan out their tail feathers. […] As Morton (1977) points out, however, the F0 of voice can also indirectly convey an impression of the size of the signaler, since F0, other things being equal, is inversely related to the mass of the vibrating membrane (vocal cords in mammals, syrinx in birds), which, in turn, is correlated with overall body mass. Also, the more massive the vibrating membrane, the more likely it is that secondary vibrations could arise, thus giving rise to an irregular or "rough" voice quality. To give the impression of being large and dangerous, then, an antagonist should produce a vocalization as rough and as low in F0 as possible. On the other hand, to seem small and non-threatening, a vocalization which is tone-like and high in F0 is called for. […]. Morton's (1977) analysis, then, has the advantage that it provides the same motivational basis for the form of these vocalizations as had previously been given to elements of visual displays, i.e., that they convey an impression of the size of the signaler. I will henceforth call this cross-species F0-function correlation "the frequency code." (Ohala, 1994, p. 330)

A perceptible increase or decrease in pitch signifies a change in the vocalizing animal's *assumed* social position. There are of course many other behaviors used to signal or establish dominance, but the "frequency code" is the primary *auditory* means for affective signaling. While most other behaviors and visual displays have species-specific significance, rising or falling F0 has been found to have cross-species generality and profound meaning for any animal within earshot, regardless of night-time obscurity, visual angle or jungle obstructions. A falling Fo implies that the vocalizer is *not* in retreat, has *not* backed down from a direct confrontation, may become a physical threat and has assumed a stance of social dominance. Conversely, a rising Fo indicates defeat, weakness, submission and an unwillingness to challenge others and signals the vocalizer's acknowledgment of nondominance.

So, why does a *rising* voice signal weakness and a *falling* voice strength (and not vice versa)? The view advocated by Morton and Ohala is that – all else being equal – a lower auditory frequency indicates a larger physical object than a higher frequency. That is as true for vocal cord resonances as for church bells: Large vibrating cavities produce low sounds; small ones produce high sounds. As a direct consequence of such acoustical physics, diverse animal species over the course of evolutionary time have learned to emit lower auditory frequencies to suggest to all listeners that they are themselves large and powerful, rather than small and weak – a hopeful ploy to scare away potential rivals without engaging in actual combat.

There is an extensive and fascinating academic literature on how and why these Fo signals have evolved, their correlations with facial expressions (smiles are correlated with higher pitch sounds) and the related, inherent sound symbolism of vowel sounds (see Morton, 1977; Bolinger, 1978; Cruttendon, 1981; Ohala, 1983, 1984, 1994; Scherer, 1995; Juslin & Laukka, 2003; Ladd, 1996; and Levelt, 1999, for further discussion). The detailed physiological mechanisms underlying the frequency code are also of interest, but the important point in the present discussion is simply that the pervasive use of high and low (rising and falling) vocalizations by diverse animal species to indicate social status is empirically well established. Deep sounds come from the big shots, and in the world of territorial struggles and sexual rivalries, everyone knows that the bullies win! So, regardless of species, even the little-shots will try to vocalize their way to power, making a play for the most beautiful frog in the pond by lowering their voices and feigning dominance.

If the frequency code were merely a peculiarity of animal communications, it could possibly be dismissed as irrelevant to human behavior in general and music in particular, but in fact the universality of such sound symbolism is known to have spilled over into human languages: Rising and falling voice intonation have related, if greatly attenuated, meanings concerning social status in human interactions. Across diverse languages, falling pitch is again used to signal social strength (commands, statements, dominance) and rising pitch to indicate weakness (questions, politeness, deference and submission): "in both speech and music, ascending contours convey uncertainty and uneasiness, and descending contours certainty and stability" (Brown, 2000, p. 289). As argued most forcefully by Ohala, the inherent meaning of pitch rises or falls is one of a very small number of cross-linguistic constants that have been found in all human languages. The patterns of syntactic constructions and the details of the pronunciation of vowels and consonants differ remarkably among the world's 6000 languages, but the basics of prosody are apparently universal.

Levelt (1999, p. 112) has noted that: "Whether desired or not, a high register universally expresses vulnerability, helplessness, or special deference. The origin of that impression may be the child's very high speech register." In this regard, we are all instinctively aware of the hidden "meaning" of voice intonation, specifically pitch height. Lexical differences, phonetic differences and grammatical differences make learning and understanding a foreign language a formidable task, but the understanding of prosody is easy. Although we may have no idea what the babbling foreigner is trying to communicate, his tone of voice will clearly indicate whether his vocalizations are commands or questions, assertions or requests! And a fundamental aspect of "tone of voice" is rising or falling pitch.

The most common statement of the frequency code in linguistics is in relation to the rising auditory frequency used in interrogatives. Although it is possible to ask questions with a falling tone of voice (usually carrying some implication about the strength or authority of the speaker), an inquiry that indicates a lack of information and a desire for an answer from another person is normally stated with a noticeable rise in the vocal pitch of the speaker. In contrast, particularly in response to an inquiry or an expression of uncertainty, answers are most frequently stated with a decisively falling melodic contour. ("Who is there?" [↑] "Me." [↓]). And when a command or assertion is expressed with a descending intonation ("You should definitely avoid that!" [↓] "Don't forget to call!" [↓]), it is not uncommon to "soften the blow" of such domineering vocalizations by using a less-daunting rising pitch contour tagged onto the end. Typically, a pitch rise following the assertion invites others to make their own comments or rebuttals or to express a contrary opinion ("n'est-ce pas?" [↑] in French, "nicht wahr?" [↑] in German, "oder?" [↑] in Swiss-German, "right?" [↑] or "okay?" [↑] in English, "deshyou?" [↑] in Japanese – all uttered with a discernible rise in auditory frequency).

Clearly, "establishing dominance" is only one of many aspects of human communication. The expression of personal "strength" or "weakness" through the use of rising or falling pitch is therefore only one part of the prosody of normal conversation, but its cross-cultural prevalence demonstrates the importance of our biological roots – extending even to the realm of language.

In its starkest form, the frequency code can be stated simply as: Falling pitch is used to signal strength, rising pitch is used to signal weakness. For both animal vocalizations and human speech, the pitch context is provided by the tonic or "natural frequency" of the individual's voice – from which relative increases or decreases can be detected by listeners (who then make their own judgments as to the validity of these vocal assertions concerning social status). Since a larger auditory framework, such as musical key, is *not* required in either animal vocalizations or human speech, the meaning of pitch movement is relative to the speaker's normal voice. It is for this reason that the frequency code can be stated simply as the *direction* of vocal pitch changes. In the context of diatonic music, however, the "meaning" of pitch changes can be deciphered only within a specific musical context. In other words, musical *key* and the location of the tonic are not "givens," but must be established within the context of the ongoing music. Normally, that is done gradually – sometimes with intended ambiguities and delays – but nearly always evolving toward a definite key within which the listener can appreciate the musical significance of any pitch movement.

Because of the importance of musical key, the question concerning why diatonic harmony has affective valence needs to be restated as: What is the minimal musical context from which pitch movement will allow the listener to hear unambiguous affective meaning? In diatonic music, the inherent affective implications of a major or minor key can be established unambiguously through the use of a resolved harmonic triad. More complex structures can of course be similarly employed, but a modal triad is the minimal structure that carries with it an explicit major or minor affective charge. Since a triad requires a pitch range of at least seven semitones (a major or minor triad in root position), a modally ambiguous triad over a range of six semitones provides a sufficient context from which a semitone rise or fall can momentarily establish a major or minor mode. It is a simple consequence of the regularities of diatonic harmony that, given this minimal context, a semitone increase *can resolve* to a minor mode, and a semitone decrease *can resolve* to a major mode, but not vice versa. Remarkably, pitch movement from *any* (nondissonant) triad that is neither inherently major nor inherently minor shows this same general pattern (Table 2.2).

Unlike the world of animal vocalizations, in music, key is all-important: An isolated tone outside of any musical context has no inherent musical meaning and even a melodic interval – rising or falling any number of semitones – can have very different musical meanings depending on the context.

For example, the affective charge of a "major third" or "minor third" (played melodically or harmonically) can be completely altered by one additional tone. If we start with the so-called minor third, A-C, it will have the affective implications of the minor mode if it is preceded or followed by E, but the same interval of a "minor third" will have the affect of the major mode if it is preceded or followed by F. Context-free intervals are that ambiguous! But as soon as a major or minor triad has been sounded, the modal ambiguity is removed, and the major or minor affect is firmly established. The composer may well then set out to transform the major or minor mode into something else with further notes, intervals or chords, but the triad itself remains as a statement of affective modality that neither isolated tones nor intervals can achieve.

The affective meaning of a triad is particularly clear for major chords – with their unambiguous stability – but, as Deryck Cooke (1959) has argued, the anguish, sadness or negative affect of minor chords is equally fixed: "There is no uncertainty in the negative affect of the minor mode. It is final and resolved" (p. 55). Unlike the inherently unresolved tension chords (and, arguably, the negative affect implied by the uncertainty itself), the finality of the negative affect of the minor mode is itself a stable ending – a sad conclusion, an undesirable outcome, an unwanted certainty that distinguishes the minor mode not only from the positive outcome of the major mode, but also from the yet-uncertain ambiguity of the amodal tension chords.

Although the musical meaning of context-free pitch movement or the musical meaning of isolated intervals is inherently ambiguous, pitch movement has explicit meaning in relation to mode, provided only that the necessary minimal context has been given. It is a noteworthy fact of diatonic music that the *direction* of tonal movement from the ambivalence of amodal tension to a major or minor triad is the same as the *direction* of pitch changes with inherent affect in animal vocalizations and language intonation. The implication of the frequency code is therefore clear: In music, as in animal vocalizations and human language, *upward pitch movement implies the mildly negative affect of social weakness, downward pitch movement implies the mildly positive affect of social strength*. When a three-tone combination shifts away from the unresolved acoustical tension of intervallic equidistance toward resolution, sound symbolism implies that we instinctively infer an affective valence from our detection of the *direction* of tonal movement: A semitone shift up is weak (submissive, retreating); a semitone shift down is strong (dominant, assertive). The fact that we feel anything at all – and do not experience such harmonic phenomena as cold "auditory information processing" – is indication that our biological beings

FIGURE 2.44. The frequency code at work in animal calls, language and music. Given an appropriately neutral starting point, rising or falling pitch has related meanings in all three realms: positive affect with descending pitch, negative affect with ascending pitch.

have been activated. When embedded in a complex musical context, the emotional response to modal harmony can occur together with goose-flesh and shivers down the spine, tears in the eyes, changes in heart rate and blood pressure, and other indications of arousal of the autonomic nervous system. These are not signs of cool-headed, detached listening, but, on the contrary, indications of emotional involvement.

Note that there is an unavoidable paradox in attempting to assign an emotional "valence" to pitch movement insofar as the contextualized, diatonic meaning of pitch rises and falls can conflict with the affect of isolated, context-free pitch changes. As has been noted by many commentators, pitch rises are "energizing" and falls are "de-energizing" – so that the emotional valence of a semitone rise can simultaneously imply a positive invigorating affect and a negative tension-to-minor-key disheartening affect! And vice versa for pitch falls – bringing the emotional ring of a loss of energy and simultaneously the harmonic transition from ambiguity to dominance! Conversely, a whole-tone rise or fall can give a unified valence. For example, a whole-tone rise from harmonic tension to a major chord brings together both the positive valence of increased energy and resolution to a major chord, while a whole-tone fall unified the loss of energy of the change in the isolated tone with the modality change from tension to minor.

Staying within the diatonic context of the minimal (semitone) changes in pitch, however, the similarity of the binary pattern of affect in response to pitch changes in the realms of animal vocalizations, human language and diatonic music (Figure 2.44) is striking and suggests an underlying evolutionary mechanism at work in the common perception of major and minor chords. It is perhaps conceivable that the positive/negative affect of falling/rising pitch in all three realms is a mere coincidence, but both ethologists and linguists have already concluded that the "coincidence" is more than chance with regard to animal and human vocalizations. Could the positive/negative affect evoked by the pitch phenomena of music have a related biological explanation?

The evolutionary roots of human behavior in general and of voice prosody in particular suggest that the basic mechanisms of both animal and human vocalizations have common origins. Since the frequency code has previously been identified as a low-level, universal "code" for signaling social status, it would be parsimonious to argue that the same pitch phenomena are at work when human beings perceive the affective meanings of harmony. The transient implication of "strength" or "weakness" will of course be generated numerous times by any melody with changing pitches, but the major and minor chords function as moments of crystallized melody – and, within the context of diatonic music, act as brief statements of movement from ambiguous tension to relative strength or weakness. These emotional focal points are notably absent in chromatic music. Their absence necessarily implies a weaker emotional response from the typical listener accustomed to the recognizable harmonic packages of sound symbolism that tonal music so generously offers.

The implications of the sound symbolism hypothesis are complex and far-reaching for discussions of the meaning of music. Suffice it to say that, even for music without lyrics and without imitation of the sounds of nature (i.e., for the inherent meaning of "absolute music," Dahlhaus, 1978/1989; Hanslick, 1891/1986), music has at least one form of biological grounding in the usage of pitch rises and falls that "mimic" the emotional meanings that countless generations of competing mammals have employed throughout evolution. Each and every rise and fall of pitch in an extended musical composition will therefore contain a miniscule implication of biological dominance or submission. Within the framework of music where the aimless meandering of pitch typical of birdsong is replaced by the context known as musical key, the usage of pitch modulation provides brief moments of biological meaning. In addition to the various ups and downs of melody (with their own sound-symbolic hints of emotional significance), whenever three-tone combinations produce the high tension of intervallic equidistance, a subsequent fall in pitch conjures up the biological twinge of "social strength," whereas a rise suggests "social weakness." Playing in parallel with these evolutionarily ancient, instinctually understood pitch signals, the musical dimensions of rhythm, timbre, lyrics and multiple layers of musical context give music great complexity. Real music will therefore virtually never be heard as a statement of biological "victory" or "defeat," but the grounding to our biological beings – the faint probes of the autonomic nervous system that we feel on hearing well-crafted music – is arguably a consequence of the frequency code woven into what might otherwise be heard as nothing more than emotionless, inherently meaningless changes in auditory frequency. In this regard, the long-standing complaints about modern atonal music are arguably on target (Pleasants, 1955). Such music is intellectual, cerebral, abstract and designed to be enjoyed as pitch patterns *without* twinges from the autonomic nervous system – *without* the "obvious" affect implied by major and minor harmonies. It purposely does *not* activate the modal emotional triggers that are normally pulled again and again in genuinely popular, tonal music. In other words, atonal music can be enjoyed as an intellectual exercise devoid of limbic probes precisely because it does *not* employ the frequency code, the instinctual meaning of which we know in our mammalian flesh.

The structural arguments concerning musical harmony in this chapter can be summarized as four triadic grids (Figure 2.45). Many secondary issues involving the effects of the upper partials (their numbers and relative strengths) remain unexplored and, as research topics, those issues may prove worth studying, particularly in pursuit of the balance of musical factors that produce an optimal emotional reaction to music. But, wherever the pursuit of "optimal" music may lead, the main conclusions of this acoustical approach to music can be adequately illustrated by assuming a relatively simple upper partial structure (F_0–F_3 with gradually decreasing amplitudes) and then examining the configuration of dissonance, tension, instability and modality on the triadic grids.

The upshot of the acoustical approach to the ancient problems of musical harmony can be stated simply: When we hear the "tension," "buildup" or "unresolved character" of harmony or, contrarily, the relative "release," "repose" or "resolution" of harmonic tension, we are not merely recalling some learned pattern of "acceptable" and "unacceptable" pitch combinations absorbed over years of intense soap-opera TV viewing and exposure to elevator music! Yes, we may have odd associations of various kinds from our individual experiences in specific musical cultures, but we also hear the pitch structure that is ringing in our ears – right now! It is not exclusively or even primarily cultural brainwashing that forms our perceptions of musical harmony, but rather acoustical patterns, the biological significance of which has deep evolutionary roots.

From the small number of harmony experiments that have been conducted using animals as subjects (e.g., Hulse et al., 1995; Brooks & Cook, 2010), it appears that only human beings have the capability of perceiving

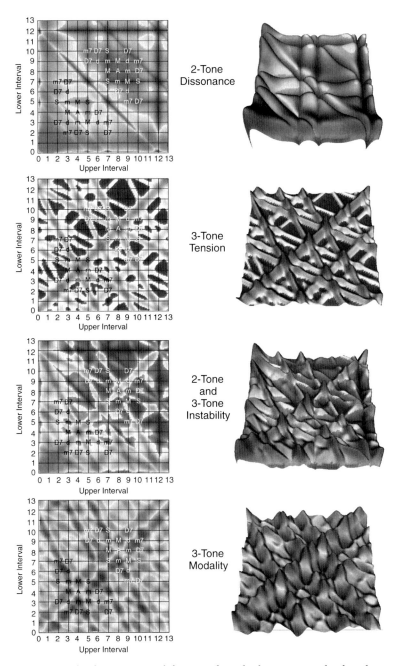

FIGURE 2.45. The four acoustical factors of triadic harmony as displayed on triadic grids. Most of the common triads lie in regions of relatively low dissonance and tension, and therefore low instability. Triads with the recognizable sonority of the major and minor chords lie in regions of strongly positive or strongly negative modality (color images are easily generated using the Seeing Harmony freeware, Cook & Hayashi, 2008).

musical harmony (as distinct from the perception of rising or falling pitch), but harmony is not unique to any specific musical culture. On the contrary, all musical traditions that use more than two scalar pitches – whether they are the pentatonic scales of many folk traditions, the 6-tone scales of Bali, the 22-tones of the Indian Ragas, or the familiar major and minor scales of diatonic music – all have implicit harmonies as soon as three different pitches have been struck. And whenever we hear a nondissonant three-tone melody or nondissonant three-tone chord, we are capable of perceiving the strange and wonderful properties of relative interval size: harmony. When there is symmetry – three-tones equidistant from one another in pitch space – we feel amodal tension that is emotionally ambiguous, unresolved and uncertain. Clearly, the tension of equal intervals is the main focus of music in the Gamelan tradition of Bali, supported by the inharmonic sonority of bells and chimes that explicitly do *not* call us back to major or minor harmonies. In the tonal music of most folk traditions, however, once the symmetry of three simultaneous or consecutive pitches is established, the symmetry can then be "broken." In creating asymmetry, neighboring intervals are altered to unequal size, we slide down from the affective uncertainty of tension and "resolve" in either of the two possible forms of inequality, major or minor, win or lose, strong or weak – and sometimes even happy or sad.

When played as either major or minor chords, the harmony is momentarily fixed, final, unambiguous and settled, but the nature of the resolution differs depending on what type of pitch movement has led to the resolution from ambiguous tension. If it is a rise in pitch, a faint sound-symbolic trigger is pulled – and we become aware of a state of nondominance – a social emotion personally experienced many times and witnessed in the interactions of others countless more times. The feelings induced by sound-symbolic pitch movement cannot be ignored, suppressed or denied any more than we can ignore, suppress or deny the pang of emotion in response to a baby's cry of unhappiness or an animal's yelp of pain. In contrast, if the unresolved harmonic tension is followed by a fall in pitch, again the sound-symbolic meaning is clear: strength, victory, dominance, contentment. We perceive the satisfactory resolution of uncertainty, the uncontested confidence of territorial dominance and the absence of conflict – a peaceful resolution of tension that only moments earlier had left us in the air.

2.5. TRADITIONAL HARMONY THEORY

Before embarking on the challenge of more complex harmonic phenomena, it is worth noting the distance we have already traveled in moving from the dyad story to the triad story. The acoustics of dyads (and the important role of upper partials) (i) tells us why the octave is the most consonant pitch interval and (ii) indicates why certain of the intervals smaller than an octave also sound pleasant. In this respect, it is clear that the nineteenth-century work of Helmholtz, Stumpf and Rayleigh in studying the effects of the upper partials is of permanent historical importance. It nonetheless needs to be said that, without including three-tone harmonic phenomena, the "scientific explanation" of music, in general, and the highly elaborated harmonic music of the Western diatonic tradition, in particular, leaves us with little insight. Three-tone musical triads do not by any means explain the full repertoire of musical tricks that real music presents to the auditory neocortex (and to the ever-sensitive autonomic nervous system), but the triads are arguably the gateway to understanding real music. They are the structural basis for the two types of pitch phenomena that give music its affective meaning: (i) tension and resolution and (ii) major and minor modality. These are the two "dimensions" of harmony that, together with the intervallic dimension of consonance/dissonance, are perceptually the most important, the most attention-grabbing, and the most salient aspects of nearly all forms of music where pitch is involved.

Music critics – particularly those once so hostile to Schoenberg, Webern and the composers of chromatic music – may have underestimated the importance (in small doses) of the violations of Renaissance harmonic sobriety that were utilized in the development of the atonal music of the early twentieth century. But critics, old and new, have never praised any musical genre for having an overall high level of consonance! On the contrary, particularly in the modern era, music that does not introduce new pitch adventures and does not flirt with the dangers of dissonance, tension and outright cacophony is likely to be criticized for being syrupy, banal, trite and simple-minded – precisely because it never wanders from the safe pathways of minimal dissonance. Relentlessly consonant music may be fine for children, who are still absorbing the basics of diatonic scales, but, for adults, something more complex is needed. Through the judicious use of dissonance and tension, the resolution of unresolved harmonies can be skirted, intimated but then denied, and generally delayed without making the music as unmusical as birdsong. But if resolved harmonies are completely eliminated in the pursuit of mathematical symmetries and the squelching of musical expectations, such truly atonal music will inevitably be as unpopular as recordings of whale vocalizations ... new and intriguing, perhaps, but no encore, thanks.

The ideas about three-tone combinations outlined in the preceding sections are, in essence, a restatement of common sense concerning musical harmony – stated without much of the terminology of traditional theory. It is therefore of interest to see how the complexities of traditional harmony theory might be reconfigured on this basis. First of all, it is evident that, if the unsettled ambiguity of the tension chords is taken to be the most salient feature of three-tone harmonies (just as dissonance is the most salient feature of two-tone intervals), then major and minor chords represent the only two possible resolutions of the unsettledness of triadic tension. In the same way that, once a dissonant interval has been played, it can be resolved into a consonance in one of two ways (an increase or decrease of pitch in either tone, leading to either unison or a consonant interval), similarly, three-tone combinations that are inherently unresolved (but not dissonant) can be resolved by pitch rises or falls. In other words, if three-tone harmonic tension is a perceptual reality, then there is inevitably a related perceptual reality in the resolution of tension through the usage of unequal intervals. Questions about the affective valence of the major and minor modes aside, the resolution of tension is inherent to the phenomenon of tension itself.

Clearly, if intervallic equidistance is the source of harmonic tension, then the existence of valleys of resolution is the inevitable reverse side of tension. Insofar as the 12-tone scale is the raw material from which harmonies are constructed, there are two and only two pitch directions to move from the symmetry of equivalent intervals (i.e., to the major and minor chords) (Figure 2.24). So, not only is it unlikely that diatonic music will evolve in a direction of unabated chromatic tension without the use of chords with resolved, asymmetrical intervals, it is also clear why, from the abundance of various church modes (the seven scales beginning on any of the white keys of the piano keyboard), only two modes have remained prevalent until the present day. While various modal *scales* remain possible and indeed in common use, only the major and minor modes are available to resolve harmonic tension.

It is relevant to note in this context that Schoenberg, one of the twentieth century's leading music theorists, predicted that the major and minor modes would disappear and go the way of the other church modes. He clearly expected that the achromatic music that he had experimented with would eventually win acceptance – despite the fact that, during his own lifetime, the listening public had so rudely failed to embrace it. In his classic textbook, *Theory of Harmony* (1911/1983), he stated that: "As for laws established by custom, however – they will eventually be disestablished. What happened to the tonality of the church modes, if not that? … We have similar phenomena in our major and minor" (pp. 28–9). Historically, he could not have been more wrong. Despite the best efforts of Schoenberg and the chromatic composers, there has been no indication of the disappearance of major and minor modes, keys or melodies throughout the twentieth or twenty-first century in either popular or classical music, and the strictly atonal music favored by Schoenberg and the chromaticists continues to have a small cult following at best.

Schoenberg's overestimation of the attractions of chromatic music can be understood as an honest miscalculation in light of the fact he – and indeed most theorists since the Renaissance – believed that pitch phenomena are built exclusively from consonances and dissonances. Given the broad historical trend toward the acceptance of gradually more and more dissonance, and away from the rigidity of "enforced consonance" in early Renaissance music theory, Schoenberg could imagine the eventual withering of preferences for all major and minor harmonies. Unfortunately, the *acoustical* reality was not as Schoenberg thought. Whatever the relative importance of interval consonance/dissonance to composer, musician and listener alike over these past few centuries, the presence of the higher-level phenomena of harmonic tension and modal resolution (i.e., symmetrical and asymmetrical harmonic triads) implies that, for there to be moments of relatively strong tension, there must also be moments of "release" that necessarily take the form of major and minor harmonies and melodies. Without the stability of the major and minor modes, the ambiguous balancing act of intervallic equivalence could not be enjoyed as a buildup of tension: Tension needs release just as release needs tension.

Listening to music, we frequently encounter pitch triads – played as melodies or chords. When we hear the unsettled up-in-the-airness of chromatic melodies or any of the tension chords, we are hearing the *symmetry* of three-tone patterns (not two-tone interval dissonance). And when, almost inevitably, the musical tension relents and the composer allows us to return to the more relaxed states of major and minor harmonies, we are hearing the *asymmetry* of three-tone patterns, patterns that resolve the triadic tension (and, again, are not simply a reduction in two-tone interval dissonance).

For anyone interested in understanding the pitch structures of music, the two- and three-tone patterns discussed here will probably now seem exasperatingly simple, crushingly boring and, in retrospect, entirely obvious. The structural arguments that underlie this view of harmony are so simple that anyone not already steeped in the ins and outs of traditional harmony theory will be thinking, "Of course, three-tone psychoacoustics has more to offer than two-tone psychoacoustics! And undoubtedly four- and five-tone relationships will bring further insights into why music has such emotional significance!" So, let us briefly address the issue of slightly more complex musical structures, in order to bring us a little closer to what might be called "real music."

Western diatonic music makes use of the acoustical properties of pitch triads to produce the psychological phenomena of tension/resolution and major/minor modality. Pre-Renaissance music – before the invention of chords consisting of three simultaneous tones – was played and presumably enjoyed without using harmonic triads, but most music in the Western tradition since then has relied heavily on three-tone effects. Under the influence of Western culture, many styles of nondiatonic music have subsequently been "harmonized" and forced into the diatonic modes – with disastrous effects on local musical traditions (and for ethnomusicologists interested in studying the original musical styles). The homogenization of all forms of music within a diatonic framework may well be an undesirable global trend, but it is also some indication of the power of the harmonic phenomena that have been incorporated into other musical traditions. For some reason, the more complex polyphonic structures employing triads and tetrads have an appeal to huge numbers of people worldwide. So, for many musicians from diverse musical cultures, there is a natural inclination to play music using traditional scales, melodies and instruments, while employing background "supportive" harmonies that are, strictly speaking, from a very different musical tradition.

The bowdlerization of folk music is in any case an unfortunate trend, but it is probably the case that the lure of harmony can no more be ignored than the lure of the big city. The added complexity that the music listener experiences in harmonic phenomena may lack the purity of rural life, but a brief return to the simplicity of isolated tones and consonant intervals can always be reintroduced amidst more complex pitch phenomena. Rather than deny the attractions of "urban harmony," the realization that harmonic sonority differs radically from interval consonance may foster greater understanding of both.

The chords of diatonic music can be classified on straightforward acoustical grounds, using the labels provided by traditional music theory, but not borrowing on the concepts of traditional harmony theory (Figure 2.46). The basic idea of such a bottom-up classification is to group the triads solely on the basis of the contributing two-tone and three-tone effects. Specifically, triads that have semitone or whole-tone intervals are necessarily rather unsettled – solely as a consequence of those dissonant intervals, whereas triads that do not have notable dissonance fall into two general classes, depending

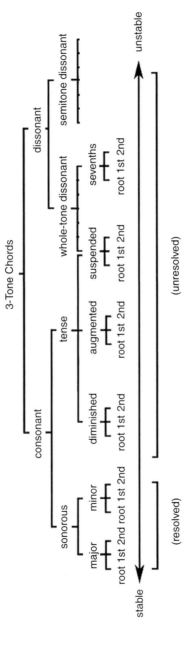

FIGURE 2.46. A classification of the triads, in which the relative "stability" of chords is influenced by both two-tone effects (consonance/dissonance) and three-tone effects (sonority/tension). Depending on the details of the upper partial structure, the relative consonance and sonority of the inversions of each chord type will vary.

on their overall tension. When three-tone tension is present, their overall instability will increase. When such tension is low (sonority is high), the triad will be stable – as found with any of the major and minor chords. In other words, stability is a function of both the interval and the triad effects.

The classification implied by consideration of both two-tone and three-tone effects aligns the triads in a series ranging from the extremely stable major and minor triads to the extremely unstable triads containing semitone dissonances. The only confusion in such a scheme is found with the suspended chords. On the one hand, these chords contain a mildly dissonant, whole-tone interval, but, particularly when played in open form, our perception of dissonance is rather weak. At the same time, they contain the tension of equivalent intervals among upper partials – and are somewhat unstable for that reason as well.

Chords containing a semitone interval are not frequently employed. Although they have their uses essentially as "passing notes" – brief moments of stark dissonance – a triad containing a semitone interval (or the major-seventh chords containing an 11-semitone interval) contains such grating dissonance that it demands immediate resolution. Insofar as it is precisely the freedom to move in various pitch directions that makes music an auditory adventure, the extreme restrictions on how a semitone dissonance *must* be dealt with makes their usage musically constraining – not necessarily unpleasant because of the stark dissonance itself, but more importantly because the semitone dissonance within a specific harmonic context must be resolved by raising or lowering one of the offending tones. At the far extreme away from such a lack of freedom are the major or minor chords, from which pitch movement of any of the three tones will lead in musically interesting directions – to be developed at will by the composer and savored by the listener.

As discussed briefly earlier (see Figure 2.37), the so-called seventh chords can be played as triads consisting of the tonic, the seventh "leading" tone and either the fifth or the third. Some of the seventh chords have a recognizably major or minor sonority, but that is not always true. In line with the three-tone acoustical model, the total tension and total modality of the seventh chords in their various inversions are as shown in Table 2.3 and Figure 2.47.

Using the concepts of tension and modality as calculated from the relative size of intervals in triads, the fundamentals of traditional harmony theory can be restated in terms of the underlying acoustical properties of chords. For anyone who has mastered the intricacies of traditional harmony theory, the urgency of exchanging one set of labels for another may seem

TABLE 2.3. *The sonority and modality of seventh chords*

Chord class	Interval structure	Empirical sonority Fukushima (2008)	Theoretical sonority		Theoretical modality	
			F_0~F_1	F_0~F_2	F_0~F_1	F_0~F_2
Generic seventh	7–3	2	0.91 (3)	1.88 (1)	1.76	1.28
	3–2	7	1.32 (8)	4.22 (9)	1.68	1.06
	2–7	9	1.12 (7)	3.74 (7)	1.55	1.57
Minor seventh	3–7	3	0.92 (4)	2.04 (2)	−1.55	0.35
	7–2	4	1.09 (6)	2.49 (3)	−1.76	−1.82
	2–3	8	1.34 (9)	4.19 (8)	−1.68	−0.98
Dominant seventh	4–6	1	0.63 (1)	2.68 (4)	−0.15	−0.11
	6–2	5	0.90 (2)	3.71 (6)	−0.16	−0.13
	2–4	6	1.03 (5)	3.14 (5)	−0.16	−1.24

Note that minor seventh and dominant seventh chords are built from seventh chords that include a major or minor third, whereas the generic seventh chords are not, and are for that reason labeled "generic." Numbers in parentheses indicate the ordinal ranking of relative sonority.

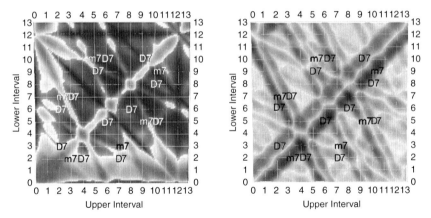

FIGURE 2.47. The total instability (left) and modality (right) of the seventh chords when played as triads. (Calculations were done with an upper partial structure of $1/n$ and including partials F_0~F_4).

minimal, but its significance lies in the fact that an acoustical description of tone combinations has an explicit scientific foundation that is lacking in traditional harmony theory.

One of the most frequently mentioned regularities of chord usage since the Renaissance is the so-called Circle of Fifths (~Cycle of Fifths)

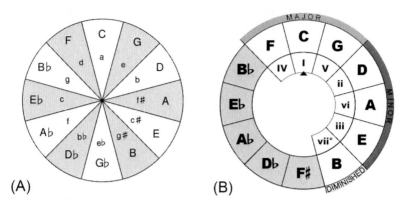

FIGURE 2.48. (A) shows the Circle of Fifths for the major (uppercase) and minor (lowercase) keys that share the same scalar tones. (B) shows the relationship between major, minor and diminished chords within a given scale (using C-major as an example).

(Figure 2.48). The Circle can be used to describe the relationships between chords, scales or keys based upon the perceptual fact that the two "closest," "most compatible" neighbors of a tone are located a fifth (seven semitones) above or below the tonic. Starting with any of the 12 arcs in the Circle of Fifths (e.g., the arc labeled C), its two nearest neighbors are F and G. If C is the tonic (I), then the nearest clockwise neighbor is G, the dominant (V), and the nearest counterclockwise neighbor is F, the subdominant (IV). By proceeding in fifths around the Circle, we eventually return to the original tonic itself, meaning that all chords are related to one another by a series of jumps of an interval of a fifth. In other words, the Circle of Fifths is a succinct summary of harmonic phenomena, with any scale or chord having two nearest neighbors in the same major/minor modality and one "parallel" neighbor (lowercase letters in Figure 2.48A) that employs the same scalar tones but has the opposite major/minor modality (due to the displacement of the tonic itself by three semitones).

The Circle of Fifths is generally understood as a shorthand mnemonic that summarizes the fact that the tonic of any major or minor key has these two "closest neighbors" – to and from which harmonic movement is "easy, natural and smooth." It is worth noting in this regard that modern textbooks on harmony theory often assert that any chord can be followed by any other chord (and it is up to the composer to prepare the listener for unusual transitions and work to justify what at first might seem to be a distant jump from one chord to the next), so strict obedience to the Circle of Fifths is hardly a fixed law of harmony. What fascinated Renaissance theorists and what is apparent to most listeners today, however, is the ease of transition between the tonic and its two nearest neighbors. Bolder harmonic leaps may be musically interesting and indeed a welcomed excitement, but the less adventuresome, comfortable transitions indicated in the Circle of Fifths totally dominate both popular and classical Western music.

Surprisingly, this mnemonic does not have an explanation in terms of traditional harmony theory. In the context of the familiar harmonic cadences of Western music, Huron (2006) has noted that

> In the chorale harmonizations by J. S. Bach, V-I cadences are 50 times more prevalent than V-vi. There does not appear to be any natural affinity between V and I apart from learned exposure. That is, there doesn't seem to be any obvious reason why V-I should sound better than some other progression (although such a reason might exist). The deceptive qualia evoked by the V-vi progression seems to reside solely in the fact that it is much less common than V-I. (p. 271)

Presumably, Renaissance musicians discovered that I-IV-V chords go well with one another – and countless tunes, ditties, sonatas and symphonies later – it may be that, today, we perceive their affinity because so much of classical and popular music repeatedly reinforces this harmonic association. But before we conclude that it is just a habit, let us examine the underlying acoustical structure of these chords.

First of all, let us review Huron's (2006) empirical work on chord transitions. Based on a large sample of classical (Baroque) and popular Western music, he found that, starting in any (major) key with a given tonic (I), there is a strong tendency for the sequence of chords to remain within the tonic-dominant-subdominant (I-IV-V) triangle (Figures 2.49A and 2.49B). Other transitions are occasionally used, and still others (not noted in the figure) are rather rare. The predominance of transitions among these three chord types is even more evident if we ignore the slight differences between classical and popular music and also the directionality of the chord transitions, and plot the incidence of sequential chord transitions (Figure 2.49C).

We find that ~85 percent of the cadences in diatonic music are among this triumvirate! Why would this be? Why are our harmonic tastes in general so hyper-conservative? On the one hand, it is true that a modulation from one key to the next is relatively "smooth" if the majority of scale tones are the same in both keys. Since the scales of nearest neighbors in the Circle of Fifths differ by only one flat or one sharp, transitions among

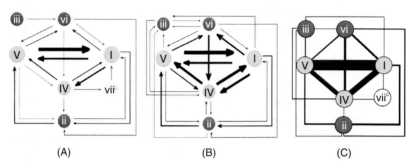

FIGURE 2.49. Huron's (2006, pp. 251, 253, with permission) tabulation of the incidence of chord transitions in (A) Baroque music and (B) modern pop music (images used with permission). The thickness of the arrows indicates the relative incidence of specific transitions. A growing diversity of cadences is evident from the increase in the number of arrows (19 vs. 24) in going from Baroque to modern popular music, although the usage of the diminished (vii) chord has decreased. (C) By collapsing (A) and (B) into one diagram, it is evident that transitions among the tonic, dominant and subdominant chords are the most common.

these neighbors are least likely to employ a tone in one scale that is not used in the other scale. That explanation may suffice for key modulation, but it clearly does not account for the smoothness of chord transitions, in general. For example, transitions from the tonic (I) to the parallel minor of the dominant (iii) (e.g., CEG to BEG) or to the parallel minor of the subdominant (ii) (e.g., CEG to DFA) are rather uncommon, but they share the exact same scalar notes as the dominant (V) and subdominant (IV) themselves. This suggests that the "nearness" of the dominant and subdominant chords to the tonic chord is not due *solely* to the commonality of the scales; there is something more to "harmonic proximity" than usage of the same scalar notes.

So, just what is "proximity?" Acoustically, it is clear that the jump from one chord to another will necessarily involve one or more rises or falls in auditory pitch. Why – among many possible chord transitions using only the scalar tones of the tonic – is the incidence of transitions among I, IV and V so domineering? Perceptually, these three chords sound "proximal," but the Circle of Fifths – and much of traditional theory – simply states what experience tells us is true, with no indication of why.

Here, the Cycle of Modes (Figure 2.40) can be employed to explain why. Starting with any major or minor triad, the number of semitone shifts that are required to arrive at another triad of the same or different major/minor/ tension modality can be easily counted. Transitions between major and minor can be achieved by changes as small as one semitone, but a transition *within the same modality* requires one full revolution around the Cycle of Modes. In other words, given a huge array of possible tonal rises or falls in each of the three pitches of a triad, the major/minor/tension modality – and therefore the affective sonority – can remain the same if the number of semitone steps is divisible by three (Figure 2.50, right).

In place of the Circle of Fifths, the harmonic "proximity" of tonic, dominant and subdominant chords can therefore be expressed by the fact that (while staying within the set of scalar tones established by the tonic) there are only two chords of the same modality that can be reached by as few as three semitone steps clockwise or counterclockwise around the Cycle of Modes (Figure 2.50, left). In this regard, *the dominant and subdominant chords are literally the two closest chords to the tonic in terms of pitch movement*. If we remove the restriction that only the scalar tones of the tonic can be used when constructing new triads (a restriction based on considerations of traditional harmony theory), then four other chords of the same modality can be reached by means of three semitone steps (e.g., given C-major as the starting point, C#-major, B-major, A-major and E♭-major can also be attained by three semitone steps). Note that, unlike the dominant and subdominant chords, these chords use one or two scalar tones that do not belong to the C-major scale. As a consequence, the perceived distance of these otherwise "proximal" major chords is relatively large (and their usage is correspondingly small). Insofar as the scale implied by a tonic chord is learned, then it might be said that these additional four chords of the same modality are *acoustically* as close as the dominant and subdominant, but they are *culturally* relatively unfamiliar in the Western idiom.

Similarly, the harmonic cadences from the common-practice period that "establish or confirm the tonality and render coherent the formal structure" (Piston, 1941/1987, p. 172) can be described as (predominantly) sequential revolutions (clockwise or counterclockwise) around the Cycle of Modes, the total number of semitone steps always being a multiple of three if the cadence is to begin and end in the same major or minor mode (Figure 2.51).

By taking both two-tone interval effects and three-tone triadic effects into account, certain basic principles of traditional harmony theory can be reproduced on an acoustical basis – the Circle of Fifths being a significant start. Music theorists would understandably be unhappy to be told that their hard-earned knowledge of diatonic music is nothing more than the Ptolemaic epicycles of an incorrect theory, but there is a far less radical compromise between traditional theory and psychoacoustics that should be considered. That is, traditional harmony theory can be considered as a summary of the

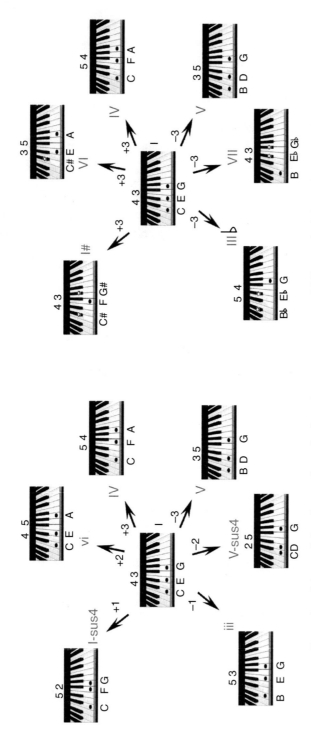

FIGURE 2.50. Examples of the modality changes brought about by semitone rises and falls. On the left are shown the chord transitions possible while using only the same scalar tones as the tonic (I). On the right are shown the possible transitions to chords within the same (major) modality with a shift of 3 semitones from the tonic. The dominant (V) and subdominant (IV) are the only two triads that stay within both the same modality and the same scalar tones.

106

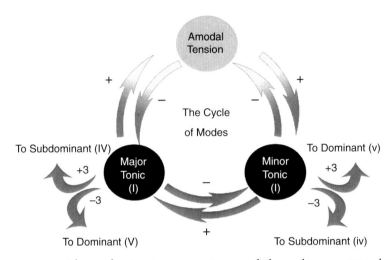

FIGURE 2.51. The predominant movements around the cycle are 3-steps clockwise from a tonic chord to its subdominant and three-steps counterclockwise to its dominant. All of the cadences (perfect/imperfect, authentic, plagal etc.) can be illustrated in similar ways in terms of the number of steps around the cycle. Each "step" in the Circle of Fifths corresponds to three steps in the Cycle of Modes.

"empirical perceptual reality" – the perceptual patterns in diatonic music that are heard and enjoyed by most listeners and that need to be explained by a science of music. Laboratory measurement of sonority, affective valence, jazziness and the like in response to isolated chords is sometimes useful for research purposes, but most of the issues of "What sounds good?" have in fact already been worked out in the marketplace of competing musical styles. Although the genres of popular music are diverse, there are far many more unrealized musical "genres" that have at best theoretical attractions and virtually no willing listeners. Traditional harmony theory – particularly as it has been developed to account for various nineteenth- and twentieth-century harmonic inventions – more or less states the case for "what sounds good" and how that can be justified within the framework of traditional theory (Goldman, 1965; Harrison, 1994; Mathieu, 1997; Persichetti, 1961; Piston, 1941/1987). But, explaining those regularities on an acoustical basis is what the science of music perception should be concerned with – not merely restating the regularities of harmonic phenomena, as if "understanding the Western idiom" were a sophisticated form of erudition that the unwashed foreigners fail to grasp. On the contrary, in the modern world the unwashed foreigners are already onboard – at around the age of 4 – as far as a fundamental appreciation of harmony is concerned.

2.6. THIS IS YOUR BRAIN ON HARMONY

The main psychological argument of this chapter has been that there is an important perceptual difference between interval dissonance and triadic tension. Both effects are rather similar in eliciting a sense of musical "unsettledness," but they are due to two distinct acoustical factors that are fully quantifiable in terms of pitch structure. The fact that they are perceptually similar, however, has led to countless mix-ups in the music psychology literature. On the one hand, interval dissonance is often referred to as "sensory dissonance," whereas all other kinds of unresolved pitch phenomena are, confusingly, referred to as "musical dissonance" – by which is meant some sense of learned instability that is acquired within a specific musical culture and that cannot be explained by sensory dissonance. Reformulating the difference between "sensory" and "musical" dissonance as the difference between two- and three-tone effects would be a great simplification, but does that theoretical argument find support in cognitive neuroscience?

In an attempt to answer this question, we have run a functional MRI experiment to measure the brain responses to interval dissonance and triadic tension (Cook et al., 2002). In that study, ten subjects listened to isolated chords and evaluated their sonority using a five-button finger pad. The chords were either (i) stable (major or minor) triads, (ii) unresolved "tension" chords (diminished or augmented triads) or (iii) highly unstable 3-tone chords that contained a semitone dissonance. All three conditions were followed by white noise stimuli, and the brain response to the auditory noise was subtracted from each of the relevant conditions. The final step in the analysis is called "exclusive masking" – whereby the activation that is unique to each condition is obtained by subtracting the activations common to the other conditions. The result of the exclusive masking is indication of what brain regions respond exclusively to (1) resolved chords, (2) unresolved chords or (3) interval dissonance and are not simply regions that respond to all kinds of auditory stimuli or all forms of musical harmony.

The main finding of interest is shown in Figure 2.52. For each of the conditions, six views of the "typical" human brain are shown, with regions of relatively strong activation indicated in black. Strictly speaking, the regions of "activation" are simply those portions of the cortex where an increase in blood flow was detected, but increased blood flow is normally found in regions that are particularly busy in processing a sensory stimulus. On the left is shown the unique activation in response to the tension chords (bilateral frontal cortex, predominantly Brodmann Areas 45, 46 and 47).

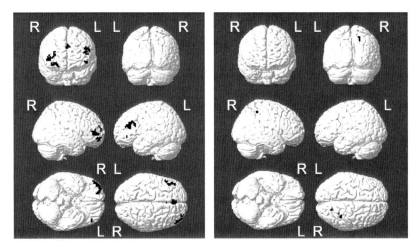

FIGURE 2.52. Cortical regions that respond to tension chords (left) and dissonant chords (right). Black regions indicate highly significant activation (uncorrected $p < 0.005$, $t = 3.25$, $k = 20$) (Cook et al., 2002).

Dissonant chords, on the right, showed activation of a small region of right parietal cortex (BA7) that is not activated in the other conditions. Activation in response to tension chords has not previously been reported by other music researchers, but other groups have also found a focus of activation in the right parietal cortex in response to semitone *dissonance* (e.g., Suzuki et al., 2008).

For researchers studying the functional neuroanatomy of the brain, the precise localization of these effects may be the main topic of interest, but a more general result is relevant here. Quite simply, the differential brain activation in response to dissonant and tension chords indicates differential processing of these two perceptually rather similar types of harmonic stimuli. When asked to give evaluations of either dissonant or tension chords along a dimension of "pleasant–unpleasant," "beautiful–ugly" or "stable–unstable," subjects typically score the major and minor chords together as pleasant, beautiful and stable, while *both* dissonant and tension chords are scored at the opposite pole. That is the *perceptual* reality, but our fMRI findings indicate the underlying neuroanatomical reality (i.e., the fact that dissonance and tension are distinct forms of pitch "unsettledness" that are processed at different locations in the brain). This finding is strong indication that a description of both types of unresolved harmony with the same label ("dissonance") is imprecise. Despite the fact that people typically report both dissonance and tension as "unsettled," the brain distinguishes between these types of pitch phenomena, and, acoustically, they are indeed distinct.

In a second fMRI study (Fujisawa & Cook, 2011), ten experimental subjects listened to two-chord cadences, each chord lasting 1.5 seconds and played as MIDI grand piano sounds. The difference between the two chords was defined in terms of the Cycle of Modes. That is, the second chord differed from the first chord by 0~3 semitone steps. Depending on the number and direction (up or down) of semitone steps, the pitch changes led to definable changes in the harmonic modality: major to minor, minor to major, major to tension and so on. There were consequently 21 different conditions (the seven varieties of pitch-change [−3, −2, −1, 0, +1, +2, +3] times the three varieties of modality [M, m, T]). The subjects were required to listen and respond with a button press indicating whether they heard a rise or fall in pitch, or no change at all. For the subjects, it is an easy task, but what the experiment was designed to detect was the change in brain activity in response to specific changes in modality within the cadences.

The most significant results concern the two conditions showing the strongest increases in brain activity. In fact, most of the chord transitions provoked only small increases in cortical blood flow (relative to white noise), but (1) a tension chord followed by a semitone rise (leading to a minor chord) or (2) a tension chord followed by a semitone fall (leading to a major chord) produced much stronger responses. What is interesting is the fact that these strong increases in blood flow involved not simply a semitone rise/fall alone nor simply a tension-to-major/minor change alone. Rather, the increases in blood flow were specifically a consequence of a combination of those factors: (i) changes from tension to resolution (major or minor) together with (ii) a semitone change in a particular direction. That is, starting from a tension chord and *lowering* one pitch by one semitone, or *raising* one pitch by one semitone produced strong hemodynamic responses. Note that these two conditions are the only conditions out of 21 that correspond unambiguously to the sound symbolism conditions (a transition from unresolved tension to strength with a fall in pitch or a transition from unresolved tension to weakness with a rise in pitch).

The specificity of the sound-symbolic effects can be seen by comparing the brain activations in the sound symbolism conditions with those obtained when the same initial and final conditions (tension-to-major) were employed, but the pitch changes were not those of sound symbolism. In other words, when the modalities of both chord transitions were the same as the sound-symbolic conditions (as shown in Figure 2.53C, tension-to-major), but the pitch movement was *not* a semitone fall to major, then

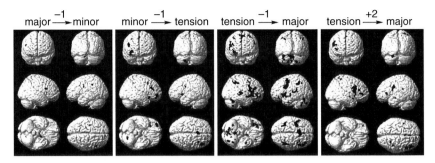

FIGURE 2.53. Having subtracted out the activation obtained in a white noise condition, the brain activity in the three conditions (ABC) where there was a semitone *fall* between the first chord and the second chord. (A) shows the negligible activation in moving from a major chord to a minor chord. (B) shows the weak activation in moving from a minor chord to a tension chord. (C) shows the massive activation in moving from a tension chord to a major chord. Note that the brain activity in a condition similar to (C) (tension to major, but where there was a whole-tone rise from the first to the second chord) is again weak (D) indicates that both the final chord and the direction of change are important factors in eliciting a brain response (all images, uncorrected $p < 0.005$, $t = 3.25$, $k = 20$) (Fujisawa & Cook, 2011).

the brain activation was notably weaker (Figure 2.53D). Clearly, the brain responded most strongly when there was a combination of falling tone and resolution to a major key.

A similar result was obtained for the only other sound-symbolic condition, the case of the minor mode. Figure 2.54 shows a comparison of the minor-chord sound symbolism condition (a semitone rise from tension to minor) (C) to the two non-sound-symbolic conditions involving semitone rises, but not ending in minor (A and B). Also shown is a comparison with the non-sound-symbolic condition in which there were similar modalities of the two chords, but a whole-tone fall from tension ending in minor (D). Again, the sound-symbolic condition (C) shows the most brain activity and the same tension-to-minor chords without the semitone rise (D) shows essentially no increase in blood flow. In comparison with Figure 2.53, it is evident that the minor-to-major cadence (B) is relatively strong; this condition showed the third strongest brain activation of the 21 different conditions, but not as strong as either of the sound-symbolic conditions.

Just one fMRI study on harmonic cadences will not provide final answers, but the results described here are some indication that the verbiage of this chapter is not merely empty speculation! On the contrary, observations of brain activity in response to specific harmonic conditions have confirmed that the strongest brain responses are the two conditions that correspond

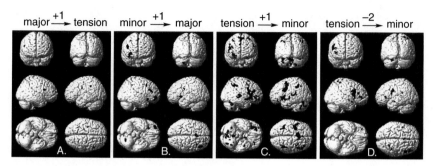

FIGURE 2.54. Again subtracting out the activation obtained in a white noise condition, the brain activity in the three conditions (ABC) where there was a semitone *rise* between the first chord and the second chord. (A) shows the weak activation in moving from a major chord to a tension chord. (B) shows the modest activation in moving from a minor chord to a major chord. (C) shows the strong activation in moving from a tension chord to a minor chord. (D) shows again that the strong activation in (C) is not simply a matter of the starting and ending chords (tension to minor), but a combination of such chords and the *direction* of tonal change (a semitone rise) (all images, uncorrected $p < 0.005$, $T = 3.25$, $k = 20$) (Fujisawa & Cook, 2011).

to harmonic manifestations of sound symbolism. That effect is confounded somewhat by the simultaneous effect of pitch rises – which, unrelated to harmonic modality, tend to produce stronger blood flow responses than pitch falls – but clearly two acoustical factors play important roles in eliciting brain responses to music: modality and the direction of pitch rises and falls.

In summary, we have found that that the brain responds most vigorously to harmonic conditions that reflect an instinctual understanding of the inherent biological meaning of pitch movement. It is noteworthy that, in music, a pitch rise or fall alone (unrelated to modality changes) was not the decisive factor, and modality change alone (unrelated to the direction of the pitch change) was not the decisive factor. Composers, take note: The kinds of musical stimulus that "excite the brain" and cause greater cerebral blood flow are specifically: (1) a semitone pitch *fall* leading from tension to a *major* harmonic resolution, or (2) a semitone pitch *rise* leading from tension to a *minor* harmonic resolution. The many other conditions with 0, +/−1, +2 or +3 semitone changes between sequential chords produced relatively weak changes in cerebral blood flow. It was specifically the +/−1 semitone changes leading from unresolved chords to resolved chords that produced increases in blood flow.

The results illustrated in Figures 2.53 and 2.54 are remarkable because the theoretical sound symbolism of animal vocalizations and speech prosody

(Figure 2.45) would predict that in a musical context there are two unambiguous, "frequency code" conditions among the 21 variants of small-scale pitch manipulations in harmonic cadences: (1) movement from tension to major by a pitch fall and (2) movement from tension to minor by a pitch rise. The third strongest brain response corresponded to the condition where there was a semitone rise from minor to major (Figure 2.54B); precisely why this condition also produced a strong brain response requires a more complex explanation in terms of the frequency code. That is, the minor-to-major cadence embodies an internal conflict between the "weakness" of the sound-symbolic rise in pitch and the "strength" of major chord resolution. The surprise inherent to the so-called Picardy cadence (the switch from minor to major at the conclusion of a minor piece) is perhaps the reason for the increased blood flow.

The only other semitone change that can lead to a resolved chord (major to minor with a semitone fall) entails a similar internal conflict, but it does not show a significant brain response (Figure 2.53A). The reason why only that semitone change leading to modal resolution did not reveal increases in blood flow is not clear, but, in general, resolution to minor chords is weaker than resolution to major chords and modulation from major to minor is a less common harmonic trick than that from the "defeated" minor to the "victorious" major. In any case, the strong emotional impact of the sound symbolism conditions moving from explicit ambiguity (tension) to favorable (major) or unfavorable (minor) resolution has been intuited by countless composers and exploited in countless musical compositions over the course of several centuries. The present fMRI study has shown brain activity that suggests why such harmonic cadences are musically so effective: They excite the brain.

Finally, by grouping together the conditions in which the final chord was either major, minor or tension (regardless of the number of semitone changes between the first and second chord), it is possible to see the locations of brain activity specific for these different modes (Figure 2.55). This is indeed the kind of brain-imaging analysis that is normally undertaken without consideration of sound symbolism and has previously shown that musical harmony more strongly activates the right hemisphere than the left (Zatorre, 2001). Moreover, the right orbitofrontal cortex is known to be involved in affective responses, in both a musical and nonmusical context. The focus of brain activity found in the right orbitofrontal cortex might therefore be considered to be a center of "musical affect."

A closer look at modality effects shows an interesting pattern (Figure 2.56). That is, the brain responses to all three harmonic modes

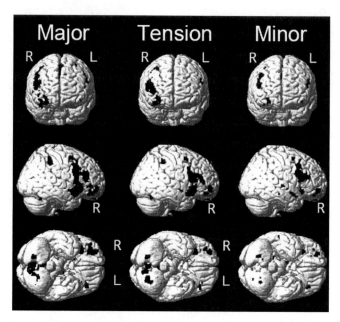

FIGURE 2.55. The brain activation for all three sets of final chords – major, tension, and minor – regardless of the preceding chord or the direction of tonal change (Fujisawa & Cook, 2011).

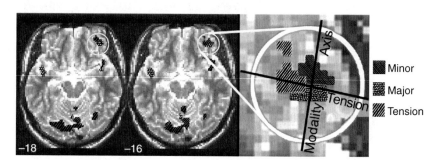

FIGURE 2.56. The harmony map in right orbitofrontal cortex, where all three modalities elicit focal activation. The dimensions of the map are tension/resolution – extending in a medial to lateral direction and negative/positive (minor/major) affect – extending in an anterior to posterior direction. The relatively strong activation of posterior sites implicates the cerebellum in the processing of harmonic mode, especially the equal-interval "balancing-act" perceived as tension (Fujisawa & Cook, 2011).

overlap in right orbitofrontal cortex. The relative positions of the three patches corresponding to major, minor and tension suggest the presence of a "cortical harmony map" with an anterior–posterior axis along which the negative–positive affect is recorded, and a medial–lateral axis along which the relative tension resolution is recorded. The possibility that, similar to so many other aspects of sensory processing, there may indeed be a two-dimensional neocortical mapping of the fundamental features of musical harmony is attractive, but further experimental work using both music and speech prosody is still needed.

2.7. WHY NOT BEFORE?

If the perception of harmony is, in essence, simply a matter of the summation of dyadic and triadic effects, why was it not fully known to Renaissance theorists? At least a partial answer to this sensible question is that a crucial piece of the puzzle of harmony was missing until Helmholtz's acoustical work on upper partials (Section 2.1) was published in the mid-1800s and their importance for music perception was not widely appreciated until much later. Hellenistic theorists (400 BC) and Renaissance musicians (AD1300–1600) did in fact have an inkling of the upper partial structure of tones from the "sympathetic vibration" of strings in synchrony with the vibrations of the fundamental frequencies plucked with the fingers, but they lacked the technological means for easy visualization, manipulation and analysis of the higher harmonics that have been available since the development of electronic equipment in the 1960s. Without an understanding of upper partials, Renaissance-era harmony theory was inevitably constructed around the patterns of tones that are actually played on musical instruments and whose identity and characteristics could be easily explored. From the perspective of the Renaissance theorist, it would be natural to believe that, even if there were higher frequency partials that tagged along with fundamental frequencies, the most salient aspects of intervals and chords would be due to the large amplitude fundamentals – with the upper partials adding only subtle flavors of timbre.

As a consequence, although the perceptual tension of certain triads (e.g., the augmented chord) would have been fully apparent to Renaissance musicians, the underlying feature of "intervallic equidistance" common to *all* of the nonmajor, nonminor "tension chords" would not be evident until the interval substructures among at least the first set of upper partials were also examined (Figure 2.22).

Similarly, the perceptual character of the major and minor chords in their various inversions was familiar to all Renaissance musicians and was an important consideration in the design of the early guitars and keyboards. But the idea that the major and minor modalities are a consequence of a certain pattern of asymmetrical intervals would not have been apparent until the upper partials were considered (Figure 2.34). The fact that the lower interval is one semitone larger than the upper interval for major chords and vice versa for minor chords is evident in four of the six major and minor triads, and this observation must have been made by countless musicians throughout the ages. But there are two "anomalies" in that simple account of modality: the first inversion of the major chord (intervals of 3–5 semitones) and the second inversion of the minor chord (intervals of 5–3 semitones) clearly do not fit the same pattern. Unless the structure among the upper partials is also considered, the seemingly inevitable conclusion is that two of the six modal chords remain anomalous and contradict a simple "difference of intervals" model. As was shown in Figure 2.34, the anomalies disappear as soon as the first upper partials are included in calculations, but this would not have been evident to Renaissance theorists (or, indeed, to many modern-day music theorists) concerned solely with fundamental frequencies.

Subsequent to the elucidation of the acoustical properties of the upper partials in the 1800s, an acoustical account of harmony would have been, in principle, immediately possible. Unfortunately, traditional harmony theory had already been developed into a truly impressive theoretical edifice – arguably, a complex Ptolemaic system, but one that works just fine, thank you! So, even with an equally thorough, strictly acoustical model of harmony in hand, it would not necessarily be the case that an acoustical description of the exact same musical phenomena that traditional theory had neatly classified and categorized would be easier to use.

Here, the astronomical analogy is truly appropriate because there are no (terrestrial) navigation problems for which the incorrect geocentric model of the universe will not suffice. Similarly, there are no (diatonic) musical phenomena that cannot be explained within the context of traditional harmony theory. But what eventually proved to be decisive in the acceptance of the heliocentric model were arguments of beauty, simplicity and parsimony. The complex epicycles of a Ptolemaic theory might be as useful – and allow daring sailors to navigate the Mediterranean Sea – but conceptually all the pieces seemed to fit more neatly with the Sun at the center of the solar system and the Earth demoted to the status of just another planet. Real-world practicality is important, but conceptual beauty is also a factor in choosing among theories.

The edifice known as traditional harmony theory is certainly not "wrong," but, lacking an acoustical foundation, it has remained a self-referential, closed system that requires more memorization than comprehension. Keys are defined in terms of scales, scales in terms of the harmonic functions of their tones, and harmony is normally described as a learned "idiom" that we somehow pick up in the first four years of life! All neat and clean, but the central "idiom" remains thoroughly unexplained. With the biological grounding provided by the frequency code, however, the regularities of traditional harmony theory can in fact be translated into fundamental psychological mechanisms that appear to have a biological basis. Importantly, the nature of those mechanisms can then be studied and tested in ways that make contact with phenomena outside of the world of music. It is for this reason that an acoustical explanation of harmony is to be preferred over the well-polished, self-consistent, but biologically ungrounded corpus developed by Renaissance theorists prior to a modern understanding of acoustics.

2.8. CONCLUSIONS

So, can it be said that the stability/instability and positive/negative affect of tone triads are "universals" of music perception? The answer is a qualified yes. Just as the traditional dissonance curve (Figure 2.8) is obtained from the cumulative *acoustical* effects of combinations of upper partials but requires a *psychological* model (the assumption of the perception of acute dissonance at ~1 semitone), the psychoacoustical model of triad perception contains both *acoustical* and *psychological* components. The acoustical components are again the cumulative effects of the upper partials, while the psychological assumption is that symmetrical three-tone combinations are heard as "tense." By adding together the effects of all three-tone combinations among upper partials, the model allows for predictions that correlate well with human perceptions on the sonority of harmonic triads – suggesting that the initial assumption concerning the perceptual tension of equal intervals is valid (as noted by Galilei in 1588 [Heilbron, 2010] and Meyer in 1956).

But, what if one were somehow raised in an environment where simultaneous two-tone intervals or simultaneous three-tone triads were not experienced? The dissonance and tension that most people raised in musical societies do perceive might then be felt less strongly, suggesting that the "universality" of both interval and harmony perception should be stated more modestly. That is, since the perception of dissonance, tension and modality is psychological in character, some early exposure in the formative years of childhood is probably required to appreciate their significance. In this regard, the music psychologists who emphasize the importance of learning, training and culture are not mistaken (pace Ball, Huron, McDermott, Patel, Sloboda and Trainor), but the perception of the affective qualities of harmony is not by any means a peculiarity of "Western culture," much less fourteenth-century northern Italy. An understanding of major and minor (and chromatic tension) requires little more than casual exposure to become an important part of our perception of the affective power of music.

The integration of rhythm, melody and harmony into music that most people would consider to be truly musical is a complex process, but triadic harmony itself is simple enough that the acoustical patterns that produce tension and the curious stability of major and minor chords can be stated in terms of the relative spacing of just three tones. Moreover, the fact that we experience emotional reactions to certain combinations of pitches can be coherently explained in terms of an ancient evolutionary story: sound symbolism. The "meanings" that *animal species* attribute to rising and falling pitch are also heard by human beings in three-tone patterns. We hear the strength of a falling pitch that is implicit in the structure of major chords and the weakness of a rising pitch that is inherent in minor chords because we can perceive the diatonic pitch context from which such harmonies arise. Although the meaning of sound symbolism is understood by a vast range of animal species simply from rising and falling pitches, the sound symbolism of harmony requires a musical context – and the minimal *neutral* context is a three-tone musical pattern containing equivalent, consonant intervals. From the ambiguous, unresolved tension that is inherent to such symmetry, movement toward asymmetry will be interpreted as a rise or fall, the emotional significance of which is instinctively known.

Relatively few animal studies have been made to determine how other species perceive the harmonic modalities that are understood by virtually all normal human beings. What has been found is that sensitivity to pitch rises and falls is similar to human sensitivities and, moreover, that animal preferences for consonance over dissonance are not unlike human preferences. But, despite some interesting efforts (e.g., Hulse et al., 1995; Brooks & Cook, 2010), the ability to distinguish between specifically major and minor chords has not been demonstrated in animal experiments – suggesting that the three-tone processing underlying harmony perception is an unusual human capability. (Note that, in the popular press, when headlines announce the "understanding of harmony" by animals, the research content is invariably an issue of interval consonance and dissonance, not three-tone harmonies.)

Without harmony, music lacks the tonal focus known as "key" and becomes as pointless and meandering as birdsong. Already by the time of the Renaissance, the importance of harmony was fully appreciated by musicians and its intricacies mapped out as formal harmony theory. There, the major mode was considered to be all-important, while the minor mode was an inexplicably euphonious variant with different emotional connotations. Everything else was "dissonance." The present chapter has outlined a contrary view: In addition to the major and minor modes, there is a third "chromatic" mode that also exhibits little sensory dissonance but has the explicit affective character of unsettled tension and ambiguity. More than anything else, it is the interplay of these three "modal" expressions of strength, weakness and uncertainty that makes music interesting – and allows harmonic cadences to toy so effectively with our emotions. The key insight that leads to this view of harmony is the realization that the human mind does not remain at the level of the perception of two-tone relations (the consonance and dissonance of intervals) but is attentive to higher-level concerns which, in the realm of musical pitch, are, to begin with, three-tone harmonies.

3

Human Seeing: Perspective

There was a simplicity to the analysis of musical harmony arising from the fact that pitch is one-dimensional: Tones are high or low or somewhere in-between. In contrast, there is an inevitable complication in studying visual depth perception due to the two-dimensional structure of the visual scene (and retinal array). Solely because of this added spatial dimension, a relatively large number of elementary configurations of visual cues must be examined in order to determine the effects of vertical and/or horizontal displacement of shapes in the visual field on our sense of 3D depth. Despite this problem, the ways in which small numbers of cues are combined can be delineated, and a coherent account of the factors underlying visual depth perception can be arrived at. As was the case for the study of harmony, it will again be necessary to go step by step through two-cue, three-cue and many-cue effects in order to build up an explanation of the overall perception of the illusory 3D space in 2D artworks.

Before addressing the question of how we see depth in static 2D pictures, let us first ask how we see 3D structure in the 3D world. Given the capability to identify individual objects at a distance from their visual appearance – and inevitably to evaluate their relevance to the seeing organism, reconstruction of the surrounding world becomes of value. A major, ongoing task is therefore to identify which of numerous visual objects are near at hand and which are in the more distant periphery. Objects that are close provide opportunities for and dangers from direct interaction, whereas more distant objects can be safely ignored or pursued later. Because of the obvious importance of such long-distance perception, it is no surprise that a variety of mechanisms have emerged in evolution – from the bat's sonar "vision" to the eagle's telescopic sensitivity. One way or another, it is important for animals to learn at short notice of the dangers and opportunities in the immediate 3D environment.

There are in fact three related answers concerning different visual skills for real-world visual depth perception. The first concerns the mere act of focusing both eyes on a specific object (ocular convergence). While we rarely think of this as "depth perception," focusing both eyes at a specific distance requires two muscular acts: stretching the optic lenses to the appropriate shape to obtain a clear image and pointing both eyes at the same location in space. Focusing at a large distance requires a rather flat lens and little convergence of the two eyes; at shorter distances a rounder lens gives focus and ocular convergence brings the same object to both fovea. The ocular effort required for eyeball and lens manipulation provides some feedback to indicate whether the object that is currently in focus is near or far.

The second type of depth perception concerns binocular stereopsis – that is, viewing an object at a sufficiently close distance that the images from the left and right eyes are slightly different. At a distance of a few meters or more, the two images are effectively the same, and the visual brain can conclude on the basis of binocular cues only that the object is "not close." But at short distances, the two eyes focused on the same object receive noticeably different images, simply because the eyes themselves are located at two slightly different, laterally displaced points in space. By comparing the two images, the brain can make approximate calculations of not only the distance to the object but also its shape. The magnitude of the disparity is inversely proportional to the distance, and the nature of the disparity between the two images provides information concerning the 3D structure of the object itself (Section 3.1).

The third type of visual computation for detecting the 3D structure of the 3D world is a related skill, but technically different from binocular stereopsis: that is, the perception of depth from movement, so-called motion parallax. In stereoscopic vision, the visual angles of the various surfaces of an object provide the information needed to identify its approximate shape and distance, whereas in motion parallax, changes in the visual scene as a consequence of self-motion provide such information. Both are important skills for deducing the 3D structure of the surrounding world, and both are skills that we share with monkeys and apes because of the similarities among our visual systems (Section 3.2).

In contrast to ocular convergence, binocular stereopsis and motion parallax, seeing the 3D shape of objects through pictorial depth cues is of a completely different character because it involves perceiving 3D structure in a static 2D picture. This type of depth perception is truly "illusory" in the sense that that there is of course no actual 3D structure in the 2D image, but it is depth perception nonetheless insofar as the 2D image provides clues that can be interpreted in terms of the structure of an imaginary 3D world. Neisser (2001, p. 79) is probably correct in maintaining that "There is no serious evidence that any animal other than *Homo sapiens* can manage" this trick of seeing 3D structure in a 2D picture "by looking into a flat picture," but experimental work with chimpanzees, gorillas and even horses suggests that they too are deceived by some of the pictorial depth cues that are used in human art (Fagot, 2000). Animals may therefore have at least some trainable talent in pictorial depth perception and can learn to distinguish different levels of illusory depth in 2D pictures. So, even though it is wise to avoid the phrase "uniquely human talent," it is clearly the case that seeing the third dimension in a 2D picture is an example of advanced visual processing and one at which human beings excel from an early age (Section 3.3).

Let us briefly examine the two "biological" depth perception skills and then consider the more "cognitive" case of pictorial depth perception.

3.1. STEREOSCOPIC VISION: TWO STATIC POINTS OF VIEW

For visual systems with two forward-looking eyes, the disparity between the two images of the left and right retina provides structural information on the protuberances and recessions in solid objects. Without any forehand knowledge about the shape of the object, the varying textures, colors and shadows on a rigid surface might suggest a 3D shape, but often one that is ambiguous. In contrast, the two different images obtained from slightly different angles in binocular vision can provide definitive information on 3D structure. That is, the small differences between the two, rather similar images indicate the presence of convex or concave deformation of what otherwise might be considered to be a uniformly flat surface (Figure 3.1).

The paired images from the left and right eye shown at the bottom of Figure 3.1 differ enough that we are aware of seeing an object from different vantage points. The magnitude of the disparity is alone indication of our distance from the object: The larger the disparity, the closer we are. The nature of the disparity, on the other hand, is used to calculate the convex/concave structure of the object. In Figure 3.1A, the left eye view of the light gray surface of the cube is notably wider than the right eye view of the same surface – clearly indicating that the left eye has a more frontal view of the light gray surface. Simultaneously, the right eye view of the dark gray surface is notably wider than the left eye view; the right eye has a more frontal view of that surface. These width disparities are consistent with the object

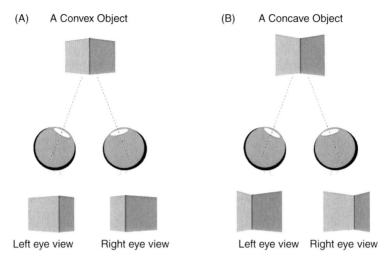

FIGURE 3.1. Perceiving 3D structure from binocular input. (A) and (B) show the two eyes and the two retinal images of convex and concave objects, respectively. Unless the object itself is 2D, a comparison of the left and right images when viewing at a close range inevitably shows differential widths of the various surfaces on the object. In (A), the left eye sees a broader expanse of the light gray surface on the left, and a noticeably narrower dark gray surface on the right than does the right eye. Having learned from early visual experience that a wider surface is a more frontal view of an object than a narrower surface, the eyes collaborate in concluding that the central vertical edge is close to us: the object must therefore be convex. In (B) again the difference in the widths of the surfaces of the object, as seen from the left and right eyes, leads to a conclusion about the 3D shape of the object: It must be concave. (Note that the relative height of the central vertical edge also differs somewhat in the concave and convex cases, but that visual cue is technically one of linear perspective, not stereopsis.) (after Purves & Lotto, 2003, p. 163, with permission).

being convex – with the central vertical edge being closest to the observer. Precisely the opposite conclusion is drawn from the cube surfaces in Figure 3.1B. Although each 2D image alone does not have any obvious 3D shape, the disparities in the surfaces indicate that it is concave and the central vertical edge is the most distant part of the concave object.

3.2. MOTION PARALLAX: TWO SEQUENTIAL POINTS OF VIEW

The limitations of stereoscopic vision for providing information about the 3D shape of individual objects are normally felt at a distance of several meters. At 10 meters, the images from the left and right eyes are effectively

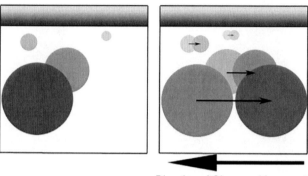

FIGURE 3.2. Perception of the distance of various objects in the visual field through motion parallax. Near objects move quickly; far objects hardly move at all. A comparison of the "before" and "after" images provides depth information that is as unambiguous as that obtained from binocular stereopsis.

identical and the visual world becomes essentially "monocular" – one picture seen by both eyes. We of course have no difficulty in perceiving the ins and outs of complex 3D shapes at this distance, so there must be other mechanisms for depth perception that we rely on when stereoscopic vision fails. The most important mechanism at this distance is motion parallax (Figure 3.2), the sense of 3D structure obtained simply by moving our heads or bodies while remaining focused on the same objects.

Similar to binocular vision, motion parallax is useful for determining the relative distance of distinct objects in the visual field. Objects that seem to move rapidly (as we ourselves move laterally) are close; slower moving objects are further away. Using only that information, we can estimate the approximate distances of the objects lying before us.

As illustrated in Figure 3.3, motion parallax is also useful for determining the inherent 3D structure of objects in a manner similar to the detection of shape through stereopsis. Although probably no one has ever been verbally instructed on the fact that the sides of objects widen or narrow depending on our lateral motion, the nature of such changes is precisely what we all learn during infancy. While crawling on the living room floor in front of chairs, tables and toys, we learn that the objects in our nearby visual world change in a regular manner depending on our own movements. If we move to the left, we necessarily see more of the left side of convex objects, and less of the right side. That is as true for toys and stuffed bears as it is for chairs, tables and buildings, so that we eventually come to understand that this visual regularity is a useful indication of convexity. Concave objects

FIGURE 3.3. The depth structure of an object deduced from motion parallax. The visual changes that occur with lateral movement provide sequential images of the external world. Comparison of any pair of such images provides information on 3D structure that is as unambiguous as the deductions from stereoscopic vision. From the changes in the relative sizes of the sides of the building, we conclude that the building is convex, with the corner edge close to us (from Hayashi et al., 2007).

show the opposite changes. In other words, we learn from experience that structural inferences can be made from sequential views of large objects, such as the cubic building in Figure 3.3.

In the real world, 3D depth is reliably computed using the disparity of binocular vision and/or the disparity obtained through self-motion. The closer the object, the more useful is stereoscopy; the further the object, the more useful is motion parallax.

3.3. PICTORIAL DEPTH PERCEPTION

The creation of pictures that produce an illusion of depth has a history as old as the cave paintings from 30,000 years ago, and, over the course of millennia, various techniques for implying depth in 2D art have been invented. The artistic principles that underlie pictorial depth perception were not, however, explicitly formulated until the Renaissance when so-called linear perspective and the depiction of cast shadows became the focus of both artistic and philosophical discussion. Since that time, many scholars have asked why the (now, well-understood) perspective techniques are so effective at deceiving the human mind. Why do we so gullibly imagine a third dimension of depth despite the fact that we know full well that the drawing, painting or photograph has only two dimensions? Why do we "look into" pictures that most animal species correctly see as flat surfaces (Hecht et al., 2003)? Clearly, in the "unreal" world of static 2D art, cues from the 3D world are no longer available, so how is it possible that we can reliably deduce depth structure from the markings on a flat surface?

An answer that first comes to mind is that perhaps we mentally construct a fictitious third dimension because, on the basis of previous experience, we know that most objects have significant 3D shapes: We have seen them from various angles, walked around them and held them in our hands. So, perhaps "the visual system builds a putative high-level 3D representation [of the visual scene]" (Papathomas, 2007, p. 84). Unfortunately, as an explanation of how we understand the 3D structure in 2D pictures, such "naïve realism" is inadequate because it does not address the perceptual process leading from 2D images to 3D understanding. It postulates, in effect, the existence of a brain module, a mental-model builder capable of recreating "objective reality" – the infallible truth corner in the cerebral cortex against which our fallible senses can compare their uncertain conclusions! When our eyes have deceived us and come to conclusions inconsistent with what we "know" to be true (so the story goes), then we have a sense of seeing a "visual illusion" – the perception of something that, on reflection, could not be real. The problem with naïve realism is that it leaves us clueless as to how this high-level module works and why other species lack it (i.e., why dogs and monkeys do *not* have a similar ability to see the 3D structure in 2D images). From the perspective of cognitive neuroscience, the postulation of an innate capability for understanding pictorial depth short-circuits the entire process of how we deduce the 3D structure from the 2D visual input. That process, of course, is the mystery that needs to be explained.

So, let us start again. If human beings do not make 3D mental models from stored knowledge, what are the visual cues in 2D pictures that are used to deduce 3D structure "on the fly"? Of course, a flat picture cannot directly show us the 3D structure of the real world, and motion parallax and binocular disparity cannot be employed to see the 3D structure. We must therefore rely on the information available solely on the 2D surface that might be interpreted as indicating depth – an "illusory" 3D structure "behind" the actual 2D surface.

The question that needs to be answered concerns what precisely are the cues that we consciously or unconsciously use in interpreting such pictures. Here, the work of artists over many centuries and from various cultures has long been far in advance of psychological research. Most importantly, rules for how to organize lines, shapes and shadows on a 2D canvas to convey a sense of depth were discovered once in Hellenistic Greece (300 BC) and once again in Renaissance Italy a millennium and a half later. From the modern standpoint and our utter familiarity with the realistic 3D scenes in 2D photographs, it is difficult to realize how revolutionary the invention of techniques for depicting believable scenes using shadows and convincing

FIGURE 3.4. Renaissance art designed to explain the importance of shadows (*The Origin of Painting*, Murillo, 1660) and a technique for perspective drawing (*Man Drawing a Reclining Woman*, from the second edition of *Dürer's Work about the Art of Drawing*, Nuremberg, 1538). Reprinted in *The Complete Woodcuts of Albrecht Dürer* (Willi Kurth, ed., New York: Dover, 1963).

perspective was. But those techniques in optical geometry transformed art from suggestive (but rather loose) story-telling to the presentation of convincing, geometrically precise depictions of realistic visual scenes.

Understandably, much of art history and much of the esoteric rancor about the discovery of perspective techniques have been concerned with the geometric transformations between the 3D world and its 2D representation (Kubovy, 1986; Damisch, 1987/1995). Indeed, this is such an important topic that some artworks have even been devoted to explaining the techniques themselves (Figure 3.4).

In discussions of the psychology of depth perception, the techniques of shadows and linear perspective are often listed as "monocular" depth cues simply because their informational content does not change with monocular or binocular viewing. But the important point is that there is no third dimension in flat pictures, only the *illusory* depth implied by the 2D cues. Unlike the binocular depth cues and the kinetic cues (motion parallax) that are utilized by many animal species, the static pictorial cues are something that human beings have a special talent for understanding.

Modern research on vision has clarified the neural mechanisms involved in binocular depth perception and motion parallax (e.g., Purves & Lotto, 2003), but progress has been slower in understanding the brain mechanisms underlying pictorial depth perception (Hoffman, 1998; Livingstone, 2002; Howard & Rogers, 2002; Solso, 1994, 2003). As illustrated in Figure 3.5 and Table 3.1, textbook discussions of the pictorial cues usually include the following: (1) relative object size, (2) relative height of objects in the visual field (~elevation), (3) occlusion (~overlap, interposition) of objects, (4) shadows and shading, (5) linear perspective, (6) transparency (~aerial perspective, atmospheric perspective) and (7) texture gradients. (Color is sometimes included, but "depth from hue" is a relatively weak cue, while

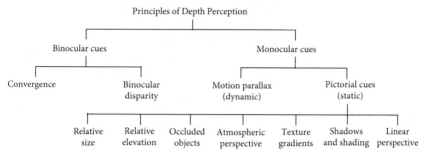

TABLE 3.1. *A typical classification of visual depth cues*

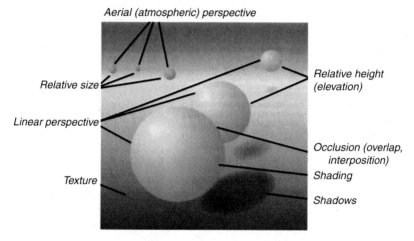

FIGURE 3.5. Illustration of the most important pictorial depth cues (after Purves & Lotto, 2003, p. 158, with permission).

"depth from the saturation or brightness of colors" more appropriately falls under "depth from shading.")

Let us examine these cues more closely.

The oldest technique for implying depth in a flat picture dates back to the cave drawings, in which relative depth was indicated simply by the overlap of objects. Aside from questions concerning the realism of the individual objects themselves, the overlap of one object by another provides unambiguous indication of relative closeness. Animals understand this, our ancestors 30,000 years ago understood this and modern "primitive"-style artists, such as Henri Rousseau and Paul Gauguin, have relied heavily on the depth cues provided by occlusion – nearly to the exclusion of other pictorial depth cues. Another ancient technique for providing the illusion of

FIGURE 3.6. An example of the aerial (atmospheric) depth perspective used in landscape painting. Distant mountains are occluded by closer mountains, and fade in intensity due to water in the air.

FIGURE 3.7. *Reconstruction of the Urbino Panel* (1993, acrylic on canvas © Ben Johnson, all rights reserved DACS, with permission) originally by Brunelleschi. Lippi's *The Annunciation* (1440, Martelli Chapel, San Lorenzo, Italy). The original Urbino Panel by Filippo Brunelleschi no longer exists, but was created around 1425 and is thought to be one of the first examples of a painting done in rigorous linear perspective.

depth is aerial perspective. As shown in Figure 3.6, aerial perspective is in fact a combination of the occlusion of landscape objects (typically mountains occluding other mountains) and the gradual fading of distant objects by atmospheric water molecules. They combine to give a strong sense of depth, mastered by Chinese landscape artists in the first millennium.

The realism of artistic painting gradually improved worldwide over the course of many millennia, but it was utterly transformed by the Renaissance discoveries of techniques for drawing geometrically precise linear perspective and optically convincing shadows (an early example of which is shown in Figure 3.7).

Following the Renaissance, various artistic genres emerged, but the emphasis remained on Realism and inevitably employed the Renaissance

FIGURE 3.8. The Neo-realism of Ben Johnson (*Tokyo Pool*, 2006, acrylic on canvas, © Ben Johnson, all rights reserved DACS, and *The Unattended Moment*, 1993, acrylic on canvas, © Ben Johnson, all rights reserved DACS, with permission).

tricks of linear perspective and shadows. Subsequent to the nineteenth century invention of photography, however, many artists rebelled against Realism and understandably seemed to find little merit in laboriously depicting scenes with an optical precision that could be achieved effortlessly with a camera click. The French Impressionists emphasized aspects of texture and color and were less precise in their use of linear perspective but retained realistic objects in realistic scenes. Often the immobile point of view of the camera lens was allowed to shift around somewhat, and the bizarre effects of more drastic changes in vantage point were explored by the so-called Cubists, such as Pablo Picasso and Georges Braque.

Early twentieth-century Surrealist painters, such as Salvador Dali and Georgio de Chirico, typically exaggerated aspects of linear perspective and/or shadows – to create slightly distorted depth scenes containing identifiable objects. Eventually it seemed that the variations on three-dimensional Realism had been sufficiently explored, and the generation of artists following the Surrealists completely rejected the "false idea" of illusory 3D depth on a 2D canvas. As scathingly documented by Tom Wolfe (1975), the new artistic ideology became "Flatism." Rather than play around with the tricks of pictorial depth and the complex geometry of linear perspective cues on 2D surfaces – previously, what most art critics would argue was the essence of human art! – the Abstractionists, such as Jackson Pollock, Mark Rothko and Piet Mondrian, returned to underlying issues of the materials themselves.

Inevitably, a subsequent rebellion against the rejection of geometric realism occurred, leading to a return to genres that emphasize the illusion of 3D depth – Neo-realism and Photorealism (Figure 3.8).

All of these artistic styles have their attractions, and personal preferences are as diverse as those for musical genres, but it should be clear from

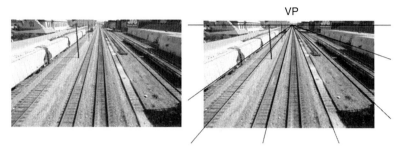

FIGURE 3.9. A picture with one vanishing point (VP). All lines of the train tracks converge at the same point on the horizon.

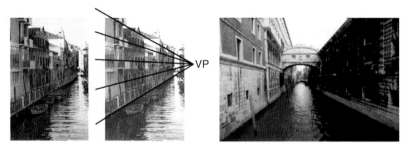

FIGURE 3.10. A vanishing point is implied, but not necessarily depicted. The canal scene on the right depicts a large concavity, implied by the centrally located vanishing point.

such familiar examples that the degree to which 3D depth is depicted and the types of depth cues employed are issues that are *central* to the artistic endeavor. At the extremes are intentionally "flat" canvases and the hyper-three-dimensionality of Surrealism and Neo-realism, but most artworks fall between these two poles as artists hone techniques for providing the illusion of depth through a specific subset of depth cues.

Of all the tricks of artistic 3D depth, the one that has had the most revolutionary effect on the art world is linear perspective. The philosophical and even metaphysical issues raised by linear perspective are of genuine interest (Kubovy, 1986; Damisch, 1987/1995), but let us begin with practical considerations. In guidebooks for artists, linear perspective is normally discussed in three stages, corresponding to the use of one, two or three "vanishing points." Examples of linear perspective with one vanishing point are shown in Figures 3.9 and 3.10. Whether explicitly drawn or not, the lines of linear perspective give the picture a focal point – and the viewer's attention is led

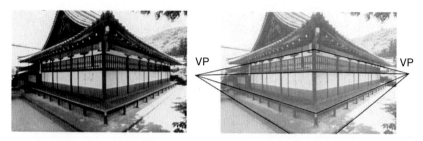

FIGURE 3.11. A picture with two implied vanishing points. The scene depicts a large convexity, a building. A line on the horizon is implied, but not depicted.

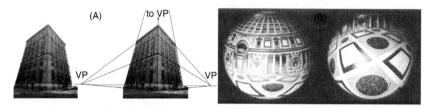

FIGURE 3.12. (A) shows a picture with three implied vanishing points and an implied horizon line. Again, the scene depicts a convexity. (B) shows two views of *The Pantheon* by Termes (with permission, 1998), illustrating depth effects obtained on a spherical "canvas."

(and indeed eyes are too; Wexler, 2005) toward the point of maximal distance in the picture, usually on the horizon.

Particularly when the scene includes architectural structures, a single vanishing point on the horizon might arise from the perspective lines on the vertical surfaces of the buildings (Figure 3.10).

Indication of a second vanishing point is often employed (Figure 3.11), with both vanishing points normally lying on the horizon. The space between vanishing points becomes a convex shape, the sides of which diminish in size as one gazes into the distance to the left or right.

A third vanishing point is less frequently employed but reflects the fact that the upper stories of tall buildings recede in apparent size as they project vertically into the air (Figure 3.12A). Perspective techniques in higher dimensionality have also been pursued (Figure 3.12B).

In the examples of Figures 3.9–3.12, there are explicit perspective lines in the pictures themselves – the train tracks, the floor and roof lines, the lines connecting windows and so on. But visual scenes with only implicit "linear perspective" will also exhibit the orderly diminution of objects – changes in relative size that imply, if somewhat approximately, changes in depth. For this reason, most experimental work on depth perception has examined pictures that do not employ the visible lines and vanishing points of the linear perspective technique. As illustrated later, in using drawings that contain only a handful of visual cues, it is possible to show how, even without explicit indications, the illusion of 3D realism can be built up from combinations of a small set of relevant cues.

Pictorial depth perception has been a favorite topic for psychologists over the past century (reviewed by Howard & Rogers, 2002; Gregory, 1997; Hoffman, 1998; Hershenson, 1999; Livingstone, 2002; Solso, 2003). Despite some real progress, the various cues underlying depth perception have remained an unorganized collection of diverse artistic techniques (e.g., Table 3.1). The terminology varies somewhat among different authors, but what has remained consistently unsatisfactory about all such lists is the fact that they are little more than a grab-bag of assorted tricks, with no conceptual links other than the fact that each technique contributes somewhat to the illusion of 3D depth in 2D paintings. Is there method to that madness?

3.4. LINEAR PERSPECTIVE

Assuming a visual system capable of object recognition (itself, a complex perceptual process, Ullman, 2000), the simplest indications of 3D depth in a visual scene are those that involve the *occlusion* of one object by another. In essence, occlusion is an example of a two-body computation. The overlap of two objects, each of which may have some inherent 3D structure, is an unambiguous indication of their relative distance from the observer. That is, it can be inferred with a probability approaching 100 percent certainty that the occluding object is closer to the observer than the occluded object. The height of the objects in the visual field, their left–right orientation and even their relative size are irrelevant to the interpretation of which object is nearer (Figure 3.13). Despite the fact that the overlap itself provides no indication of the metric involved, the interposition of one object between the observer and a second object is decisive indication that the interposing object is closer. So, we may have no idea whether the objects depicted in Figure 3.13 are the size of ping-pong balls or the moons of Jupiter, but we know for certain which one is nearer to us.

In contrast, the relative distances of nonoccluding objects of the same size are fully ambiguous (Figure 3.14). We might imagine that either sphere is closer (and physically smaller) or further away (and larger), but we have no visual information to guide our imagination, no grounds to support or deny a preconceived notion about differential depth, and indeed no strong

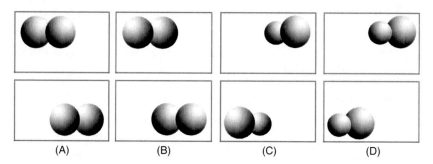

FIGURE 3.13. Occlusion (overlap) is unambiguous indication of relative distance regardless of the vertical or horizontal position of the objects in the visual field or their relative size. The relative wholeness ("Prägnanz") of the two intersecting objects suffices to indicate relative closeness.

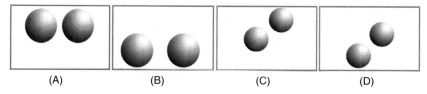

FIGURE 3.14. The relative distances of two nonoverlapping objects of the same size are uncertain, regardless of relative height in the visual field.

sense that one interpretation is correct or inevitable. Without further information, even differences in the relative height of similar-size objects give little support for a differential depth interpretation of the two objects.

If, in addition to differential height, we add differential size (in the retinal image), the depth story becomes somewhat more complex (Figure 3.15, upper row). We still cannot be certain that one of the two objects is closer than the other, but a "default" interpretation that the larger of two similar objects is closer is likely because we know from experience that objects loom larger as they approach and become apparently smaller as they recede. That depth interpretation may turn out to be incorrect, based on further cues (Figure 3.15, lower row), but, without such clues, the seemingly larger one may be closer than the seemingly smaller one. Maybe so, but it is not necessarily so.

The presence of a horizon line in the background adds another factor (Figure 3.16), but, even if we assume that the gray region is a surface, above which the objects somehow float, we nonetheless have no additional information to conclude which of the spheres is closer. Perhaps the larger is closer, as we might have concluded from the diagrams in the upper part of

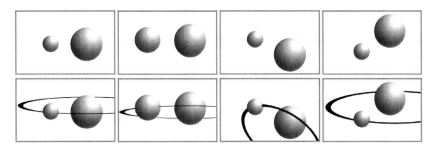

FIGURE 3.15. No definitive indication of differential depth is obtained from relative size, but the larger object is more easily interpreted as closer than further away (upper row), unless further cues indicate otherwise (lower row).

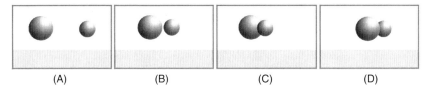

FIGURE 3.16. Even with an apparent horizon line, the relative distances of the spheres from the observer are uncertain until they overlap.

Figure 3.15, but we can imagine, without contradiction, that the apparently smaller object is considerably smaller than the larger object but is closer to us than the larger object. On the other hand, the occlusion of either the small or the large object (Figures 3.16C and 3.16D) conclusively resolves the relative depth issue and eliminates all speculation about the possible significance of relative size as indicating closeness. If occlusion is apparent, the presence of a horizon line is irrelevant and the overlap decisively indicates relative distance.

Thus far, our progress in understanding pictorial depth perception from simple drawings has been a little bit laborious! What is certain is that the "two-body computation" demanded by the overlap of two objects is a definitive clue concerning relative depth, whereas other two-body pictorial depth cues – relative size and relative height – leave the interpretation of depth uncertain. So, we might ask, is this uncertainty simply because of the ambiguous nature of these floating spheres? Or would familiar objects bring a greater sense of 3D structure?

Figure 3.17 shows several drawings containing familiar objects – light bulbs of similar shape, but different size and drawn at different relative positions in the visual field. The light bulbs themselves are caricatures

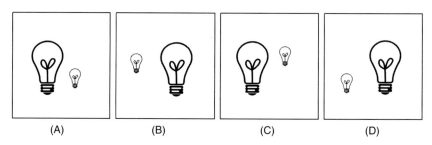

FIGURE 3.17. Pictures containing two familiar objects. The "default" interpretation is that the larger light bulb is closer, but it is possible to perceive each picture either as (i) a large light bulb and a small light bulb or (ii) a near light bulb and a far light bulb. The familiarity of the objects plays no role.

with no indication of inherent 3D structure, but we bring to such pictures some general knowledge: We know that light bulbs are roundish convex objects, generally less than one meter but more than one centimeter tall. With only these two visual objects in the scene and no visual context aside from the picture frame itself, two quite different interpretations are again possible: (i) Each picture could be a scene with a large and a small light bulb lying at similar distances from the observer, or (ii) they could be two light bulbs of similar size, one of which is considerably closer than the other. The uncertainty here is similar to that seen in the upper row of Figure 3.16 depicting spheres.

Simple diagrams such as these are of course far from being natural scenes and, therefore, lack the many additional visual cues that would most likely indicate which interpretation – differential size or differential depth – is correct. From the perspective of a biologically evolved visual system, the answer to the question of which interpretation is correct will be a judgment about the naturalistic probabilities (Purves & Lotto, 2003; Howe & Purves, 2005) of finding two such objects in either a big/small or a near/far relation – with full consideration of the visual context in which these two objects are found. Without such context, the judgment concerning their relative size or relative distance from us is uncertain ... until more information is supplied.

The mere presence of a horizon in Figure 3.16 did not strongly affect our perception of the distances between us and each of the two spheres, and the grayish region in the picture did not clearly distinguish between a picture with objects above a nearby table-top or objects hovering above a distant plain. What about the effect of the *relative height* of objects from such a horizon line?

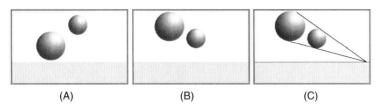

FIGURE 3.18. Different size spheres at different heights relative to a horizon line.

In Figures 3.18A and 3.18B, spheres of different size *and* different relative height are shown together with an apparent horizon. We can again imagine that either sphere is closer or further away, but the depth relations in Figure 3.18B are somewhat less ambiguous than those in Figure 3.18A. Why? In Figure 3.18A, given no visual information concerning their absolute sizes, a default interpretation that the two spheres are in fact similar in physical size despite their different apparent size is of course conceivable, but, if that is the case, then the alignment of the spheres bears no relationship with the position of the horizon: The visual "context" provided by the horizon plays no role in our interpretation of the structure of the 3D scene. The relative sizes of the spheres in Figure 3.18B, however, are related to the vertical location of the horizon line, if we assume that the spheres are of comparable size, but lying at different distances. By drawing explicit perspective lines, as in Figure 3.18C, the depth relationship becomes unambiguous. Without the drawn perspective lines, it is the rather weak contextual clue that suggests a slightly stronger depth interpretation for Figure 3.18B than for Figure 3.18A. In fact, the 3D structure of Figure 3.18B is still ambiguous, but we are inclined to believe that, even without drawing the perspective lines in Figure 3.18C, the smaller sphere may be more distant than the larger sphere because an interpretation of differential depth has one additional bit of supporting evidence that is not found in Figure 3.18A.

Note that the numbers of visible objects in the scenes in Figures 3.18A and 3.18B with large and small spheres are identical, and yet the implication of depth is different for Figure 3.18A and Figure 3.18B. Try it! While staring at ("looking into") each of the scenes, one at a time, try to imagine them as depicting one sphere as closer than the other, and then imagine the reverse situation. Similar to understanding the structure of the Necker cube (Figure 2.18), most people can do this with a little bit of mental effort, but – depending on rather subtle differences – the depth relations will be seen with somewhat more or somewhat less difficulty. What are the factors that facilitate or inhibit the imagination here?

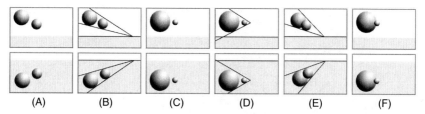

FIGURE 3.19. The horizon line in (A) suggests differential depth for the two objects since it provides a third cue that is consistent with linear perspective. The depth effect becomes unambiguous when perspective lines that converge at the horizon are also shown (B). In (C), perspective lines that would connect the spheres do not converge on the horizon (D), so that the differential retinal size, even in the presence of a horizon line, has only weak implication of differential depth for the two objects; they now appear to be large and small spheres floating in space unrelated to the distant horizon. In (E) and (F), possible differential depth interpretations due to relative size, height or perspective orientation are trumped by the more powerful occlusion cues. The upper and lower rows again indicate the irrelevance of viewing angle for determining the depth relationship between the two objects.

In Figure 3.19A, the relative ease with which we see the large sphere as close and the small sphere as further away is based on an implicit assumption that the two objects of similar shape but different retinal size are of similar physical size. That assumption may prove to be correct or incorrect, depending on further cues, but, particularly if the alignment of the two objects is consistent with linear perspective (Figure 3.19A and 3.19B), an interpretation of differential depth will be slightly favored over an interpretation of differential size. In contrast, if the alignment of the two objects of different apparent size is *inconsistent* with linear perspective (Figure 3.19C and 3.19D), then the implication of differential depth will be weaker or entirely absent.

The depth interpretation of two context-free objects is unambiguous only when occlusion is apparent (Figure 3.19E and 3.19F). The addition of further cues in a scene without occlusion will eventually disambiguate the possible interpretations, with the weight of the evidence normally supporting one interpretation over all others, so that we are not left vacillating between contradictory interpretations. Such summation of multiple cues is typically referred to as "contextual" effects and explained as a consequence of "higher-level" cognition. That may well be true, but before jumping to the vacuity of "contextual" explanations, some tentative conclusions can be drawn from a comparison of two-cue versus three-cue configurations.

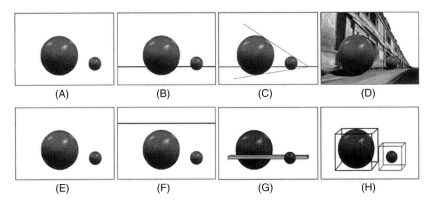

FIGURE 3.20. "Contextual" effects. Although there are weak implications of depth solely from the relative size of two objects (A&E), additional cues consistent with linear perspective indicate the greater likelihood of a "differential depth" interpretation (B~D), whereas other cues indicate a "differential size" (F~H) interpretation of the same objects.

We have already seen that, simply by adding a third visual cue (e.g., a horizon line), the strength of the inference concerning the depth versus size interpretations of two objects can be altered. Starting again with two spheres of different apparent size (Figure 3.20A), when a horizon line that is consistent with linear perspective is drawn (Figure 3.20B), a depth interpretation becomes favored – not decisive, but favored – over a differential size interpretation. Adding perspective lines that connect corresponding regions of the objects and converging on the horizon line (Figure 3.20C) and/or multiple perspective cues (Figure 3.20D) further strengthens the depth interpretation. In contrast, if the horizon line is *inconsistent* with the possible linear perspective implied by the objects themselves (Figure 3.20F), if occlusion cues are *inconsistent* with the differential depth interpretation (Figure 3.20G) or if a more complex visual context not suggestive of alignment of the objects in-depth (Figure 3.20H) are drawn, the weight of evidence from the multiple cues argues decisively against the differential depth interpretation.

Although the depth interpretation of two nonoverlapping objects is ambiguous, and the depth interpretation of two nonoverlapping objects with a horizon line may be suggestive, the situation changes drastically when a third object is present. In Figure 3.21A, again light bulbs are depicted – now with a third light bulb added. The potential ambiguity in interpreting this scene is between "three light bulbs of different sizes (small, medium and large)" or "three light bulbs at various distances (near, intermediate and

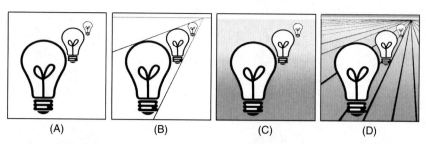

FIGURE 3.21. Pictures of three objects. Are they three objects of different sizes or three objects at different depths? With clear indication of linear perspective lines (B), a texture gradient (C) or both (D), an interpretation of depth is strongly favored, but even without any further visual information, an interpretation of 3D depth is favored in (A) simply if the three objects are aligned in accordance with linear perspective.

far)." Again, both interpretations are possible – conceivable to the human imagination – but there is already a strong bias for choosing the 3D depth interpretation as more likely. Why is that?

The answer is simple, but profound. Even without explicitly drawn perspective lines, a background gradient or additional contextual cues, the human visual system cannot help but to infer the undrawn perspective lines that potentially join homologous points on the light bulbs. Adding contextual information may further bias the interpretation in favor of 3D depth (Figures 3.21B through 3.21D), but a leap from ambiguity (two equally probable interpretations) to an "obvious" sense of 3D depth is brought about simply by the appropriate alignment of *three* objects. Even without the definitive information provided by occlusion, the implication of alignment *in depth* is perceived (already in Figure 3.21A and strengthened by further contextual cues). At long last, we have come to an insight concerning pictorial depth perception that is not already fully understood by 6-year-old children!

If the same three objects are *not* aligned to suggest (undrawn) perspective lines (Figure 3.22), an interpretation of 3D depth is not entirely impossible: Maybe there are light bulb-shaped balloons floating in 3D space before us! And with some effort we can imagine precisely that. But, from the point of view of a visual system that has evolved primarily to disambiguate inherently ambiguous visual information, the absence of additional evidence in the scenes in Figure 3.23 leaves the depth interpretation uncertain. It is as easy to imagine that three different size objects lie at more or less the same plane of depth in the visual scene, as to imagine three similar size objects arranged in a complex depth configuration. We cannot disambiguate.

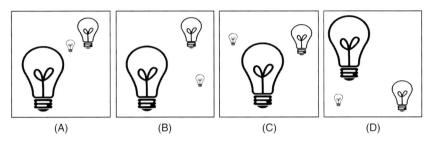

FIGURE 3.22. When the same three objects are not linearly aligned, the suggestion of depth is weak. The likelihood of a 3D interpretation differs depending primarily on the vertical elevation of the big and small objects, but in no case is the 3D interpretation of three unaligned objects natural, easy or inevitable.

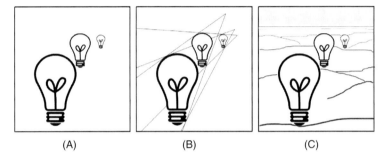

FIGURE 3.23. Three unaligned objects do not indicate unambiguous linear perspective, but, with the various implied vanishing points lying in the same quadrant of the visual scene (B), a depth interpretation is still strong (A) or, with additional cues, absolutely convincing (C).

In contrast, the three aligned objects in Figure 3.21A are more easily imagined as being aligned "in depth," while perceiving them as three different-sized objects at the same depth requires some additional mental effort.

Note that the *alignment* or *misalignment* of the three objects is the crucial factor for facilitating or inhibiting the interpretation of 3D depth. The arrangements in Figure 3.22 do not favor the depth interpretation, but the slight misalignments in Figure 3.23 are not so decisively negative. We can easily imagine a 3D scene in which the three objects might be arranged in depth to produce these configurations. In contrast to the display of the objects in Figure 3.22, the *approximate* alignment in Figure 3.23 is sufficient to suggest 3D depth. Lacking the draftsman's precision or additional contextual cues, the interpretation of depth is perhaps less than conclusive, but the approximate alignment in Figure 3.23A and 3.23C is still more suggestive of 3D depth than similar objects pasted at random on a 2D canvas.

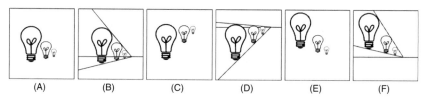

FIGURE 3.24. Again three objects aligned to suggest depth. The orientation is not critical, but the (approximate) alignment is.

But if the three objects are precisely aligned in accord with linear perspective (Figures 3.21A, 3.24A, 3.24C and 3.24E), it is difficult *not* to infer their alignment in depth. Note that their relative orientation (horizontal or oblique) is not crucial. In either case, the implied linearity favors an interpretation of objects arranged in 3D, not 2D, space.

Having suggested that the alignment of *three* objects is decisive for perceiving depth, it is important to consider why the "alignment" of *two* objects is somehow *not* decisive. The argument boils down to one of "redundant information." Two similar objects of different size are necessarily and inevitably "aligned" in the sense that we can always connect homologous points with perspective lines that indicate a vanishing point and horizon. Unlike configurations of three or more objects, two objects can never be "unaligned," so that the implication of a vanishing point is not strong and the imagined "alignment" in depth is not convincing. In other words, the alignment itself for *two* objects provides no additional information. This means that the possible inference of depth from *two* objects that are *consistent with* linear perspective (but not including any further perspective cues) is inherently weak (Figure 3.25A and 3.25B, upper row) and open to alternative interpretations.

In contrast, the presence of a third (nonoccluding) object that is aligned with the other objects such that they are all consistent with linear perspective provides redundant information, all indicating the same depth interpretation – the same vanishing point and a unique horizon at the same depth. In other words, a scene with three aligned objects is already an extremely low-probability configuration if they are objects of different sizes (Figure 3.25C and 3.25D, upper row). With or without perspective lines converging on a common horizon line, it is unlikely that three objects of different sizes but at the same distance from the observer would be aligned by chance in such a linear array. The fact that they are aligned in these 2D images (Figure 3.25C and 3.25D, upper row), such that perspective lines, if drawn, connect corresponding regions of all three objects (Figure 3.25C&D, lower row), indicates

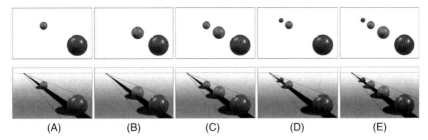

FIGURE 3.25. Pictures in the upper row illustrate the effects of small numbers of visual objects on the perception of illusory 3D depth. Pictures in the lower row illustrate how further visual cues (horizon line, shadows, perspective lines, surface gradients, etc.) make the depth relationships explicit. In the upper row, the depth relationships of the two objects in (A) and (B) are inherently ambiguous; they can never be "unaligned" and therefore can be interpreted either as "objects of different size but at similar depth" *or* "objects of similar size but at different depth." Although similar "differential depth" or "differential size" interpretations are possible for three or more aligned objects, as in (C– E), the "differential depth" interpretation is already strongly favored over the "differential size" interpretation, even without the additional depth cues shown in the lower row. Why? (See the back cover for a color version of this figure.)

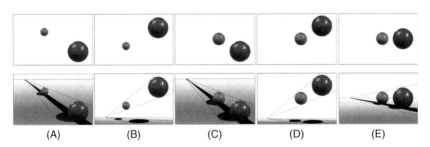

FIGURE 3.26. Two nonoccluding objects of different size (upper row) are necessarily "aligned" in depth and can always be interpreted as consistent with the laws of linear perspective (lower row), regardless of their placement in the 2D image.

that the higher-probability interpretation is one of "differential depth" as distinct from a literal (retinal) "differential size" interpretation. The alignment of additional objects (Figure 3.25E) provides further evidence of depth structure, but the qualitative jump from ambiguity to illusory depth lies in the difference between two and three objects.

Figures 3.26 and 3.27 further illustrate the significance of the alignment of three or more objects – and indeed the nonsignificance of the alignment of two objects. Two objects of different size are never "unaligned" – which

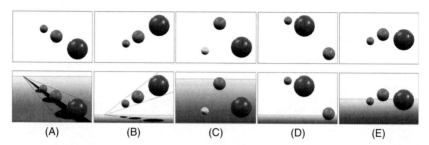

FIGURE 3.27. Three nonoccluding objects of different size (upper row) are either aligned (and consistent with one unique interpretation of depth, as in (A) and (B)) or unaligned (and inherently ambiguous with regard to a depth interpretation, as in (C) and (D)). Note that the geometrically imprecise, but approximate alignment in (E) enhances a possible in-depth interpretation. (See the back cover for a color version of this figure.)

makes the inference of their "linear perspective" depth relation inherently unconvincing (Figure 3.26, upper row), until contextual cues are included (Figure 3.26, lower row). But three objects can be aligned or unaligned – which makes an inference of depth significant when they are aligned (Figure 3.27A and 3.27B) or approximately aligned (Figure 3.27E).

Psychological experiments in which subjects are asked to report their subjective impression of depth in such scenes support these arguments (Cook et al., 2008a), but it is important to point out that they are *not* delicate effects reliant on the statistical analysis of staged laboratory results! These examples simply illustrate that there is a notable difference in the likely interpretations of the depth of pictures between scenes with *two* identifiable objects and those with *three*. Of course, photographs of natural scenes and "real art" normally have many more visual cues to help us disambiguate the possible interpretations, but, reduced to simple stimuli with a small number of visual cues, it is apparent that *three* aligned, nonoverlapping objects *can* signify 3D depth, whereas *two* nonoverlapping objects alone are inherently *ambiguous*.

The conclusion that follows from this examination of simple drawings is that there are implications of three-object configurations that are not found with two-object configurations. Needless to say, although three aligned, nonoccluding objects provide prima facie evidence in favor of a differential depth interpretation (Figure 3.28A), it is of course possible that further cues will indicate whether an inference of differential depth (Figures 3.28B through 3.28D) or one of differential size (Figure 3.28E and 3.28F) is correct. Drawing a vanishing point, horizon line and/or perspective lines (Figure 3.28B), adding further objects aligned in linear perspective

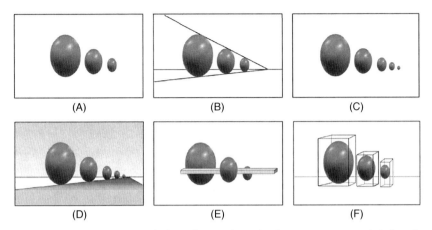

FIGURE 3.28. Three nonoccluding objects aligned in linear perspective (A) already provide inferential evidence of differential depth, rather than differential size, simply as a consequence of their alignment (implied convergence of perspective lines that join corresponding sites on each object). Additional cues in accord with linear perspective (B~D) further strengthen the depth interpretation, whereas cues inconsistent with linear perspective strongly imply that the objects are of different intrinsic sizes (E and F) despite the fact that their alignment remains unchanged.

(Figure 3.28C) or incorporating other disambiguating contextual cues (Figures 3.28D through 3.28F) will favor one interpretation over the other. But, provided that we are dealing with nonoverlapping objects, it appears that the linear alignment of *three* objects is the simplest configuration providing inferential evidence of 3D depth on a 2D canvas. Nonoverlapping two-cue configurations, in contrast, are inherently ambiguous and a definitive interpretation of the 3D structure of the visual scene awaits further information.

So, can this effect be summarized simply by saying that "The total number of visible objects determines the strength of the illusion of depth?" Is it simply that the more objects in sight, the firmer our sense of 3D structure? Experiments indicate that the crucial factor is not in fact the number of objects, but rather the number of objects that are aligned in linear perspective (Figure 3.29). On the one hand, it is found that the number of spheres aligned in a manner consistent with linear perspective produces notable increases in the sense of 3D depth. That alone is hardly surprising (and can be interpreted as a "total information" effect), but it is noteworthy that the sense of depth reported by experimental subjects is statistically identical for the paired pictures in Figure 3.29 (A&B, C&D, E&F and G&H). For each pair, the number of levels of depth (as depicted by spheres of different sizes)

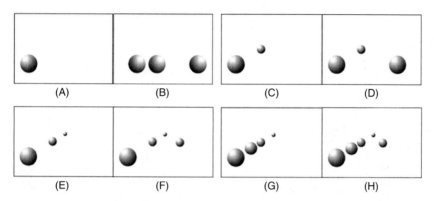

FIGURE 3.29. Stimuli used in an experiment to measure the sense of depth in simple pictures (Cook et al., 2008a). The depth increases with an increase in cues indicating different depth levels; the number of spheres itself is not the crucial factor.

is the same (one-, two-, three- or four-levels), but the actual number of objects is variable. In other words, the number of objects is less important than the number of object sizes in each picture. At the same time, the sense of 3D depth increases from AB to CD to EF to GH. That is, the number of *different* depth markers (spheres of different sizes consistent with linear perspective) strongly influences the sense of depth.

A related effect is obtained using the pictures shown in Figure 3.30 as experimental stimuli. Although the number of objects remains the same (three), the sense of depth reported by subjects is significantly stronger for pictures in Figures 3.30B and 3.30C than for Figure 3.30A (and slightly stronger for Figure 3.30B than Figure 3.30C). Clearly, even when the perspective and horizon lines are not depicted, there is something in the alignment of three objects (of different retinal size) that we perceive in judging their 3D depth.

The two-body depth cues of relative size and relative elevation (Table 3.1) are, on their own, inherently ambiguous. Similar to the question of the inherent "meaning" of two-tone pitch intervals, the depth "meaning" of (nonoverlapping) large and small objects lying anywhere in the visual field can become clear only when a third cue is present. The minimal third cue in a visual scene is a horizon line at an appropriate elevation relative to the different sizes of two objects. A stronger cue is a third object of appropriate size, since it provides indication of not only the vertical height of a horizon line but also the lateral position of one or more vanishing point(s) (Figures 3.30E and 3.30F).

Stated in its most general form, the perception of 3D depth in 2D pictures appears to be strongly influenced by "triadic visual processes"

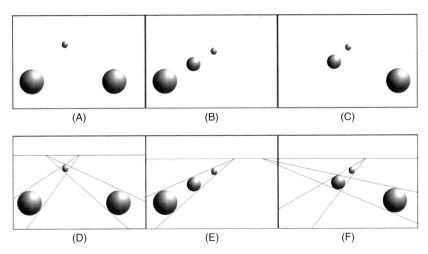

FIGURE 3.30. Scenes with three objects. When three levels of depth are depicted (B and C), the subjective sense of 3D depth is unmistakably stronger than when only two levels of depth are depicted (A). In (A) and (C), there are, respectively, two and three independent indications of the vertical location of the horizon, as shown in (D) and (F). In (B), there are three redundant indications of the location of a vanishing point – as calculated from any two spheres (E). The redundancy of the perspective lines reinforces our certainty about the vertical height of a unique horizon line, providing a stronger indication of depth in (B) than (C) and in (C) than in (A) (Cook et al., 2008a).

[i.e., undertaking an inferential mental operation to calculate the (illusory) depth of a scene involving at least three visual components]. When three objects are aligned to suggest 3D depth, the human brain leaps to the conclusion of 3D depth. In contrast, in animal studies using 2D visual stimuli with numerous (countless!) indications of 3D depth, the suggestion of 3D structure is not readily taken up by the animal brain, and – even with highly evolved object recognition capabilities at work – apparently only shapes of various sizes distributed over the 2D plane are seen: There is no imagined dimension of depth insofar as the "relation between relations" (i.e., three-body configurations) are not processed. The question of precisely what that "leap of faith" into imaginary realms is for the human mind will be addressed in Section 3.8.

3.5. SHADOWS AND SHADING

To the adult mind, shadows are obviously not material "things," but for each of us individually that understanding was not always so certain (Figure 3.31).

FIGURE 3.31. A photograph showing young children pondering the puzzle of why gravel can hide the pavement, but can't hide the shadow (Casati, 2003, with permission).

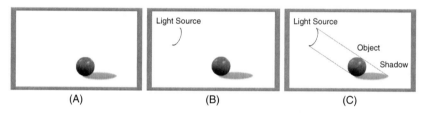

FIGURE 3.32. An object and its cast shadow (A) imply the presence of a light source (sometimes, but not often visible in the visual scene) (B). The physics of light emission implies the linearity of the three-body relationship among the light source, the opaque object and its shadow (C).

The meaning – and, in a sense, the lack of meaning – of shadows is something that we learn. Although shadows have a real visual presence, their utter dependence on an opaque object and a source of light that are spatially separate from the shadows themselves makes them less "real" than most other visual stimuli. It takes some years to figure out what shadows are. The children in Figure 3.32 are in the process of learning that a shadow cannot be treated as if it were an independent object on a par with the gravel that they have used in attempting to hide the shadow. From an examination of the behavior of shadows, they will eventually realize that the form and location of shadows can be reliably traced back to the opaque object and the light source that generated them, but the shadow itself is quite ephemeral. That is a noteworthy insight.

An artistic appreciation of the importance of shadows in explicating the 3D shapes on a 2D canvas was also learned, and late to develop. Although the techniques needed for realistic depiction of individual objects were understood already some 30 millennia ago – as seen in the cave drawings of lions, bison and horses – there are few indications of shadows in painted artworks until much later. Hints of geometrically correct shadows can be seen on Hellenistic vases (ca. 400 BC), but it comes as a surprise to most students of art history that nearly all pre-Renaissance art from around the world included no shadows whatsoever or, at best, shadows with unlikely shapes and orientations.

An understanding of shading (as distinct from "cast shadows") was obtained earlier by Giotto and other pre-Renaissance painters (Hills, 1987; Gombrich, 1995). By definition, shading is the darker tones on unlit regions of an object, whereas a cast shadow is the unlit region cast by one object onto an underlying surface or onto neighboring objects. Technically, they are similar in requiring the artist to use darker colors in drawing the unlit portions of objects, but the depiction of realistic shading can be achieved with a more approximate technique than that required for depicting shadows. Unlike shading, a cast shadow makes explicit the orientation of a light source relative to the opaque object, whereas shading suggests, at best, only the quadrant of the visual array from which the light source shines (above or below, left or right).

In either case, both shading and shadows, as visual cues, are inherently insubstantial, come and go with the weather and are defined in physics by what they "are not" – an absence of light. Unsurprisingly, they are not understood by young children or animals and are among the most difficult techniques for aspiring artists to master. In much of Medieval art, *shading* was used to depict semirealistic solidity for 3D objects – clothing and faces, in particular – but an understanding of the optical principles underlying the drawing of *cast shadows* did not become firm until the Renaissance.

The cognitive question that needs to be addressed is: What is the nature of the information that shadows (and, to a lesser extent, shading) provide for our interpretation of the 3D world? The answer is known and quite simple. The significance of real-world cast shadows lies in the fact that they provide geometrically precise indication of the spatial configuration of *three* visual cues: (i) a physical object, (ii) its shadow falling on a separate surface and (iii) the light source that produced the shadow (Figure 3.32).

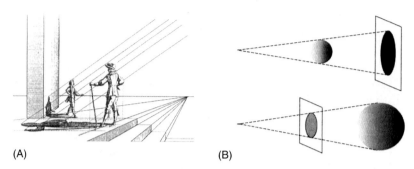

FIGURE 3.33. (A) Demonstration of the linear geometry of both perspective and shadows (after Jean Dubreuil's *Practical Perspective*, London: Bowles & Carver, originally published in 1642, p. 140) (available at http://books.google.com/books). Light rays run parallel; perspective lines converge. (B) Illustration of Leibniz's argument that "shadows are perspective" and differ only with regard to the location of the projection plane (Casati, 2003, with permission).

Various authors have commented on the specifically *triadic* nature of shadow information. Casati (2003, p. 62) notes that "Shadow is useful because it makes visible an alignment that we could not otherwise see. Three points lie along the same line: the light source, the [object's] tip, and the tip of the [object's] shadow." Similarly, Baxandall (1995, p. 42) states that "The gross form of any particular shadow is due to a particular relation between three principal terms – a positioned light source, a positioned and shaped solid, and a positioned and shaped support or receiving surface." In effect, the mere presence of a cast shadow provides unambiguous information about the 3D structure of a small portion of the visual scene (Figure 3.33). The same optical linearity that pertains to the shadow/object/light source in the real world can be maintained on a 2D canvas and conveys a sense of 3D structure, insofar as the rigid geometry of shadows is accurately depicted in the picture.

The principles of optical linearity that Macaccio, Leonardo and other Renaissance painters mastered are well illustrated in a drawing by Dubreuil (Figure 3.33A). There he has drawn the parallel light rays that give rise to realistic cast shadows from the Sun and the converging perspective lines that give rise to realistic diminution of objects with distance. Together, these two triadic geometrical "tricks" give paintings and drawings a structural realism that other pictorial depth cues do not convey.

Since the Renaissance, most of the discussion of depth perception has focused on the optical laws that underlie these artistic techniques. The perception itself – the intuitive understanding of the 3D structure in a painting constructed following the laws of linear perspective – is "obvious," but the relationship between the 3D world and the 2D projection on the retina or on the image plane is geometrically complex. Among many Renaissance thinkers, Leibniz was impressed by the geometrical similarities between linear perspective and shadows – noting that they differ only with regard to where one places the image plane: "The doctrine of shadows is nothing more than a reversed perspective" (quoted in Casati, 2003, p. 188) – as illustrated in Figure 3.33B. Note that this similarity is distinct from the *perceptual* triads inherent to both perspective and shadows and is perhaps better labeled as a *conceptual* triad. In either case, it is the inherent linear alignment that makes the geometry "rigid" and gives the 2D image an unambiguous 3D structure.

Both the linearity of shadows and the linearity of perspective can be illustrated with an extremely small number of relevant cues, but to maintain the illusion of 3D depth over the entire canvas, the shadows and perspective cues from many objects need to be employed in a self-consistent manner. A shadow falling to the left next to a shadow falling to the right or perspective lines that indicate both a high and a low horizon line will work *against* the viewer's intuitive ability to obtain a coherent spatial understanding of the visual scene. It is for this reason that the great Renaissance painters sketched out the entire scene and used rulers and guidelines that were later painted over in order to construct a unified picture allowing for only one correct interpretation.

Point light sources (such as candles), reflecting surfaces (such as mirrors and ponds), semitransparent solids (such as windows and vases) and complicated objects casting shadows on complicated surfaces (Figure 3.34) provided Renaissance artists with countless geometrical themes to perfect their optical techniques. Eventually, the enthusiasm for optical realism generated a backlash and a deemphasis on such geometric precision in subsequent generations of artists. But without plunging into the controversy about what constitutes "great" art, it is clear that an understanding of both types of three-body inference that underlie the realistic depiction of 3D depth – perspective and shadows – was achieved during the period known as the European Renaissance.

Cast shadows, typically falling beneath the object itself, are joined by lines connecting (i) the object, (ii) its shadow and (iii) the light source. In the 3D world *cast shadows* provide (i) several parallel, linear indications of the precise position of the light source relative to the illuminated object(s) and (ii) some indication of the topology of the surface onto which the cast shadows fall. Artificially contrived scenes (such as objects illuminated on a stage with multiple light sources) can obscure the linear relationships,

FIGURE 3.34. The Renaissance obsession with geometry is on display in Niceron's *La Perspective Curieuse* (1638, image used with permission). It seems that the geometric principles of realistic perspective and shadows are so easy that even children can understand!

but the physics of light emission implies an unambiguous, default interpretation of three such visual cues (Figure 3.35) – an interpretation that all normal children eventually come to understand.

The mere shift of a cast shadow from directly beneath a sphere (Figure 3.36A) to a position away from the sphere itself can simultaneously move the sphere forward in the scene (Figure 3.36B). Note that a similar shift when *three* objects are aligned again moves one sphere forward, but less convincingly, as we stubbornly want to believe that the chance alignment of the three spheres themselves on the 2D surface of the painting are indicative of their linear perspective alignment in 3D space – despite the fact that the shadows tell us that they are not all resting on a common surface (Figure 3.36C&D).

In contrast to cast shadows, the spatial implications of *shading* are weaker – suggesting the convex/concave shape of the 3D object itself – but

Human Seeing

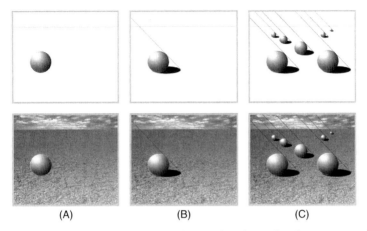

FIGURE 3.35. Without a shadow (A), the depth relationship between an object and its environment is uncertain. By adding a shadow (B), the object becomes grounded, and its position relative to the light source and the surface on which the shadow is cast becomes unambiguous. Adding further objects with their shadows (C) provides further information on the depth structure of the 3D scene.

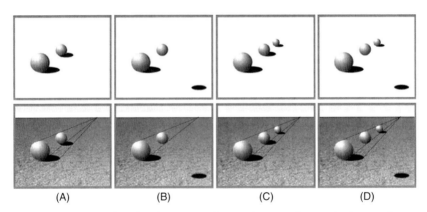

FIGURE 3.36. Note that the alignment of two spheres with cast shadows indicates a specific depth configuration (A), but that interpretation of the scene is radically altered by an aberrant shadow indicating the forward position of the second sphere (B). With a third sphere in the linear array (C), the depth structure is similar to that in (A), but a change in the position of the cast shadow of the second sphere produces a conflict between the depth alignment of the spheres and the nonlinearity of the shadows.

indicating only approximately the direction from which the object is illuminated. Indeed, shading alone – without depiction of the light source or cast shadows – indicates a 3D structure that is inherently reversible. The convexities and concavities in Figure 3.37A, for example, are normally perceived

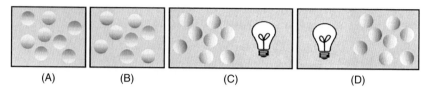

FIGURE 3.37. (A) shows some Ramachandran-style concavities and convexities. Most people perceive five as sticking in and three as sticking out, solely on the basis of the tacit assumption that the light source is from above. (B) shows the same figure rotated 180 degrees; now five are sticking out and three in. (C) and (D) show that, even cartoon light bulbs that we perceive as unrealistic and flat, nonetheless are effective in determining the convexity or concavity of the bumps.

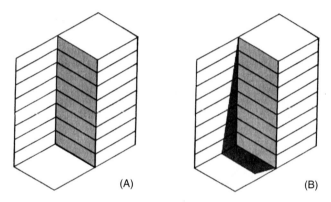

FIGURE 3.38. The Mach illusion. The 3D depth structure in (A) is reversible, but the cast shadow in (B) makes only one interpretation possible.

as sticking in or out under the implicit assumption that the illumination is from above. But, since the directionality of gravity does not change, their convexity/concavity is instantly inverted simply by 180 degree rotation of the same figure (Figure 3.37B). Note that even the position of an unconvincing drawing of a light bulb can deceive us into seeing illusory convexity or concavity of these textured circles (Figure 3.37C and 3.37D).

Shading alone gives ambiguous indication of 3D structure because the directionality of the light is not clear. For this reason, the famous Mach illusion in Figure 3.38A can be seen as a folded card illuminated from either the left or the right (with remarkably different 3D structure, depending on the assumption about the location of the light source). With the addition of a cast shadow, however, the directionality of the light becomes fixed, and the reversibility of the figure is eliminated (Figure 3.38B): The light source now lies unambiguously to the right.

Manipulation of shading, shadows and the alignment of objects is a source of endless fascination, psychological experimentation and artistic delight (Hills, 1987; Gombrich, 1960/2000; Stoichita, 1997), but the rather elementary argument of this chapter is simply that the psychological effects of these visual cues are realized in combinations of three. Artworks that contain only three such cues would hardly be valued as art, but multiple, mutually consistent depictions of perspective, shadows and shading through triads of visual cues ultimately provide a convincing illusion of 3D depth on a canvas that we know full well to be flat.

3.6. HISTORICAL PERSPECTIVE ON SHADOWS

The importance of shadows in various genres of art has been documented by Hills (1987), Baxandall (1995), Stoichita (1997) and Gombrich (1996/2000), and is often mentioned as a small part of the story of the invention of linear perspective (Kemp, 1997). The complete absence of cast shadows in Medieval art and in most styles of contemporary Oriental art is noteworthy. Since both of the revolutionary techniques of pictorial depth perception (linear perspective and the linearity of shadows) were introduced into painting in Europe in the fourteenth through sixteenth centuries, it is of interest to consider what the relationship between these techniques might have been in the minds of Renaissance artists. Although it is conceivable that the discovery of the geometry of shadows and the discovery of the geometry of linear perspective could have been made totally independently in distinct artistic cultures, historically that appears not to have been the case. Why? And is there an explanation for why the principles underlying triadic musical harmony were also invented in the same region at the same time?

No obvious answer was forthcoming in Chapter 2 as to why triadic chords were first introduced in Renaissance Europe, but a plausible answer to the shadow puzzle can be found in the field of Renaissance astronomy. As unlikely as that may seem, it is precisely the triadic relationship implied by *shadows* that led to the elucidation of the geometry and dynamics of the Sun–Earth–Moon system (Figure 3.39) (Casati, 2003). The phases of the Moon – or any externally illuminated sphere, for that matter – provide unambiguous information concerning the direction of the source of the illumination. That astronomical geometry was in fact already debated in Hellenistic Greece, but an understanding of its significance for the heliocentric structure of the solar system awaited the invention of the telescope and observation of the phases of planets.

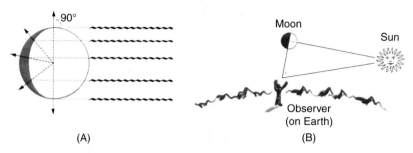

FIGURE 3.39. (A) The shading on a sphere gives geometrical information on the location of a distant light source. (B) The lunar shadows of the Sun–Earth–Moon triad provided the astronomical paradigm that eventually led to an understanding of the heliocentric solar system (redrawn after Casati, 2003, with permission).

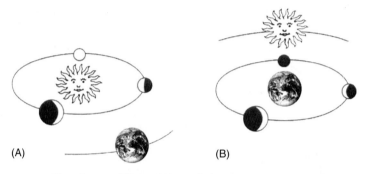

FIGURE 3.40. The phases of Venus (A) in a heliocentric universe and (B) in a geocentric universe. Galileo's insight into the geometry of the solar system came from the changing phases of Venus, a planet that lies closer to the Sun than the Earth (see Casati, 2003, and Heilbron, 2010, for discussion).

Specifically, using a primitive telescope (with merely sixfold magnification!) that revealed the phases of Venus, Galileo discovered that the changing brightness of Venus could be explained only if Venus revolved around the Sun (Heilbron, 2010) (Figure 3.40). If, instead, Venus revolved around the Earth, there would be phases where it reflects no light toward the Earth. Therefore, in *not* observing phases similar to those displayed by the Moon, Galileo deduced that Venus revolved around the Sun, not the Earth. Many astronomical details concerning the rotation and revolution of heavenly bodies remained to be sorted out before the idea of heliocentricity was widely accepted, but, ultimately, the concept of a geocentric universe was doomed by the triadic geometry of *shadows* (Figure 3.39B).

Others have offered plausible scenarios about the possible causes and effects of discoveries in the arts and sciences (Shlain, 1991), but the triadic conceptual link between linear perspective and shadows has apparently not been investigated. While firm conclusions remain difficult to draw without careful examination of Renaissance documents, it is interesting to consider the possibility that the remarkable fermentation of human culture during the European Renaissance might be traced back to a new awareness of triadic relations. This is not to say that triadic relations were not *perceived* until then, but the topic of the "geometry of the universe" became an explicit focus of both academic and popular attention during the Renaissance. Among artists, the concrete triad of object, light and shadow was a topic of active debate and conscious deliberation. While the human ability to perceive the depth in flat paintings preceded the promulgation of the artistic "laws" of linear perspective and shadows by some 30 or 40 millennia, it appears that Renaissance scholars succeeded in bringing those ideas to conscious attention – ultimately formalizing those ideas into systematic, triadic theories that brought coherency not only to astronomy, but also to the visual arts. Were the triadic insights of harmony also involved?

3.7. A RECLASSIFICATION OF DEPTH CUES

The present chapter has been an attempt to answer basic questions about pictorial depth perception, with particular emphasis on the significance of two- and three-object depth cues. Disregarding the relatively weak depth effects of hue, there are seven varieties of pictorial cue that are typically listed as the techniques artists employ to create the illusion of 3D depth on a 2D canvas (Table 3.1). All of those cues can be reclassified into one of two types of cognitive operation that are two- and three-body computations (or combinations of same). The strongest two-body cues are stereopsis, motion parallax and occlusion. In a static, monocular context, when it is apparent which of two overlapping objects in the retinal image has a more "complete" form, their relative depth is immediately known. In contrast, relative size and relative elevation are two-body effects that provide only tentative answers. It may be evident which of two objects is larger (occupies more retinal area) and which object lies higher/lower in the retinal image, but their relationship in depth remains uncertain.

In contrast, three-body cues allow one to draw inferences concerning the likelihood of the geometrical alignment of objects in 3D space. Can the lower regions of three objects be joined by a straight line? Can the upper regions of the same three objects also be joined by a straight line? The inference of linear depth is already answered if a straight line connecting the three points is visible – as in the pavement connecting the base of neighboring buildings – but the same inference can be drawn by imagining such a line. If a linear connection can be imagined, then we infer a depth relation. Of course, a similar depth relation might be inferred from the imagined perspective lines connecting only two objects, but the crucial point is that there is *never* a case where two objects of similar shape but different size *cannot* be so connected. Two "aligned" points therefore tell us nothing, whereas a configuration of *three* points linearly aligned by chance is unlikely, so that we infer a causal mechanism underlying the alignment – an inherent structural relationship among the points that appear to be "in line." Moreover, three aligned points connecting the lower portions *and* three aligned points connecting the upper portions of the same three objects is such an extremely low-probability configuration of three different-sized objects at the same distance from the viewer that the natural conclusion – the default interpretation for the human mind – is one of depth. Again, it may in fact be simply an unlikely coincidence that has happened by chance – and further visual cues will make that fact clear. But the alignment of three cues is alone circumstantial evidence of a nonrandom alignment in depth.

Naturally, all but the most drastically simplified drawings will have multiple – often innumerable – pictorial depth cues, but, even in complex scenes, the local cues contributing to the perception of illusory depth can be reduced to these two types of two- and three-body inferences. Stated negatively, there is no need to invoke qualitative, aesthetic or cultural arguments concerning the visual "context" that leads to pictorial depth perception. As Purves and Lotto (2003) have argued with regard to other aspects of perception, the empirical probabilities of various configurations of visual cues provide strong evidence concerning their correct interpretation: "The visual system is not organized to generate a veridical representation of the physical world, but rather is a statistical reflection of visual history" (p. 227). The alignment of three visual cues is experienced countless times in a normal lifetime, often reflecting an alignment in 3D space. Taking that linearity seriously and making inferences about the real world is what the human mind is good at.

So, are we in a position to reclassify all of the various pictorial depth cues in light of the two- and three-cue distinction? Not yet. Clearly, relative size, relative height and occlusion are two-component comparisons, and the case for considering shadows and linear perspective to be essentially three-component perceptions has been outlined earlier, but where do aerial (atmospheric) perspective and texture gradients fit in?

Aerial perspective can be understood as an effect of multiple two-body occlusion cues. That is, the suggestive depth effects elicited by the fading

Human Seeing 159

TABLE 3.2. *A reclassification of depth cues based on the number of visual cues (compare with the classification in Table 3.1)*

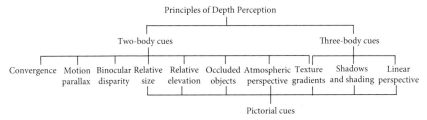

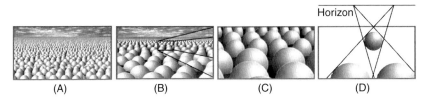

FIGURE 3.41. A ground texture (A) can be viewed as the summation of multiple 2-body effects (occlusion, relative height, relative size) (B and C). Multiple three-body effects (linear perspective, shading and shadows) become invisible in texture gradients (D).

intensity of distant objects can be attributed to a multiplicity of two-body occlusion effects. Although individually too small to detect without microscopic analysis, the occlusion of distant objects by water vapor in the air is responsible for so-called atmospheric effects. Neither the relative size of objects, relative elevation nor the three-cue effects of linear perspective or shading/shadows plays a role.

Texture gradients are, by their very nature, multi-cue indications of depth, but they can arguably be reduced to underlying two- and three-body effects. Textured surfaces consisting of granular components can be constructed by defining the relative size of the granules, frequency of occlusion and their density over a predetermined area (Figure 3.41A through 3.41D). As such, a variety of two-cue effects (occlusion, relative size and elevation) contributes to the perception of depth. Multiple three-cue effects (linear perspective, shading and shadow) are also present, although often obscured or too minute to be seen within the expanse of a textured surface.

By considering texture gradients and aerial perspective to be inherently multiple two- and three-body effects, the seven principal types of pictorial depth cue that artists have exploited for millennia to provide the illusion of 3D depth in 2D drawings and paintings can be classified as in Table 3.2.

Implications of 3D depth from hue and from semantic knowledge about familiar objects fall outside of this classification, but it appears that the most salient pictorial depth cues have a simple psychophysical basis that can be described in terms of the number of visual cues contributing to the perception of relative depth.

Historically, the simpler two-body depth cue effects were understood and employed prior to the three-body effects. Already in the cave drawings from the Paleolithic era, the occlusion of smaller and therefore seemingly more distant animals by larger, seemingly closer animals produces an illusion of surprising realism. Precisely how much larger or smaller different objects should be portrayed to obtain optical realism over an entire visual scene and precisely where their shadows should fall on an underlying surface were problems that were not understood for many millennia. But the intuition of two-body relations suffices for producing an approximate illusion of 3D depth.

Much later, once an understanding of the phenomenology of light itself and the optical laws underlying depth portrayal had been discovered, visual art attained a 3D realism that is still enjoyed many centuries later. Together with the exaggeration of perspective (e.g., Surrealism), its partial disruption (e.g., Cubism), complete negation (e.g., Flatism) and the inversion of the illusion of 3D depth (e.g., Reverspectivism, Section 3.9), we have all become experts in evaluating the 3D structure suggested in 2D pictures. Those same visual depth cues are certainly seen, but hardly noticed by animal species. Here too the stark unusualness of the human mind is evident.

3.8. "PERSPECTIVE AS SYMBOLIC FORM"

Although the real world sometimes provides us with perspective lines that connect objects, sometimes provides a clear horizon line, and occasionally even a vanishing point on the horizon, such explicit indications of linear perspective are not, in fact, frequently present. In artworks depicting modern urban environments, portions of perspective lines on sidewalks, streets, railway lines and buildings are not unusual, but – even without the perspective lines – three approximately-aligned trees, bushes, flowers or light bulbs, for that matter, elicit a depth interpretation from the human mind. Similarly, depiction of a spotlight in a foggy or smoky environment may reveal the trajectory of light rays that connect a light source, an object and its shadow, but even without such explicitness we easily interpret a 2D scene as having 3D structure from the ephemeral cues provided by shadows and shading. What is it that motivates this slightly overly enthusiastic willingness to infer depth in pictures that we know to be physically flat? Why do we insist on "looking into" a picture to see the illusion of 3D depth on a 2D surface (Hecht et al., 2003)?

An answer to these questions was given in the title of a short, but notorious book by Erwin Panofsky (1927/1997): *Perspective as Symbolic Form* (an abstruse 46-page essay, demanding an 18-page translator's preface and 124-pages of notes, diagrams and addenda). The ins-and-outs of that essay, and the academic debate it kindled (e.g., Damisch, 1987/1995), are fascinating, but not as important as the basic argument already apparent in the title: Perspective, Panofsky argues, is "symbolic." How can that be? Clearly, the perception of depth facilitated by linear perspective cues has nothing to do with language or the representation of phenomena through mathematical symbols. Panofsky's point, however, was that perspective is "symbolic" in the sense that it allows the viewer to enter into an imaginary realm with its own logic and internal coherency. The third dimension in a 2D drawing exists only insofar as the observer is willing and able to infer depth relationships among the objects provided on the 2D canvas. Since the retinal image itself is 2D, reading it as 3D requires the viewer to discern the fact that the literal "surface structure" can be interpreted in terms of a "deep structure" that is, in fact, fictitious. "The material surface upon which the individual figures or objects are drawn or painted or carved is thus negated" (Panofsky, 1927/1997, p. 27) and only an imaginary world remains.

The "symbolism" here is not the pairing of concepts with arbitrary written symbols, but rather the symbolic leap is the intentional interpretation of inherently-unreal 2D visual cues as if they had meaning concerning a third spatial dimension in which the viewer is willing to be momentarily immersed. As Panofsky states it: Linear perspective introduces "something entirely new: a realm where [the events drawn in the artwork] become a direct experience of the viewer [and] erupt into his own, apparently natural, visual space and so permit him to 'internalize' [the scene]. Perspective, ultimately, opens art to the realm of the psychological, in the highest sense" (1927/1997, p. 72).

Interpreting a 2D scene as a "real" 3D world is not inevitable. Most animal species do not see a 2D picture as a "still" image of a 3D world; they see shapes on a flat surface – with no implications concerning the 3D world. Even for normal people, pre-Renaissance art can be interpreted as 3D scenes, but a picture with grossly incorrect linear perspective remains an "unreal story" hanging on a wall – there to be "decoded," rather than experienced as a virtual visual reality. If one perceives the differential areas of shapes on a 2D surface as indicative of their actual physical bulk, then the image does not undergo manipulations in an imaginary third dimension of space. If, however, one inhibits – effectively suppresses – the literal 2D interpretation and does not succumb to a facile evaluation of the differential retinal size of visual objects, then an alternative interpretation of the same 2D image becomes possible in the form of an imagined 3D space. One then enters into a world that owes its coherency to an interpretation of the same visual cues – now "seen" through the regularities of linear perspective as a possible 3D fantasy space.

Complex, multi-cue, visual scenes painted in accordance with linear perspective provide the viewer with ample hints on how to conceive of the scene as having believable 3D depth. Similar to the symbolic world constructed through spoken language, where the "literal" interpretation of speech sounds – onomatopoeia, rising and falling pitch, long and short words and phrases – is largely *irrelevant* to the meaning of language, the symbolic world constructed through linear perspective has an internal metric that is not implied by – and, in fact, is often contradictory to – the "literal" interpretation provided by two-body cues. The link between the symbolic worlds of language and pictorial depth perception is apparent from the fact that both worlds require the speaker/viewer to enter willingly into an imaginary realm and to suppress possible superficial interpretations. The quality of the sounds of speech are not attended to when we contemplate the meaning of language and, in visual art, the shapes of objects on the 2D surface are also ignored, in favor of an imagined 3D interpretation. But the scene takes on 3D spatial structure only if the normally *invisible* perspective lines and light trajectories are drawn in the mind's eye to connect the visible objects in the 2D image.

In this view, the formulation of the "laws" of linear perspective was the establishment of artistic guidelines such that the art viewer does not inadvertently interpret the static 2D image as indicative of a 2D world. Prior to the delineation of those laws, various pictorial tricks had been utilized to create an approximate sense of spatial depth – and clearly people worldwide were already capable of interpreting geometrically-imperfect drawings as depicting 3D scenes. But, although they could interpret imperfectly depicted pictorial scenes as representing a 3D reality, they could not "live" in them: They were not automatically suspended in an imaginary world, but rather necessarily fell out of the spell whenever geometrically-incorrect indications of depth came into awareness.

In order to enforce the illusion of 3D realism and to make the 2D scene a believable 3D reality, the geometrical rigor of linear optics was necessary.

Before the Renaissance, the pictorial depth cue of two-body occlusion provided information about relative depth that both people and animals readily use in interpreting 2D pictures. But occlusion is inherently nonmetrical: interposition alone does not indicate the relative distances between the observer and the occluding/occluded objects. Moreover, without occlusion, relative size and elevation do not provide reliable (absolute or relative) information on the depth of the objects in a 2D scene. In contrast, three-cue configurations provide relative depth information and, given the presence of familiar objects of known size, absolute information on 3D depth. Using the techniques of shading, shadows and linear perspective, there arose the possibility for the viewer of art to maintain a self-consistent, coherent understanding of the visual scene, as if the viewer himself had been momentarily placed within that particular 3D world.

Once the laws of linear perspective had been formulated, artists delighted in creating geometrically-precise scenes, often including polyhedral solids and other irrelevant objects that emphasized the geometry with fanatical precision (e.g., Figure 3.34). After that particular itch had been scratched to satiation, artists in the following centuries purposely strayed from the "correctness" of linear perspective. For some of the hard-line defenders of Realism, modern art that lacks comprehensible depth barely deserves the label "art," while to others (Gablick, 1977) art comes completely into its own symbolic realm only once the remnants of realism are gone. In either case, it is clear that much of abstract, modern art is a conscious departure from geometrical realism in search of other effects.

Panofsky's point that perspective itself is symbolic suggests that – quite aside from the issue of the realistic depiction of objects, all art that maintains the illusion of 3D structure on a 2D surface pulls us into an imaginary realm. Story-telling, realistic objects and even object recognition are of secondary importance, but the illusion of a coherent depth structure is crucial. In this respect, the scorn that Tom Wolfe (1975), in particular, has poured on modern abstract art for no longer being concerned with the experience of 3D space depicted on a 2D surface is well-directed. Insofar as visual art denies the uniquely human dimension of "looking into" the 2D display to see more than a monkey might see (Hecht et al., 2003), it lacks the essential illusion that makes human art human. It is specifically the jump from 2D shapes on a flat canvas (seen by the ape brain with a clarity comparable to ours) to 2D shapes that imply 3D structure (perceived by all normal human beings who have had minimal exposure to the culture of visual art) that is the step into the visual symbolic realm.

3.9. VARIATIONS ON THE ILLUSION OF DEPTH

Particularly since the advent of photography, the remarkable sense of realistic 3D depth obtained in paintings created with linear perspective is no longer startling and it is easy to be "unimpressed" with the human capability to see illusory depth in flat pictures. Of course, we can perceive the 3D structure "inside of" 2D snapshots! And it can be a recurring surprise to witness that our pet cats and dogs haven't an inkling that it might be possible to "look into" photographs to see their favorite toys, friends and enemies. With visual systems that are comparable to ours for object recognition, animals do not pick up on static pictorial depth cues, but instead rely on binocular depth cues and/or motion parallax for all of their real-world 3D navigational needs. They have no comprehension of visual symbolism and indeed no need for the pictorial depth illusion that underlies our insatiable consumption of picture postcards, photographs and fine-art paintings. For real-world, 3D navigational purposes, we are much like our animal friends, but when we are not on the run we can abstract useful structural information from static pictures that animal species see as flat wallpaper.

Following soon after the invention of linear perspective, some artists used the techniques of perspective to produce not simply the illusion of 3D reality, but illusions that were meant to be discovered by the viewer as delusions (*trompe l'oeil*). The variety of such visual illusions is fascinating and endless, but it is particularly worth noting the sustained popularity of such "trick art" (e.g., Figure 3.42). Oftentimes shunned as "kitsch" or "graphic" (as distinct from "fine"), trick art has a fascination and staying power that even the connoisseurs of the classical masterworks must envy. Visual illusions that have neither the historical reputation nor the price-tag nor the political provocation that would draw crowds for other reasons consistently draw the museum-going public in huge numbers. People motivated by nothing more sophisticated than the desire to look at something unusual – to be visually surprised and momentarily dumbfounded – enjoy well-crafted "trick art" variations on realistic linear perspective.

A recent art style that the critics ignore, but the viewing public highly values was invented in 1991 by the London artist, Patrick Hughes, in the development of what has come to be known as reverse perspective (or reverspective) (Figure 3.43).[1] What is of interest about reverse perspective

---

[1] Hughes himself makes reference to his early works from the 1960s as the precursors of the reverse perspective illusion, but his earliest depth-inverted painting on a 3D canvas was created in 1989 and is more accurately described as a form of conceptual art.

Human Seeing 165

FIGURE 3.42. The ever-popular visual illusions of William Hogarth (*Perspectical Absurdities*, engraved frontispiece from J. Kirby's *Dr. Brook Taylor's Method of Perspective Made Easy in Both Theory and Practice*, Ipswich, 1754, used with permission) and a simple drawing containing deceptive linear perspective and shadows.

paintings is that they produce an illusory movement within the picture itself (when viewed in person by a mobile observer). Although there are precedents in the intaglios and cameos of Renaissance Italy, the illusion of the hollow mask (Hill & Johnston, 2007), the folded card illusion of Mach (Figure 3.38) and early experiments by Hughes himself (Slyce, 1998), the creation of

> In that work (*The Point of Infinity*), the reverse perspective technique of depth inversion was used, but the characteristic illusion of false movement in the picture itself was weak or absent. As we have shown experimentally (Cook et al., 2008b), the illusion does not occur with a single depth inversion, as employed in Hughes's first reverspectives, but requires *multiple* inversions. Hughes was in fact quick to realize the need for multiple depth reversals – as seen in *The Present* (1991) and in *all* of his reverspectives since 1992, without exception. As is true for most artistic discoveries, there is some dispute about who was the first to invent reverse perspective, but Slyce (1998) has documented the gradual evolution of Hughes's work – from conceptual "oxymoronic" 2D paintings concerned primarily with linear perspective and shadows, through single-inversion 3D reverspectives, to double-inversion reverspectives (that, without shadows and without a continuous floor or ceiling surface, also do *not* induce the false movement illusion) to, ultimately, the mature, multiple depth-inversion technique that the viewing public, at any rate, fully appreciates. The stages through which Hughes developed the reverse perspective illusion clearly illustrate the focused, trial-and-error process of artistic discovery. Others since the Renaissance may have stumbled onto the same visual effect, but only Hughes has had the tenacity to turn it into a genre of art that psychologists, if not necessarily the art critics, highly value.

FIGURE 3.43. Reverse perspectives by Patrick Hughes (*The Present*, 1991, private collection and *Faith Moves Mountains*, 1997, private collection, with permission).

static artworks that consistently produce illusory motion through the use of conflicting visual cues was a true discovery in visual perception.

The illusory "trick" underlying the reverse perspective illusion (as well as the hollow mask illusion) is the use of inferential depth cues (shadows, shading and linear perspective) to play against the observer's learned expectations about motion parallax when viewing 3D structures (Cook et al., 2008b). Unfortunately, the visual effect that makes depth inversion unusual cannot be seen in 2D reproductions, because the painting itself must be a 3D relief structure (typically 200 × 100 × 30 cm in size). The unavoidable three-dimensionality of the reverse perspective illusion has meant that, with one lone exception (Livingstone, 2002, p. 106), the most powerful and arguably the most psychologically profound of all of the known visual illusions is not even mentioned in the textbooks on visual psychology!

The reason why reverspective paintings produce illusory movement is that they embody an inherent contradiction between pictorial depth cues (that indicate an unmistakable 3D structure "in the painting") and motion parallax cues (that indicate an unmistakable, but *contrary*, depth structure of the 3D canvas). Walking in front of a reverspective, the 2-body computations underlying motion parallax are pitted against the 3-body depth cues of shadows and linear perspective – forcing two mutually-contradictory interpretations of the 3D shapes within the picture. (As Hughes has commented, "Your eyes tell you one thing, your feet tell you something different.") Unable to ignore either your mind or your body, these normally-reliable, but contradictory indications of 3D structure force the brain to resolve the paradox by projecting movement onto the static painting. In other words, the only way to perceive a reverse perspective painting *without* concluding that one's own previous experience of solid objects is wrong or that one's own previous experience in understanding pictorial depth is wrong, is to "see" the painted objects in the depth inverted scene as moving "of their own accord."

As a consequence, even knowing with certainty that the painting on the wall has *not* moved, observers of reverse perspectives consistently report that they see the painting itself rotate, wiggle and move spontaneously. That "knowingly incorrect" interpretation is the only solution that does not violate the viewer's hard-earned, experience-based understanding of motion parallax and also does not violate the viewer's hard-earned understanding of the meaning of linear perspective, shading and cast shadows. Interestingly, a small minority of viewers finds the contradiction between what-they-see and what-they-know to be so discomforting that they shield their eyes and refuse to look any longer at the painting. ("Do not wake me from my preconceived notions!")

But, curators, take note! Most museum-goers with only a modicum of curiosity about visual psychology find reverspectives to be fascinating and typically invest lengthy periods of time examining the reverspective artwork – time and energy that would not be spent on perusing the Mona Lisa, much less a Maplethorpe installation.

The contradictions between different types of depth perception provided by reverse perspectives allow for many opportunities for psychological experimentation. A straight-forward experiment which indicates that the conflict between pictorial cues and motion parallax is indeed responsible for the illusion involves the use of prism goggles that reverse the visual field, left-to-right and right-to-left. When observers wear prism goggles, the reverse perspective illusion simply disappears (Cook et al., 2002). What has changed while wearing the goggles is specifically the *learned* connection between self-movement and changes in the visual field. The normal motion parallax effects (illustrated in Figure 3.3) are then left-right reversed, so that the visual field changes associated with movement to the left now occur with movement to the right, and vice versa. Given enough time, one can "acclimatize" to this reversal, but the effects of prism goggles remain highly counter-intuitive for at least many hours. During that time while wearing prism goggles, movement in front of a reverse perspective (with the reversal produced by depth inversion in the reverse perspective artwork itself) gives the observer the impression of seemingly-normal motion parallax – a "double reversal" that eliminates the illusory effect. Contrarily, when viewing a normal-perspective 3D "shadowbox" of the exact same scene, subjects wearing the prism goggles report seeing the reverse perspective illusion (Figure 3.44).

A second experiment (Cook et al., 2008b) measured the relative strength of shadows and perspective lines in producing the illusion. If subjects are shown reverse perspective paintings that differ with regard to shading/

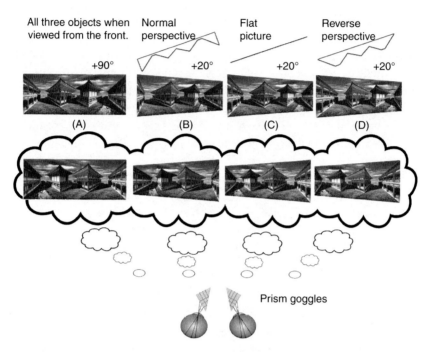

FIGURE 3.44. When viewed from the front (90 degrees, A), all three pictures appear the same, but, with a 20 degree rotation of the pictures (B, C, and D), distinct changes in the relative sizes of the faces of the buildings can be seen. In normal perspective (B), the changes are those expected from normal motion parallax. In the case of a usual 2D picture (C), there occurs a slight narrowing of all objects in the picture – as to be expected for any flat picture. In the case of a reverse perspective painting (D), however, the changes are counterintuitive – contrary to what one would expect when moving to the left in front of a 3D building. This counterintuitive effect is the source of the illusion. In the row of pictures presented within the thought bubbles, all of the pictures are left-right reversed due to the prism goggles. Here, the normal perspective painting becomes counterintuitive (showing changes suggestive of movement to the right). The reverse perspective now presents motion parallax effects that are seemingly normal (details in Cook et al., 2001, 2008b).

shadows and linear perspective cues (Figure 3.45) at a distance of, say, ten meters, the typical illusory effect is normally experienced. But, as subjects move slowly toward the paintings, depth perception through binocular stereopsis eventually reveals the actual structure of the relief canvas (at typically 1~3 meters) – and the illusion evaporates. The precise distance at which the shape of the canvas becomes evident depends on the effectiveness of the various pictorial depth cues. What is found experimentally is that

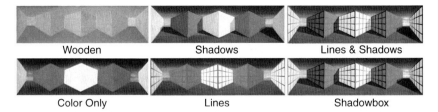

FIGURE 3.45. Six pictures used as stimuli in both a behavioral and an fMRI experiment. All six were "relief" paintings (140 × 42 × 20 cm), five being reverse perspectives and the one on the bottom right being a normal perspective "shadowbox" of the same dimensions, but with the "far" regions in the painting physically receding and the "near" regions protruding. Note the varying pictorial depth cues. Except for the "wooden" stimulus consisting of unpainted plywood surfaces, all of the stimuli were painted in bright colors; the "color only" stimulus contained no shadows or perspective lines. In the behavioral experiment, the strength of pictorial depth cues was determined by measuring the distance at which such cues can no longer compete with stereopsis and motion parallax cues (and where the illusion fails).

FIGURE 3.46. Several views of a reverse perspective (*House I*, 1996) by Roy Lichtenstein at the National Gallery Sculpture Garden in Washington DC. On the left is the concave "front view" which produces the reverspective illusion. In the middle is the convex back view, which appears as a normal house. On the right is a side view, which reveals the inverted construction.

the illusion is maintained up to much smaller distances for the paintings containing more numerous linear perspective and shading/shadow cues – again indicating that the illusory effects are a consequence of the struggle between (two-body) motion parallax cues and the surprisingly powerful (three-body) pictorial depth cues, until stereoptic cues at short distances provide definitive information on the 3D structure.

The reverse perspective effect has been mimicked by both well-known artists, such as Roy Lichtenstein (Figure 3.46), and other pretenders (Figures 3.45 and 3.47), but has not yet been hailed in the art world as the astounding discovery that it is. The problem is of course that the illusory

FIGURE 3.47. Exaggerated linear perspective and dark shadows produce unmistakable 3D structure – that isn't there (*Fuan Jingu*, Cook, 2011).

effect cannot be explained in the conventional terminology of aesthetics, but must be addressed in the slightly awkward terminology of vision science – and specifically in terms of the conflict between two- and three-visual-cue inferences about 3D structure.

Although the unexpected placement of shadows or perspective lines to produce an illusion might be dismissed as a minor variation within the genre of "trompe l'oeil," the popularity of the reverse perspective illusion, in particular, is clearly a consequence of the unusualness of the visual experience that it brings. For most viewers, the illusory effect is powerful – a wrenching feeling in the pit of the stomach that is a direct consequence of the irresolvable conflict between motion parallax (a two-body calculation) and shadows/shading/linear perspective (three-body visual effects). Caught between the rock of indubitable motion parallax and the hard place of unquestionable pictorial depth perception, the viewer "sees" the artwork as twisting-and-turning in counterintuitive ways as he moves back-and-forth, looking at the painting. The typical viewer delights in an illusory effect that is not experienced in normal life, but, unfortunately, the same inexplicable effects when felt by art critics tend to reveal a lack of understanding of visual psychology (and specifically of the known processes that lead to the illusion of false movement). As a consequence, just as the brain is quick to project the internal contradiction onto the artwork itself and blame it for "moving on its own," the art critic is quick to blame the artist for trickery: "Fine art should not confuse us!"

The upshot is that, just as M. C. Escher lived most of his life without critical acclaim ("thirty years of poverty" and "able to keep going only thanks to the help of parents on both sides," [Escher, 1986, p. 144] while creating, ultimately, hugely popular artwork that generates 3D confusions on a 2D surface), Hughes is destined to antagonize the psychologically naïve, uncomprehending critics of his 3D/4D confusional artworks. The apparent lack of interest in reverse perspective among the art elite has not, in fact, prevented Hughes from exhibiting his work to enthusiastic audiences

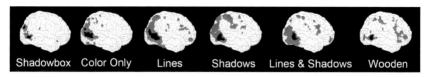

FIGURE 3.48. The brain activation obtained by viewing the six rotating pictures shown in Figure 3.45. The activation shown is that remaining after subtracting the activation due to static viewing of the same pictures. The foci of strongest activation (black) correspond to area MT (bilaterally symmetrical, only the right hemisphere shown here). Note the weakness of the activation using the "shadowbox" (normal perspective) stimulus, where the same picture is presented, but no illusion is experienced (Hayashi et al., 2007).

worldwide, but it does raise embarrassing questions about what contemporary art critics are in search of, if not new types of visual experience.

From the standpoint of human visual perception and specifically with regard to the psychology of pictorial depth cues, Hughes's work is unprecedented, profound and certain to outlast the transient notoriety of artwork that can be fully explained as variations on familiar techniques, materials and political posturing. But just what does reverse perspective tell us about the psychology of depth perception? A functional MRI study has provided some answers.

3.10. THIS IS YOUR BRAIN ON REVERSE PERSPECTIVE

In order to understand the brain response to reverse perspective paintings, we measured cerebral blood flow when subjects viewed the stimuli shown in Figure 3.45 where the nature of the pictorial depth cues was carefully controlled. The task for the 14 subjects was simply to observe the illusory movement in the picture when the artworks were in view and evaluate the strength of the illusion. The stimuli were presented statically for 30 seconds and then dynamically (rotating back and forth at 0.5 Hz) for 30 seconds (and repeated 5 times). Analysis was done by subtracting the brain activity during the static viewing from that of the dynamic viewing to determine what parts of the brain are active during the experience of the illusion (seen only during self-motion or rotation of the painting itself). It comes as no surprise that the most strongly activated region of the brain for all stimuli was a portion of visual cortex (MT), known to be active during the perception of motion (the black regions in Figure 3.48).

There were also indications concerning the specific regions of visual cortex that process linear perspective vs. shadows (dark gray), but one

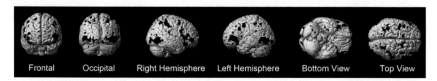

FIGURE 3.49. The brain activation for the unpainted "wooden" reverse perspective (Hayashi et al., 2007). Note the significant frontal activation in both hemispheres.

finding was of particular interest. That is, the reverse perspective picture with the *fewest* pictorial depth cues (Figure 3.45, upper left, the unpainted "wooden" object) nonetheless (i) produced the reverse perspective illusion (as reported by the subjects) and also (ii) showed the strongest activation of the frontal lobes (areas BA46 and BA47) (Figure 3.48, far right, and Figure 3.49).

Even without the abundant pictorial depth cues that are typically found in a reverse perspective painting, the illusion can indeed be experienced provided only that the painting is far enough away from the viewer that binocular depth cues do not reveal the actual structure of the object. In the fMRI experiment, the paintings were placed at a distance of 2.3 meters and observed via a mirror, while the subject's head was immobilized in the fMRI apparatus. Even in such a strange condition, the illusion was visible to all subjects, but, in order to see the illusion in an *unpainted* structure, typically some "mental effort" is required. The effort is similar to that needed to reverse the Necker cube or to see an imbedded figure in 2D trick art. The question in the present context is: "What precisely is the mental effort that is being made?" From a comparison with the other pictures in the experiment, it is clear that the undrawn shadows and/or undrawn linear perspective lines must be imagined – in order to perceive the concave shapes as convex buildings. In a word, the subjects appear to be *imagining* the pictorial depth cues that give convex three-dimensionality to a concave surface. As seen in Figures 3.48 and 3.49, the posterior visual cortex was relatively inactive – there being neither shadows nor linear perspective lines to view, but frontal cortex was active bilaterally.

Anatomically, BA47 is known to be one of the *first* regions of frontal neocortex at which nerve tracts from the three main sensory modalities (touch, vision and hearing) converge (see Chapter 4). It is absent in the chimpanzee brain and is clearly a region of association cortex involved in "high-level" cognition. Both BA46 and BA47 have previously been found to be active in a variety of cognitive operations – sometimes referred to as "reality monitoring" or simply "imagination," and play a role in cognitive functions that are not tied to any particular sensory modality. In the visual task of our fMRI experiment, these regions appear to be involved in imagining the perspective lines, shadows and shading needed for perceiving the concave shapes as convex. Without such imagination, the artwork is simply a rotating box, but with the aid of (real or imagined) shading/shadows and perspective lines, the concave shapes become convex "buildings" that twist-and-turn in a direction opposite to the physical object itself. These same brain regions were shown in Chapter 2 to be active when hearing the rather unfamiliar tension chords – three-tone auditory configurations. It would appear that the commonality between the visual and auditory conditions is that these portions of recently evolved prefrontal cortex are engaged in 3-body cognitive processes.

3.11. CONCLUSIONS

The idea that the underlying mechanism of the perception of 3D depth in 2D pictures entails an understanding of the alignment of, at a minimum, three visual cues cannot alone explain the fascination that visual art holds for most people, but the complexities of the illusion of pictorial depth *begin* at this rather simple level. "Real art" – outside of the psychology laboratory – compounds the effects of the alignment of three visual cues many-fold to enhance the sense of 3D structure in 2D paintings, drawings and photographs, and effectively hides the triadic nature of pictorial depth perception behind a plethora of cues.

Whichever of the artistic tricks that are used to depict pictorial depth, the illusion of depth in flat pictures is the cognitive dimension that has come to utterly dominate most genres of visual art (fine, trick, great and otherwise). Similarly, music employs countless pitch triads in chords and melodies to produce a complex narrative of "emotions" that are every bit as fictitious as the "depth" in flat paintings, but it is one of the principal attractions of music for most people. When reduced to their basics, it is apparent that both art and music have discernible two- and three-component processes at their core – with the three-component processes leading into the "higher" dimensions that the human mind finds alluring. Intimations of both visual and auditory illusions can be detected already in simple triadic configurations, but they become overwhelming, convincing and meaningful in complex musical or artistic compositions. For reasons still to be elucidated in experimental cognitive neuroscience, the same triads that provide pictures and music with, respectively, depth and affective modality for human observers are seen and heard by animal species, but not perceived. As a consequence, they do not experience the "illusion" of depth in a flat painting and do not feel the "illusion" of emotion in auditory vibrations – the illusory effects that make art and music "real" for us!

4

Human Work: Tools and Handedness

In order to understand the origins of the triadic talents discussed in the two previous chapters, it is essential that we examine the evolutionary history of *Homo sapiens*. Although art and music are two of our modern obsessions, unambiguous indications of such "high culture" date back little more than 40,000 years. In contrast, the origins of another type of characteristically human behavior go back several million years – and are as concrete as … rock – the primitive stone tools of the stone age. Although the construction and use of stone tools may seem rather unsophisticated to us today, the beginnings of human ways of thinking arose there.

The beauty of stone tools is of course that they last for indefinitely long periods of time. Language too undoubtedly evolved well before painting and instrumental music, but there are few empirical facts concerning the evolution of spoken language, and indications of written language date back only several thousand years. Vocal signaling by our early ancestors must have begun much, much earlier, but today we are in the difficult situation of knowing only the modern versions of human language – with nothing remaining of "simple languages" comparable to the ancient relics of tool use and toolmaking. Fortunately, the long trail of fossilized bones and stone tools provides some unambiguous evidence concerning the earliest origins of our higher cognitive talents.

In this chapter, I summarize several key discoveries about toolmaking and tool use and show that the human capabilities for both triadic *perception* and triadic *cognition* began more than two million years ago. Although the "threeness" of early tool use is not often mentioned in the evolutionary literature, the jump from two-cue to three-cue perception can be delineated in several different tool-technology contexts and indicates why the seemingly modest jump from (i) maintaining two objects "in mind" to (ii) contemplating the relations among three objects was a crucial perceptual

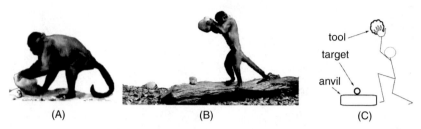

FIGURE 4.1. (A and B) Tool use by capuchin monkeys. Primitive as it may be, the task of lifting a heavy stone to crack the shell of an edible palm nut requires considerable hand–eye motor coordination to propel the stone directly onto the nut. (C) The conceptual innovation is the monkey's triadic understanding that a suitable "anvil" – the base on which the nut must be placed to crack its shell with a hammer stone – is essential. The same action standing on gravel or the soft earth fails to produce an edible morsel.

innovation leading initially to tool use and then to tool making. The story of the human fascination with tools is, I maintain, very much the story of how our ancestors learned to think "in threes."

4.1. STONES AS TOOLS

The earliest stone tools employed by human beings (Schick & Toth, 1993) were probably similar to those used by modern-day capuchin monkeys to crack open the shells of palm nuts containing an edible seed (Figure 4.1). The "tool" is simply a stone, but to become useful as a tool it must be a stone of sufficient size, weight and rigidity to be suitable for the job. The stone itself need not be altered in any way, but a new way of thinking is required. Specifically, the conceptual insight displayed already by the capuchin is an understanding that, in order to crack open the external shell of a palm nut, the nut needs to be on a hard surface when struck with a large stone. The same action with the same tool, but performed on soft earth will not crack the shell. So, for success, the monkey must bring both the nut and the hammer stone to the location of a flat rock that will serve as an anvil. That is the triad. Quite unusual among all animal species, the capuchin monkey understands that the action of opening the palm nut requires that three distinct objects – tool, target and anvil – must be used together, as it goes about the motor task.

The perceptual triad of tool, target and material "context" is arguably the most fundamental triad inherent to tool use of all kinds. More primitive examples of using material objects, but not dependent on a specific context, can be undertaken by relatively many animal species. Untrained chimpanzees use sticks to fish out the termites from termite nests and crumple leaves to make sponges for soaking up water. Similarly, the sea otter is known to dive under water to fetch a mollusk and a rock, and then to return to a dry perch to pound the mollusk with the rock. But, unlike the capuchin's situation, the entire perch, the rocky shoreline, is an anvil and does not require consideration as a distinct, third component in the tool-use task. Untrained seagulls, as well, drop clams and mussels onto paved surfaces in order to crack them open, and must have in mind both the shellfish and the physical location of the pavement – a cognitive dyad – to achieve success. But, for the seagull, there is no external, third object – a tool – involved. The capuchin, in contrast, must maintain three distinct perceptual items in mind.

To conflate these various types of clever animal behavior is to miss an important point. All examples of using sticks and stones and anvils might be described as "a series of planned actions within an appropriate context to achieve a specific goal," but the capuchin's mental state – with three rather than two concrete objects under active consideration – is one step more complex than that of either the otter or the seagull. Innumerable "contextual" factors will of course make these behaviors possible or impossible – wind and rain, competing animals and objects scattered on the anvil surface – but the focus of attention for the capuchin is specifically on three material entities and how they relate to one another.

4.2. TOOLMAKING AND HANDEDNESS

Undoubtedly, hominid tool-use had such humble beginnings, but already some 2.5 million years ago (mya), our Australopithecine ancestors were using stone tools not simply to obtain food but also to make other tools. The simplest of those tools were the "cores" and "flakes" obtained by pounding one rock against another (Figure 4.2).

To the nonspecialist, the historical evidence concerning stone tools is bewilderingly simple, even uninteresting, but, to the paleontologist, the markings on primitive tools are often unambiguous and of deep significance for the cognitive operations that they reveal. From an examination of early stone tools, three types are known to be contained "within" the rock itself – and require only that the rock be fashioned into a more appropriate shape. By striking a "core" stone with a separate "hammer" stone, the desired implement – to begin with, a "flake" with a cutting edge (Figure 4.2) – can be made. The next step is to strike the same core several times to obtain a slightly larger "knife" – that is, the remaining core

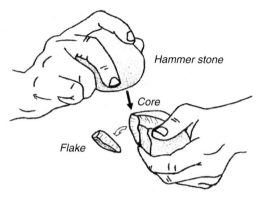

FIGURE 4.2. Toolmaking. A hammer stone is used to remove flakes of appropriate size and shape to be used as cutting knives from a core (after Wynn, 1996, with permission).

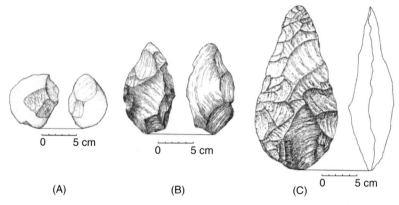

FIGURE 4.3. Stone tools with one sharp edge (A), two sharp edges (the so-called biface) (B) and well-chiseled hand-axes with symmetrical form (C). The difference between (A) and (C) is the difference between 6 and 60 precision strikes on the core (after Wynn, 1996, with permission).

with a sharp-edge and a smooth surface that can be easily held (Figure 4.3A). And the next step is to fashion the core into other shapes, typically, "hand-axes," for specialized uses (Figure 4.3B and 4.3C). This is primitive stone toolmaking.

The capuchin monkey is intelligent enough to *use* a tool, but the conceptual revolution of *making* a tool is profound, for when a capuchin breaks open the shell of an edible nut, the pay-off is immediate – and indeed eaten on the spot. In contrast, in the act of making a stone tool, the reward is necessarily delayed until the newly created tool itself can be used in a completely separate task. Whether the toolmaking task is the creation of flakes (later to be used as blades) or the creation of a sharpened edge on a core (later to be used as a knife), the toolmaker with hammer stone in hand must keep three inherently unrewarding objects in mind. In this regard, both capuchin-style tool use and Australopithecine-style toolmaking are triadic, but, as a consequence of the temporal delay in obtaining a reward through toolmaking, the true motivation underlying tool creation is less obvious: The anticipated benefits of the tool being created must also be kept "in mind" without any rewarding gustatory feedback.

Clearly, the difference in the motor expertise needed for pulling the twigs off a branch or crumpling a leaf to use as a sponge versus striking one stone against another is huge. On the one hand, there is very little "technique" and virtually no danger of failure for those simpler tasks. In contrast, modern-day experiments demonstrate that stone toolmaking requires a considerable learning process to attain the manual skill. The stone core must be struck repeatedly at an appropriate angle and with sufficient force to remove flakes of appropriate size and shape. The unforgiving medium of stone implies that – in addition to the insight of imagining the stone tool latent within the core – the process of learning how to knap and chisel the core into a suitable object was necessarily slow and laborious.

In fact, observations on young capuchin monkeys indicate that the "simple" act of cracking open a palm nut is learned from their elders over the course of *several years*! A hammer stone that is too small will fail to crack the nut, the stone that is not elevated high-enough will produce insufficient momentum, a nut placed on the soft earth will make crushing the shell impossible and even a nut placed on the anvil will not produce the desired result if the stone does not strike it from directly above. The young capuchin learns these facts by imitation, trial and error and relentless practice until the relations among all three components are properly understood, behavior is coordinated and a nutritional snack can be consistently obtained. A similarly prolonged learning process was undoubtedly necessary for hominids making the earliest stone-tools.

Stone toolmaking was a cognitively unprecedented behavior and required holding a triad of objects in short-term memory. Such activity alone is a remarkable feat and indicative of extremely unusual animal intelligence. But, as bold and imaginative as that stone-age invention was, it leaves a deep, unresolved mystery that many scholars have commented on. That is, almost as astounding as toolmaking itself is the fact that there were no changes in toolmaking technology over the course of the subsequent *one million years*:

> From the perspective of the typological study of stone tools, nothing much changed in Europe and Africa between 1.4 million years ago and 300,000 years ago. Hominids made the same type of tools they always had – hand-axes and cleavers, and a range of flake tools where the emphasis continued to be on the characteristics of the edge. In some sites flakes dominated in percentages, and in others bifaces dominated, but there were never any "new" tool types. (Coolidge & Wynn, 2009, p. 155)

The revolution of making and using tools (and thereby being able to feast on animal carcasses) undoubtedly transformed the lives of these hominids, but stone tool technology was simply not taken any further. Having invented tools out of the raw materials that most animal species had trampled on for eons, *our ancestors then spent one thousand millennia pounding rocks!* (And you thought you've had a rough life?) How can we understand their ability to invent tools and yet their inability to develop anything more advanced? What is the psychology underlying the quantum leap to stone tools and yet the failure to go beyond that for such a long time?

At least one part of the answer lies in the difficulty of the cognitive leap from tool use to toolmaking. In progressing from the capuchin-like chore of using an external object to hominid tool creation, hominids were forced to develop a favored hand. Unlike the *bimanual* stone-throwing needed to crack open a shell, stone flaking and core knapping imply that the hands will undertake coordinated, but *different* tasks – holding the core with one hand and striking it with the other (Figure 4.4A). Above all else, the training of *one* hand/arm (unilateral motor cortex) would be important to achieve control over the directed-impulse needed to separate a flake from a core, with the other hand immobilizing the core and necessarily engaged in a very different form of motor control.

We know from common experience that the precision of manual skills improves with practice. For crumpling leaves or breaking off twigs, practice is unimportant, but for the control of a trajectory to strike one object against another at a specific angle and with a specific force, repetition and gradual honing of skills would be essential. The capuchin learns its bimanual tool-use skill without distinguishing between the roles of the right and left hands, but the creation of hand-axes (Figure 4.3C) entails imparting *several dozen* precision blows by one hand to the core immobilized by the other hand. It requires handedness.

As every sportsman and weekend carpenter knows, a *unimanual* skill is best learned by first deciding on a favored hand, and then mastering the

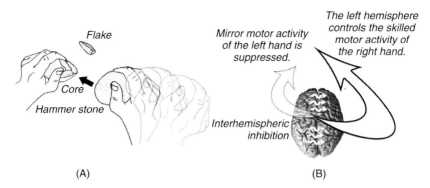

FIGURE 4.4. (A) The roles of both hands in creating stone flakes from a core. Immobolization of the core requires calculations in static coordinate space, whereas the dynamics of the right hand require vector calculations in momentum space. (B) Unimanual training implies interhemispheric inhibition (wavy arrows) in order to suppress unwanted "mirror movements" of the other hand.

motor skill through repetition. Particularly for a survival skill such as stone toolmaking, the nonchalance of alternating between the left hand and the right hand would not have been a luxury that our early ancestors could afford. To begin with, either hand might be used to wield the hammer stone, but, for each individual toolmaker, practice with the same muscles (and the same region of contralateral motor cortex) would be important to obtain improvements in performance. Independent training of both hands in the same skill (i.e., ambidexterity) would slow down the training process twofold and would simply not be useful. On the contrary, insofar as the nondominant hand is also learning the new, if less complex, skill of core immobilization, the learning of both skills by both hands would be notably inefficient. Differential training of the hands was essential.

In brief, the step from symmetrical to asymmetrical manual activity is noteworthy because it implies a rudimentary form of lateralized cerebral specialization (Frost, 1980) (Figure 4.4B). The training of a motor skill by the right hand (for example) would mean training the neural circuitry of the neocortex of the left hemisphere that controls the right-sided musculature responsible for the trajectory of the hammer stone. The left hand's stabilization function is not irrelevant, but the motor task of immobilizing the core is static, relatively unskilled and, above all else, *different* from the more complex, dynamic skill undertaken by the right hand.

Individually, early hominid toolmakers probably favored the left or right hand in about equal numbers, as is common for most mammalian species given tasks that can only be performed with a unilateral hand or paw. But, eventually over the course of a million years or so, a species preference for the right hand must have emerged and become genetically engraved (Crow, 2002). Whatever the underlying genetics of handedness, it is likely that there would have been advantages to unilateral handedness for a community of hominids insofar as a uniformly right-handed group of toolmakers could avoid left–right confusions among themselves. Any motor skill could then be learned by mere imitation (without mirror reflection) – provided that all members of the group had the same handedness. And indeed, stone relics themselves give clear evidence of the reality of right-handedness already about 1.8 mya:

> Most right-handers use the left hand to hold the stone core while pieces are removed by a hammer stone in the right hand. The hammering imparts a slight twisting movement to the core, and the chips that fly off have a slight, but definite twist, allowing one to tell the handedness of the knapper.... Our earliest ancestors, then, like us, were mostly right-handed. (McManus, 2002, p. 213)

It is therefore likely that, as a consequence of a million years of core knapping, the dominance of one hemisphere for fine motor control (i.e., neocortical functional asymmetry and a species-wide bias for right-handedness) eventually emerged. In other words, the lateralized cerebral specialization characteristic of the human brain has its earliest, identifiable roots in simple stone-toolmaking.

The importance of hand preference in the modern world is of course minimal; for most tasks, either hand will suffice, and even ambidexterity can sometimes be advantageous. So, it is important not to equate handedness, and specifically right-handedness, with the special cognition of *Homo sapiens*. Particularly in light of the known, if small, behavioral asymmetries in many other animal species – it is clear that motor asymmetries themselves do not correlate with intelligence. Nevertheless, the evolutionary significance of handedness during the emergence of toolmaking competence some two million years ago cannot be so easily dismissed. Tim Crow (2002), in particular, has pursued the implications of handedness and cerebral lateralization and made a strong case for believing that one crucial speciation event that brought divergence of *Homo sapiens* from other early hominids is inextricably linked with the hemispheric specialization associated with the motor control of, to begin with, the hands. The Crow scenario has profound implications for psychopathology and the various abnormalities of lateralization found in schizophrenics (Crow, 2000), but the more basic argument of the present chapter is simply that the two hands (hemispheres) were *unequally* involved during the first million years of toolmaking.

It appears that the simultaneous training of *different* motor skills by the two hands was important for successful stone-toolmaking.

Lacking any empirical evidence suggesting that language preceded stone tool use, we have every reason to believe that the hemispheric division of labor of the human brain began with toolmaking, and only many millennia later generalized to unilateral dominance for other motor skills, such as speech – probably through the generalization of hand gestures to vocal modes of communication (Corballis, 2009). The factor common to toolmaking and speech is of course that they are both extremely "fine" motor skills for which the millisecond sequentialization of motor events is crucial. Although speech and hand movements involve very different muscular systems, it would be a major convenience to have one cerebral hemisphere specialized for sequencing both types of goal-directed motor function ("praxis," Corballis, 1991). For the time-sensitive coordination of the striate muscles of the arm, hand and fingers, the inevitable delays of messages being sent long-distance across the corpus callosum would be awkward at best. Particularly for the control of the midline muscles involved in speech, the coordination of motor commands coming from both hemispheres could become tricky. In other words, the sharing of motor skills between the cerebral hemispheres would provide no obvious advantages – and it could be a problem to have two "equal partners" competing for control over the midline muscles of the lips, mouth and tongue (a known source of pathological stuttering, Fox et al., 1996). Clearly, if neither the left nor the right hemisphere were trained to the level of true motor expertise and the issue of dominance remained disputed, confusion would reign. For this reason, the unilateral "dominance" of one cerebral hemisphere for executive control was probably advantageous – while relatively unskilled "supporting tasks" not requiring careful sequential ordering of muscular activity were relegated to the other hand (hemisphere). A solution to this control problem was apparently discovered prior to 1.8 million years ago.

Whatever the mechanism underlying the initial asymmetry of handedness may have been, the *suppression* of synchronous bimanual actions (Figure 4.4B) would have been a necessary feature of the behavioral asymmetry: The activity of the hands needs to be independent. Once the decoupling of manual motor activity was accomplished, one hemisphere (for obscure reasons, the *left* hemisphere – possibly due to asymmetries emerging much earlier in the evolution of vertebrates, MacNeilage, 2008, p. 202) became the "dominant" executive controller of striate muscle – the main toolmaker and tool user for more than 90 percent of all human beings. It is therefore no surprise that, in the modern brain, the mere visual perception of tools and other man-made artifacts activates the left hemisphere (posterior regions BA 18, 19, 36 and anterior regions BA 45, 46), whereas the perception of objects from nature elicits responses symmetrically from cortical regions in both hemispheres (Perani et al., 1995). So, without disparaging the various supporting skills of the right hemisphere (left hand), it is clear that, for most people, the left hemisphere (right hand) was and still is the CEO capable of skilled manipulation of material objects.

4.3. THE DIVISION OF LABOR BETWEEN THE CEREBRAL HEMISPHERES

Despite the fact that the cerebral hemispheres are structurally very similar to one another and, not incidentally, despite the fact that these two "brains" experience essentially the same world throughout a lifetime, their electrical, blood flow and metabolic activity is often reliably dissimilar – asymmetrical to a degree not found in animal species. Periodically, attempts at debunking the importance of hemispheric functional asymmetries in human beings have been made (e.g., Efron, 1990), but modern brain-imaging research clearly demonstrates that the two cerebral hemispheres do not respond in the same way to the same world. Precisely why functional asymmetry is prevalent in human beings, but notably weaker in most other species is not certain, but two factors are thought to contribute: (i) innate differences due to microscopic differences in neuronal circuitry and/or neurotransmitters and (ii) differences due to dynamic hemispheric interactions across the corpus callosum (see *The Parallel Brain*, Zaidel & Iacoboni, 2003, for a thorough discussion). Both factors are likely to lead to small hemispheric asymmetries in other species, as well, but the explosive evolutionary growth of the corpus callosum in the primate line indicates that left–right interactions are particularly strong there, especially in the human brain, which contains several hundred million callosal fibers.

From the interactionist perspective on cerebral specialization, the most important anatomical facts concern the corpus callosum: (i) It is the largest fiber tract in the human brain and contains enough fibers to connect bidirectionally every cortical column in each cerebral hemisphere with its contralateral counterpart; and (ii) the density of callosal connections is highest in association neocortex and lowest or absent in primary sensory and motor cortex. Since lateralized functions in the modern human brain (i) employ association cortex and (ii) are strongest for the "unique" human talents (language, toolmaking/handedness), callosal involvement is likely. In other words, on anatomical grounds alone, the corpus callosum is poised to play a role in producing and/or maintaining the asymmetries of higher cognitive functions (Cook, 1986; McGilchrist, 2009). As such, it is a likely factor contributing to the uniqueness of the human mind (Gazzaniga, 2000).

Those who favor the view of innate hemispheric differences leading to laterality have argued that even small perceptual asymmetries may "snowball" into large-scale cognitive asymmetries (Hellige, 1993). For example, Richard Ivry and Lynn Robertson (1998) have advocated the idea that an innate difference in processing high- and low-frequency information (in both the visual and auditory modalities) results in differences concerning the visual or auditory features that are emphasized by the hemispheres. Given an unexplained but plausible starting "innate" asymmetry of that kind, an interesting story then unfolds, but hemispheric interactions are, from the outset, excluded as possible causal factors. Indeed, in their concluding chapter, Ivry and Robertson (1998, p. 276) themselves note that the "lack of consideration of interhemispheric communication is an obvious weakness [in most theories of human brain function. It is, however, an inevitable] consequence of the fact that the nature of callosal function remains mysterious." Still uncertain, perhaps, but arguably far too important simply to be overlooked.

The presence of small, innate, left–right differences in neuronal circuitry or small, innate neurotransmitter asymmetries is of course likely, but there is abundant evidence from many species suggesting that callosal communications accentuate differences in the information processing of the left and right hemispheres (Bianki, 1993). Given the nature of neuronal functions in general, there is a small number of possible roles that callosal fibers (and, more generally, long-range corticocortical connectivity) can play. Functionally, neurons can be either excitatory or inhibitory, and their axons can terminate on neurons that are themselves either excitatory or inhibitory. Structurally, callosal fibers can terminate either diffusely (one-to-many termination) or topographically (one-to-one), giving four main patterns of possible interhemispheric communications depending on excitatory/inhibitory and diffuse/topographic factors. Indeed, in the laterality literature, each of these four kinds of callosal activity has been defended as the "most important" form of interhemispheric information processing: (i) topographical excitation (the "carbon-copy" transmittal of a pattern of activity to the contralateral hemisphere) allowing both hemispheres to process information from the entire visual field (Sperry, 1968); (ii) diffuse excitation, by which a cortical region in one hemisphere alerts the other hemisphere of possibly relevant information (Guiard, 1980); (iii) diffuse

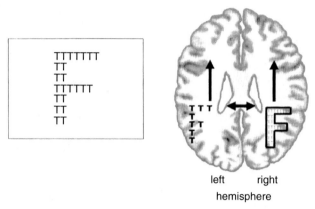

FIGURE 4.5. Model of whole-part perception in relation to the cerebral hemispheres. When presented centrally with the stimulus shown on the left, the perceptions of the two hemispheres differ (from Schulte & Muller-Oehring, 2010, with permission).

inhibition, through which one hemisphere becomes the dominant processor (Kinsbourne, 1982) and (iv) topographical inhibition, through which complementary regions of processing (e.g., figure-ground) are activated (Cook, 1984).

Arguably, the 200 million fibers of the human corpus callosum perform all these functions depending on permutations of these factors at different cortical regions. In recent reviews of callosal functions (Bloom & Hynd, 2005; McGilchrist, 2009; Glickstein & Berlucchi, 2010), the relevance of both inhibitory and excitatory mechanisms has been duly emphasized. Not only is it true that "there appears to be no structure or chemical constituent that is present in one hemisphere and not the other" (Galaburda, 1995, p. 52), but several generations of neuropsychologists have found evidence in both normal and clinical populations for functional asymmetries (McGilchrist, 2009). On the basis of experimental work in normal subjects, Schulte and Muller-Oehring (2010) conclude that "the corpus callosum mediates lateralized high-order cognition such as whole-part perceptual integration. Specifically, anterior callosal integrity appears to mediate local-global interference (callosal inhibition), whereas posterior callosal integrity seems to mediate local-global facilitation (callosal cooperation)" (p. 183) (Figure 4.5).

Unfortunately, the current uncertainty on the relative importance of innate factors versus callosal interaction in producing human laterality means that firm conclusions cannot be stated concerning the modern brain, much less the hominid brain several million years ago. What is known, however, is that (i) somewhat different kinds of cognitive strategies have emerged in the cerebral hemispheres (as revealed most clearly by unilateral brain damage and brain imaging with normal subjects, but sometimes in remarkably low-tech experiments; see, for example, Section 7.7) and (ii) the precise duplication of cortical information bilaterally does *not* occur. Neural network simulations indicate that functional hemispheric asymmetries can be achieved either through callosal inhibition (i.e., one- or two-synapse, direct corticocortical effects) or by inhibition at the level of the brainstem (Cook, 1999, 2002; Reggia et al., 2001; Shkuro et al., 2000). Both types of inhibitory mechanisms may be at work, but what is certain is that, behaviorally, *without an intact corpus callosum* (i.e., in young infants whose callosal fibers are developmentally immature and individuals with a congenital absence of the corpus callosum), people exhibit mirror movements of the arms and hands and generally a lack of bimanual coordination (Jeeves, 1994). So, whatever the detailed neurology of the normal development of hemispheric relations may be, we can conclude that callosal neurons are involved in the *suppression* of bilaterally *symmetrical* motor activity.

In an evolutionary context, the suppression of left-hand mirror movements (callosal inhibition of the right hemisphere by the left) would have been necessary for early toolmaking. Only one hand (and the motor cortex of only one hemisphere) needed to be trained in the dynamic skills of stone-toolmaking. But, in order to prevent replication of that motor activity and in order to prevent disruption of the training of the somewhat different skills of the contralateral hand, the dissociation of the motor activity driven by the cortex of one hemisphere from the contralateral motor activity was required (Figure 4.4B). At this level of rather basic motor competence, we are indeed "lopsided apes" (Corballis, 1991) – and probably have been for a thousand millennia (McManus, 2002) – with the corpus callosum playing an important role in allowing for the necessary motor asymmetry.

Whatever the case may be for callosal functions and hemispheric specialization, the asymmetrical use of the hands alone does not explain the cognitive revolution that toolmaking represents. We must also ask: What other kinds of changes occurred in the human brain that allowed us to make tools?

4.4. BRAIN SIZE

In the latter half of the nineteenth century, much of the debate concerning human evolution focused on resolving the contradiction between the apparent biological *continuity* of all animal species and the seeming *discontinuity* of specifically human intelligence. Although differences in brain size were evident to all, it was unclear how *qualitative* changes in cognition and behavior could emerge from a *quantitative* increase in brain volume. Even a threefold increase in neuron number – from a few billion to many billions – has no explanatory power in and of itself, and many commentators have been left with little more than the hopeful assertion that the wonders of the human mind simply must be a consequence of our "big brains" (*Big Brain*, Lynch & Granger, 2008). In modern times, the paradox has been slightly exacerbated by the known extinction of the big-brained Neanderthals and the seemingly modest intelligence of the even bigger-brained dolphins.

So, clearly, "more neurons" is not the entire explanation, but a large increase in brain size did in fact occur in the primate line over evolutionary time. Geographical differences and possible "species" differences among various hominid groups continue to be debated by specialists, but the overall trend from brains that were similar in size to the modern chimpanzee brain to hominid brains that were two-, three- and eventually almost fourfold greater is the principal finding in the fossil record. As seen in Figure 4.6, what is of greater interest than subtle regional or species distinctions is the apparent spurt in brain size quite *early* in hominid evolution – precisely during the era of simple stone-toolmaking. Why was there an explosion of brain size between 2.5 million and 0.5 million years ago? It is easy to speculate about the possible impact of many other factors over this 2-million-year stretch – changes in the organization of society, the use of fire, cooking and so on, but what we *know* for an empirical certainty is that, throughout this period, *Homo erectus* and *Homo habilis* were busy making stone tools – *simple* stone tools. Why on Earth was a bigger brain needed? As Oppenheimer (2003, p. 23) has emphasized: In order to understand the mystery of human evolution, we need an answer to "the question of what new behavior drove that rapid growth 2.5 million years ago."

The answer to the enigma of brain expansion during the early stone age lies in an understanding of the changes in the *relative* size of various regions of the cerebral neocortex. It is true that the hominid brain expanded remarkably during that era, but we also know from modern neuroanatomy that there are large disparities in the relative growth of cortical regions. In a word, the most outstanding fact of human brain evolution is the relative expansion of so-called association cortex. As illustrated in Figure 4.7, by first expanding a chimpanzee brain fourfold and then comparing individual cortical regions with the human case, it has been possible to calculate the percentage changes entailed in moving from the chimpanzee brain to

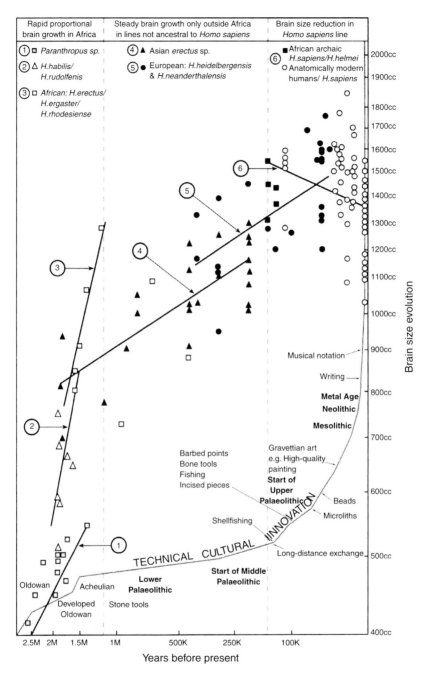

FIGURE 4.6. Changes in hominid brain size in relation to various cultural innovations (after Oppenheimer, 2003, p. 17, with permission). Note that most of the growth in brain size from 400 to 1200 cc occurred prior to 1 mya – during the period of development of the simplest of stone tools.

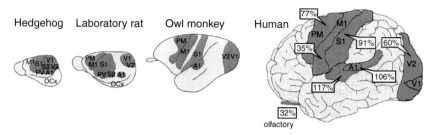

FIGURE 4.7. The absolute and relative increases in association cortex are evident from a comparison of the hedgehog, rat and monkey brains (redrawn after Striedter, 2005, p. 205, with permission) with the human brain (percentages from Deacon, 1997, p. 217, with permission). The percentages denote the difference between the actual size of cortical regions and the sizes expected from an ape brain expanded to human size. Note that much of the association cortex in the hedgehog and rat is paleocortex (olfactory cortex, OCx).

the modern human brain. We can then address the question: If the overall volume of the chimpanzee brain is adjusted to that of the human brain, which cortical areas have increased in size and which have decreased? How are we *different* from big-brained chimps? The answers are not necessarily what we might expect.

To begin with, we find that the human brain is less concerned with smell: Our olfactory cortex is only 32 percent the size of the (expanded) chimpanzee's. But there are also surprising decreases in other cortical regions. Particularly noteworthy is the fact that, although tool usage is a human specialty and made possible by hand–eye coordination, there are relative *decreases* in both motor cortex (to a mere 35 percent) and visual cortex (60 percent) in moving from a chimpanzee brain to a human brain! Even premotor cortex – the region that coordinates the motor commands eventually sent to the skeletal musculature for toolmaking – has not shown expansion relative to other regions (a decrease to 77 percent). Primary and secondary auditory cortex have grown slightly (106 percent and 117 percent) – suggesting the growing importance of auditory information processing, but what has shown tremendous growth is the polymodal association cortex (the light gray areas in Figure 4.7). So, it is a truism to say that *Homo sapiens* has a large brain, but the greatest increases in cortical hardware have come in the form of "association" cortex – and not in the changes to purely visual, somatosensory or motor cortex.

In summary, the broad outlines of the evolutionary changes in brain anatomy clearly indicate two major neurological developments: functional asymmetry and cortical expansion. The asymmetry is a likely consequence of right-handed stone-toolmaking behavior over the course of 2 million years and can be understood as indicating the importance of fine-grained, differential motor control of the hands. But the emergence of lateral asymmetry does not explain the fourfold expansion of the human brain or the explosive increase in size of the associational areas. With or without the unprecedented handedness of the early stone-toolmakers, the massive enlargement in the human brain also requires an explanation.

Specialists in brain evolution have frequently noted that the ballooning of the human brain did not come without costs (e.g., Allman, 2000; Striedter, 2005). Not to even mention the increased danger of giving birth to babies with big heads, the human brain in particular is a demanding drain on nutrients of all kinds. If there were no evolutionary advantage for hominids to have more association neocortex, cerebral expansion would not have occurred again and again over the course of 2 million years. In fact, the expansion of brain size has subsequently halted – or even mildly reversed over the past 50,000 years – but massive increases in association cortex were apparently useful *during the early stone age*. That indisputable finding in the fossil record is strong indication that – well before the emergence of symbolic language, art, music, agriculture, wheels, writing and the many other complexities of human culture – big-brained hominids were having their own behavioral successes. In the jargon of evolutionary biology, big brains were "selected for" during a span of 100,000 generations of our ancestors. What, we must ask, was so important about association cortex? An answer more precise than "we got smarter" would be useful.

So, before considering the further evolution of tools themselves, let us consider the fundamental nature of the perceptions entailed in toolmaking during the time when the simple implements of Figure 4.3 were being produced. Similar to the small-brained capuchin, our ancestors eventually must have learned to discern among different types of rocks and to choose those that could be fashioned into tools. That was a visual task or perhaps a visual and tactile task undertaken while walking through riverbeds in search of hard, shiny, igneous rocks. With appropriate rocks in hand, our toolmaking ancestors learned that a single blow could achieve the desired sharp-edged flake. Like the capuchin using a stone "anvil" on which to pound one rock against another, the visual-and-tactile motor task of obtaining a sharp-edged "tool" demanded hand–eye coordination. But, beyond stone flakes, it is noteworthy that the creation of a large, symmetrical hand-axe was *not* the product of a lucky, hit-or-miss, one-off effort in banging two rocks together. On the contrary, it was the product of persistent, careful concentration, requiring that a core be struck with precision *several dozen times in succession.* In the case of hand-axes (Figure 4.3C), multiple, nonidentical blows to the core would have been required in order to obtain a tool with a specific shape, symmetries, sharp edges and possibly a point.

The progression from simple stone flakes to hand-axes may seem like a minor development, but consider the mental state of the hominid intent on making a hand-axe. The focused effort to produce this stone tool meant that a rather haphazard approach – which might suffice for flake production – would end in failure. In making a hand-axe, every blow to the core in a series of perhaps *60 repetitions* would be crucial for success. One overly enthusiastic blow could destroy the core and a misdirected blow directly onto a sharpened edge could ruin a day's work, whereas *faint-hearted* taps on the core would make the task endless. It is within this real-world context that we need to consider the experience of toolmaking by our early ancestors. To repeat, their brains were twice the size of chimpanzees (not yet three or four times), and regions of posterior neocortical association cortex that are barely discernible in the chimp brain were blossoming, but apparently not yet preoccupied with linguistic concerns. Why, we must ask, would an expansion of *association* cortex facilitate specifically stone-toolmaking? What, precisely, is association cortex?

4.5. TRIMODAL CORTICAL REGIONS

Surprisingly concrete answers to questions about the nature of association cortex come from the musings of nineteenth-century philosophers and the findings of twentieth-century anatomists. Both were concerned with the convergence of visual, tactile and auditory sensory information at the neocortical level, concluding that, rather than a *quantitative* increase in brain size (i.e., simply the presence of more neuronal hardware), the crucial *qualitative* feature that distinguishes the human brain is the presence of neocortex where the three main sensory modalities come together. In other words, it is not brain size (or neuron number) but the processing of diverse types of sensory information within the same region of (by definition) association cortex that allows for a certain type of higher-cognition. Beyond the evolving capabilities of unimodal visual, unimodal tactile and unimodal auditory processing, the convergence of two or three types of information might lead to insights not apparent in one modality alone.

Such speculation is easily said, but how do the modalities of sight, sound and touch in fact work together? "Setting the weights" of the synapses in a neural network for any combination of visual/tactile/auditory input would clearly be a nontrivial chore for a brain that is not already well-versed in cross-modal interactions. As Steven Mithen (1996) has discussed in detail, it is difficult for modern human beings with well-integrated sensory modalities to appreciate the problem of weighing the relative importance of information that arrives through different sensory channels. Today, we effortlessly use information in one modality to compare against information in other modalities and easily draw connections among sights, sounds and touch sensations, but such cross-modal integration was not always the case. More pointedly, we know from the study of patients who have had the auditory or visual modality suddenly "switched on" after a lifetime of deaf or blind existence that the melding of different forms of sensory information is not, to begin with, easy. On the contrary, the counterintuitive surprise from clinical neurology is that the addition of a new sensory modality is not immediately perceived as a welcomed "expansion of awareness" but rather is more commonly a source of confusion than insight (Sacks, 1985). Conclusions that were once certain on the basis of auditory or tactile information alone suddenly become uncertain in light of simultaneous visual patterns, or vice versa.

The initial confusion created by adding a sensory modality does not mean that ultimately multimodal processes would be less reliable than unimodal processes, but learning when and where to rely on different types of information from one modality rather than another would necessarily take time. How much evolutionary time is, of course, the big question – and not easily answered. Of interest is the fact that, in the world of artificial neural networks, the *unsupervised* learning (which evolution necessarily is) of appropriate synaptic settings in a network containing only two input signals and only four neurons can require several hundred (!) iterations (simply to calculate the XOR-function [see Section 7.5], Rumelhart et al., 1986, p. 318). In an evolutionary context, if each failure to compute the correct result because of an inappropriate weighting of visual, auditory or tactile information led to the death of the organism, then clearly it would be a long, slow process of learning how to rely on sensory information arriving at multimodal association cortex.

Two early examples of the successful evolution of *cross-modal* processing are well known and must have evolved around the time of the bifurcation of apes and hominids. The first is bipedal locomotion. Most obviously, standing on two legs requires fine-grained proprioceptive perception that quadrupeds need not worry about. A sense of two-legged balance requires attention to the tactile sensations of the strains and stresses pulling on our bones and muscles and the inner-ear proprioceptive sense of vertical orientation. But, without visual input, standing or walking is a chore requiring unrelenting attention to the proprioceptive sense of balance in order not to stumble and fall. Together with the monitoring of visual changes, however, the bipedal stance and locomotion becomes trivial.

The second example of cross-modal processing is inherent to hand–eye coordination. Extending a finger to make contact with an object in sight entails only visuomotor coordination; visual information alone will suffice to guide the seen hand to the seen object. In contrast, grasping an object using the appositional thumb requires not only visual perception, but also tactile feedback in order for the object to be taken into possession. Object-manipulation capabilities require more than contact, so that coordination between sight and touch is needed for grasping: "Simple!" we may think, but grasping is a primate talent that has not been mastered by most other animal species. Clearly, the ability to walk on two feet and grasp with the hands alone does not make an animal human, but both capabilities require cross-modal sensory processing that did not emerge among mammals until the primates evolved.

Evolutionary scenarios of this kind necessarily remain rather speculative, but the gross anatomy of multimodal information processing in modern animal brains is in fact well understood. Histochemical tracing of the connections stepwise from primary sensory areas to secondary, tertiary and quaternary cortex has revealed the precise locations of the convergence (in the primate brain) of visual, auditory and tactile information (e.g., Jones & Powell, 1970; Seltzer & Pandya, 1994). Although the anatomical connections alone do not answer questions about cognitive mechanisms, they leave no doubt about the areas where multimodal processing occurs. Specifically, the regions of convergence of the three main sensory modalities are regions where, in the modern human brain, higher cognitive functions are undertaken (Figure 4.8).

The implication of the stepwise convergence of the sensory modalities in primate evolution is that previously unknown forms of bimodal and then trimodal association would have occurred in cortical areas that are not devoted exclusively to one sensory modality. Of course, within-modality dyadic and triadic associations among different sensory features (e.g., color, shape and texture in vision; pitch, loudness and timbre in hearing; smoothness, temperature and rigidity in touch) would undoubtedly have evolved earlier, but the much later convergence of the different sensory modalities in cortical regions not already committed to a specific modality would have allowed for computational results that are potentially different from those within a given modality.

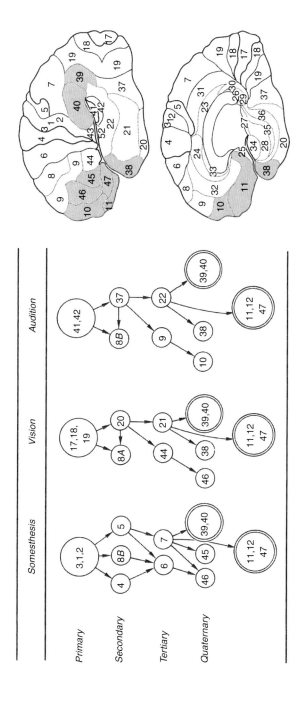

FIGURE 4.8. The flow of sensory information from primary through quaternary areas for each of the sensory systems is known from anatomical studies (data from Jones & Powell, 1970). Interestingly, trimodal convergence first occurs at regions of association cortex (in the human brain) known to be involved in language processing (BA 39, 40) and at prefrontal regions involved in conditional associations (BA 11, 12, 47).

So, let us again return to questions concerning the evolution of cognition in stone-age man. The archeological evidence indicates that our ancestors were pounding rocks for the better part of 2 million years *before* they developed the more complex hafted tools consisting of several components (bows and arrows, fish-hooks and lines, animal traps). Curiously, their brains doubled, trebled and then quadrupled beyond the chimpanzee dimensions during that time of simple stone hammering. Noteworthy also is the fact that much of the expansion of the brain occurred while the human larynx had *not* yet descended to its modern position (which eventually made the production of modern speech sounds possible). In other words, the evolutionary emergence of association cortex seems to be *unrelated* to the subsequent developed of language capabilities. On the contrary, it occurred simultaneously with seemingly very *modest* developments in toolmaking. What was going on, we must ask, during the stone age that favored a huge increase in association cortex and that, only much later, supported the explosive development of many other human cognitive talents?

Was the act of pounding stones really all that important?

The answer appears to be yes. Clearly, the hand–eye coordination needed to produce stone "flakes" involved tactile and visual sensory processing together, but, in holding a stone and imparting *multiple* blows using another stone, the *sound* of each blow would give the toolmaker an unmistakable indication of when his efforts were being fruitful. The striking of stone against stone gives an auditory vibration, rich in higher harmonics that would indicate whether either object was too soft to be useful. Plink! would indicate igneous rock against igneous rock, whereas Thunk! would indicate some softness and Thud! would tell the toolmaker that one or both rocks are useless sandstone. Moreover, the sound of igneous rock hitting igneous rock differs with the directness of the blow; a glancing blow sounds different from a direct hit, and both will differ from the sound obtained when the core is on a firm versus a soft, uneven or gravel base. It is therefore likely that using such auditory information would have played an important role in early stone-toolmaking.

In a modern context, the Ping! of a well-struck nail will be followed by a similar trajectory of the hammer, again and again until the task is completed, whereas the absence of Ping! will warn of a misplaced blow and demand realignment of wood, hammer and/or nail. In other words, *hearing* – auditory perception – provides feedback that is immediately useful for guiding the hand–eye (tactile–visual) coordination. The same is *not* true for sewing, pottery, leathercraft and toolmaking using softer materials, but striking stone against stone would necessarily engage the auditory modality because relevant information is available there. It is, moreover, information that arrives on average 40 milliseconds before visual information (Libet, 2004) and, as such, is a valuable feedback signal. As Coolidge and Wynn (2009, p. 94) comment: "Tool cognition is more than motor memory.... [T]he knapper relies on visual, aural and tactile patterns – how the surfaces and angles appear, what the blows sound like, and how they feel." I suggest that the cognition involved in the earliest stone-toolmaking is inherently triadic and, moreover, that no earlier form of hominid behavior required the coordination of tactile, visual and auditory cues.

Note that auditory information processing among modern monkeys, apes and gorillas is predominantly a matter of detecting the distant sounds of approaching predators and raising an alarm to compatriots. For orientation in a complex and dangerous jungle environment, hearing is of course relevant, but, beyond its usefulness in long-distance warning, most primates exhibit little interest in auditory information processing in comparison with the time and energy they spend on visual and tactile phenomena. Chimpanzees, in particular, are highly vocal in expressing pleasure and displeasure, but they are not known for subtle auditory prowess. That fact is demonstrated in specialist texts on primate cognition, where "hearing," "auditory perception" and "listening experiments" are generally absent from the index.

Indeed, the indifference of rhesus monkeys to most auditory stimuli was the fatal flaw in a recent case of scientific misconduct at Harvard (Bartlett, 2010). Expecting monkeys to habituate to auditory tone stimuli (A-B-A) and then to respond differently to simple changes (A-B-B) was an understandable human bias in the design of a monkey experiment: a simple task to illustrate a simple point about learning, but requiring that the experimental animals are attuned to the auditory stimuli! Some commentators have sensationalized details about where the experiments were run, but the only lasting significance of the experiments is the fact that monkeys are rather inattentive to inherently meaningless (for them) auditory tones. When suitable gustatory motivation is included in the design, cross-modal associations can of course be trained in many species (Calvert et al., 2004), but no primate is naturally and spontaneously as attentive to auditory cues as is *Homo sapiens*.

In summary, the main point of this section on brain evolution is that stone-age, stone-wielding toolmakers were the first hominids to exploit three-way sensory associations that included not only vision and touch but audition as well. To begin with, the posterior polymodal sensory association areas would have been actively engaged in the work of stone-toolmaking (the inferior parietal lobe, BA39 and BA40 in the modern human brain, corresponding to a small sliver of cortex in the superior temporal sulcus of the monkey brain). In strong support of this hypothesis is the fact that areas of posterior association neocortex expanded *first* (2~3 mya) – well *before* the enlargement of frontal cortex (1~2 mya) (Holloway, 1995). Ralph Holloway (1999) has been one of the more cautious interpreters of the cognitive significance of fossilized hominid skulls, but he has been explicit here:

> The fact that the amount of primary visual striate cortex is relatively reduced in humans is tantamount to saying that the relative amount of parietal lobe association cortex is larger. The comparative evidence for this is indisputable. (p. 90)

Moreover, on the basis of brain-imaging experiments, Dietrich Stout and colleagues have reported on the cortical regions activated before and after naïve human subjects are taught the stone-knapping technique. Stout and Chaminade (2007) conclude that

> Initial tool use training is a lengthy process involving modification of bodily representations in the rostral part of the intraparietal sulcus [=posterior association cortex].... This suggests that it may be acquisition of the necessary sensorimotor capabilities, rather than executive capacities for strategic planning, that represents the critical bottleneck in the initial development of complex tool-use and toolmaking abilities. (p. 1098)

Behaviorally, our ancestors were busy refining their techniques for making simple stone tools over this period of nearly 3 million years. Cognitively, they were concentrating on (i) the sight of their own hands holding and manipulating rocks, (ii) the feel of the weight, texture and momentum of those rocks and (iii) the sounds of the rocks striking rocks and other objects. Neurologically, they were learning how to evaluate the relative importance of sensory input from visual, somatosensory and auditory cortex in the newly evolving regions of posterior association cortex. In brief, the first stage in the evolution of the specifically hominid brain involved growth in *posterior* and, much later, *frontal* association cortex (Figure 4.9).

The neuroscience of cross-modal neocortical processing is in fact a complex issue (Ghazanfar & Schroeder, 2006). Not only are there significant species differences, but modern studies "suggest that little cortex is truly unimodal.... Thus, it may be more relevant to characterize cortical areas not by their dominant modality but by the relative weights and roles of different types of inputs" (Kaas & Collins, 2003, p. 285). Nonetheless, even if there is some modality mixing in primary sensory cortical areas

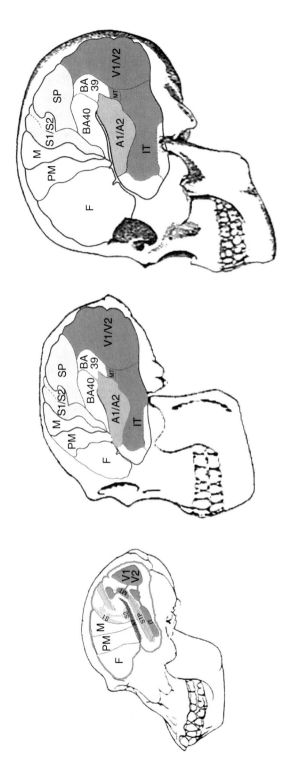

FIGURE 4.9. The major sensory and multisensory regions of the brains of the chimpanzee (A), *Homo erectus* (B), and *Homo sapiens* (C). Dark gray regions are predominantly visual, gray regions are predominantly tactile and lighter gray regions are predominantly auditory. The superior parietal area (SP) is both visual and proprioceptive; inferotemporal cortex (IT) is both visual and auditory. The brain of *Homo erectus* shows enlargement of posterior association cortex (BA 39/40); whereas that of early *Homo sapiens* shows a subsequent enlargement of frontal regions (F).

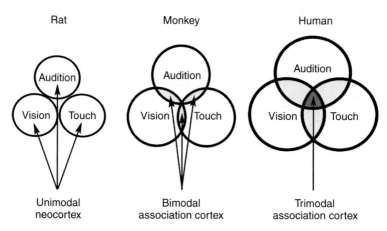

FIGURE 4.10. The evolution of the brain has entailed not only quantitative increases in neocortical areas but also the emergence of association neocortex (unimodal, particularly evident in mammals), bimodal (particularly evident in primates) and trimodal neocortex (particularly evident in *Homo sapiens*).

(possibly playing a role in various forms of synesthesia), it is abundantly clear from classical neuroanatomy that regions of polysensory association neocortex have increased hugely in the evolution of the primate brain (Figures 4.8 and 4.9). The upshot is that particularly the human brain is capable of not only a quantitative increase in "associations" due to an increase in overall brain size but also a qualitatively unprecedented mixing of visual, auditory and tactile information in the supramarginal and angular gyri, areas that barely exist in the ape brain (Ramachandran, 2011, p. 22).

The evolutionary trend in the mammalian line can be characterized as in Figure 4.10, where both the increase in unimodal cortex and the gradual appearance of bi- and, later, trimodal cortex are depicted. Note that the evolutionary process leading to the emergence of polymodal neocortex can be stated simply as the natural selection of brains with a greater expanse of neocortex, but, more important than raw size, the overlap of the different sensory modalities is an inevitable result of the physical constraints of cortical expansion within a skull of limited dimensions. Quantitatively, we have big brains, but qualitatively we have polymodal brains.

I submit that the puzzle of the 2-million-year stretch of stone-age toolmaking with surprisingly few innovations (a time of "almost unimaginable monotony," Jelinek, 1977, p. 28) can be understood as an era of *polymodal neuronal software* innovations. The hardware needed for associations among the sensory modalities was in place, but the difficult balancing act of weighing auditory, visual and tactile information took eons to perfect.

In other words, it is no mere coincidence that chiseled hand-axes – those unmistakably "crafted" tools of the stone-age – emerged when they did: *not* soon after apes became bipedal, *not* while the brain was still of chimpanzee dimensions and lacking trimodal association areas and *not* after cave art, long-distance trading, music and burial of the dead had appeared on the scene. The hand-axes emerged *together with* the convergence of auditory/visual/tactile sensory information in association cortex *because* it is only when all three modalities are used to fashion tools that stone-toolmaking becomes efficient.

As Merlin Donald (2001, p. 164) has also commented, "The human brain got a great deal larger, but its expansion was not an indiscriminate, across-the-board increase in size. It expanded in a very specific way, led by the most abstract regions of the cortex, the so-called tertiary, or association areas." Absolutely! But before we discuss the issue of "abstract" thought, there is the very concrete issue of modality convergence – using visual, tactile and auditory sensations together in the guidance of motor behavior. The ability for such polymodal processing is where the true significance of early toolmaking lies – and from which abstract thought eventually evolved.

Ramachandran (2011) has also emphasized the importance of this region of newly evolved polymodal sensory cortex in trying to explain human uniqueness: "The inferior parietal lobe – strategically located between the touch, vision and hearing parts of the brain – evolved originally for cross-modal abstraction" (p. 178). But the essential point here is not the "abstraction" so much as the concreteness of the polymodal associations. *Before* the emergence of abstract thought (however that might be defined), there emerged the concrete cognition of harnessing touch, vision and hearing in the construction of stone tools.

One other skilled, manual behavior of our ancestors undoubtedly played a significant role in our evolution: fire making. At some early stage, hominid geniuses must have noticed the sparks that fly when rocks strike one another and then thought how to use the sparks in creating fires. Whether for the purposes of cooking, keeping warm or scaring off the wild beasts, the ability to control fire would have been a discovery of paramount importance (Wrangham, 2009). That discovery is most likely to have been made during the era of intense stone-toolmaking 1~2 mya, but, unfortunately, there are few historical traces of fire-making capabilities until hearths and fireplaces became a part of semipermanent encampments (250,000 years ago). In the present context, the important point about making fire by striking a flint stone is that it (i) was a motor skill, comparable in difficulty to stone-toolmaking and also (ii) required unilateral training of the dominant hand (and hemisphere). Being essentially a task involving the striking of one stone by another to produce a spark, (iii) a triad of sensory associations would undoubtedly have played a major role in learning how to make fire. Here also, auditory perception – in conjunction with visual and tactile perception – would have been useful in ways distinct from the auditory warning signals of approaching predators.

Other scenarios in which simultaneous visual, auditory and/or tactile information is relevant to human behavior are of course possible, but the huge expanse of time in which our early ancestors made stone tools (but apparently little else) suggests that the refinement of stone-tool technology was crucial for human evolution. This was cutting-edge technology some 2 million years ago! And any stone-toolmaker who attended to and made appropriate use of all three of the main sensory modalities would have accomplished toolmaking tasks more efficiently – and would have been "selected for."

Stated conversely, it seems unlikely that toolmaking and fire making would have been invented by our ancient ancestors if they had been deaf or were as apparently disinterested in the auditory realm as are modern-day chimpanzees. The work behind the knapping of a stone to obtain a useful tool required more than hand–eye coordination; it also required coordination with and concentrated listening to auditory cues to know when progress was being made. It is therefore likely that, generation by generation, individuals who had undergone brain changes that included an increase in trimodal association cortex would have fared well in stone-age society insofar as they had the "brain power" to associate visual, tactile and auditory cues for the efficient making of tools. They would have been the most productive individuals in the high-tech work of fashioning stone tools and creating fire and would have enjoyed an enhanced social status in primitive hominid society, eventually leading to a proliferation of descendants with similarly advanced trimodal brain power.

A related idea about the evolution of human cognition has previously been outlined by Steven Mithen in *The Prehistory of the Mind* (1996). As a paleontologist, he has been concerned primarily with the cognitive *contents* of various human intelligences and how they relate to the fossil record. These include the "technical intelligence" underlying toolmaking, the "biological intelligence" behind the exploitation of surrounding flora and fauna and the "social intelligence" making human societies possible. Mithen's key insight is that initially each form of intelligence would have developed within a very narrow range of applications – truly a cognitive Swiss Army knife, where each cognitive tool had a limited number of specific uses – and only later were the cognitive barriers among the different realms overcome. For example, we know that the working of stone to make useful tools and weapons was developed 2 mya but was not immediately generalized to the working of wood, ivory and bones. Although the similarities between those hard materials and stone seem obvious to us today, wood, ivory and bones were initially considered within the closed framework of biological cognition, and not within the framework of technical implements.

Mithen provides many plausible examples of the apparent lack of cross-domain thinking among early hominids and the eventual dissolution of cognitive barriers that allowed for similar behaviors in different domains. His general hypothesis is persuasively consistent with the fossil record, but consideration of the neurological substrate on which cognition occurs suggests a slightly different classification of intelligences related to the independence and eventual convergence of the sensory modalities.

So, without disputing Mithen's distinctions among different intelligences, the broad strokes of brain evolution would indicate different types of cognition that depend on the different types of sensory information employed. Whether modern human talents are described as due to a core "human intelligence" (a flexibility of cognition that we, but not other species, have mastered) or as the "multiple intelligences" of various domains, the neuroanatomical reality is that we differ from other species primarily in the fact that we have relatively large expanses of polymodal association cortex where a variety of algorithms have evolved for dealing with "three-way relations." As related earlier, a coherent evolutionary scenario can be constructed around the idea that stone-toolmaking was facilitated by trimodal sensory processing. Importantly, in this view there is *no* need to postulate an inexplicable mutation of neuronal circuitry that miraculously propels humanity into unprecedented cognitive realms. On the contrary, the *quantitative* expansion of neocortex within a finite skull alone would have inevitably led to the neocortical overlap of the sensory modalities – and eventually to trimodal cortical regions. More than anything else, the existence of trimodal cortex made qualitative changes in sensory processing possible, ultimately leading to new cognitive and behavioral capabilities.

It should already be obvious that triadic associations are not a single, unique trick, but rather a diverse set of computations – various ways in which the relations among three independent sources of information can be evaluated. From the perspective of cognitive neuroscience, the gradual process of learning complex behaviors based on triadic relations necessarily has its origins at the neuronal level (Figure 4.11). Initially, behavior as mundane as

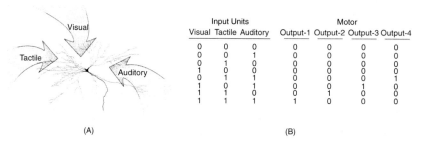

FIGURE 4.11. (A) Trimodal information processing at the single-neuron level (after Stein & Meredith, 1993, p. 234, with permission). (B) The Truth Table for various forms of trimodal information processing. In each case, the neuron should fire only in response to a certain combination of sensory inputs.

stone-toolmaking must have been important, but there are countless other behaviors where the coordination of the different modalities would eventually have become useful. With trimodal information reaching regions of association neocortex, appropriate motor behavior would be computed by relative synaptic weighting. The so-called Truth Table for several possible trimodal processes is shown in Figure 4.11, illustrating which set of inputs is required for a specific motor output. Expressed as simple neural networks with on/off inputs and outputs (see Section 7.3), the computations are indeed trivial, but adjusting the synaptic weights for this kind of neural network was the painstaking, trial-and-error chore undertaken by our ancestors over the first 2 million years of toolmaking. As an unsupervised learning process, the evolution of suitable networks that could guide survival behavior was far from trivial.

However the algorithms for multimodal information processing may have evolved, we now live much of our lives in a cognitive universe that transcends the world of Pavlovian stimulus–response pairs (i.e., those types of one-to-one dyadic associations that essentially fill the lives of animal species). While animals excel in various sensory skills – and are often faster and more precise in their motor responses than we can ever be – our higher-level, cross-modal cognition has given us a deeper and broader understanding of the natural world, an understanding that has allowed us a much wider range of behaviors than are available to other species.

4.6. HAFTED TOOLS

The fossil record concerning tool use shows four notable stages. The first stage began more than 2.5 mya with the advent of the hammer stone and

TABLE 4.1. *The numbers of components found in the artifacts from Middle and Upper Paleolithic Europe*

	1	2	3	More than 3
Middle Paleolithic	Wooden spears, clothing, scrapers, knives, etc., if unhafted	Stone-tipped spears, scrapers, knives, etc., if hafted		
Upper Paleolithic	Needles, ivory spears, bone/antler polishers, etc., if unhafted, grinding stones	Bone/antler polishers, etc., if hafted.	Arrow (shaft/tip/flights)	
		Bows (single or composite), traps		Ladders, multipiece clothing, necklaces, nets

After Dennell, 1983, p. 83.

the ability to produce flakes with one sharp edge (Figures 4.2 and 4.3A); the second stage was the use of the hammer stone to produce biface cutting stones created by multiple strikes to suitable cores (Figure 4.3B); the third stage was the period when hand-axes were created (Figure 4.3C). The fourth stage beginning some 250,000 years ago was that of so-called hafted tools – tools comprised of two or more parts.

For nearly 2 million years, our ancestors were clever enough to make and use tools – and thereby to survive in African environments that were, at least periodically, extremely harsh. But they were not clever enough to develop tools consisting of two or more parts until the so-called Middle Paleolithic period (250,000 years ago) when the recursive nature of complex tool construction was mastered (Table 4.1). Since the Upper Paleolithic period (150,000 years ago), tools have become increasingly complex – to such an extent that it is irrelevant to count the number of parts involved. The important conceptual leap was the jump from a one-piece tool to a two-piece tool! That is the jump into a new form of triadic processing, for the hafting of one object to another – for example, an arrowhead onto an arrow shaft – necessarily required a third component, either a piece of hemp to tie it on or sticky tree sap to glue it in place. Here again, a triadic invention entailed the manipulation of three kinds of materials to produce useful tools. Once this new form of triadic cognition had become available, recursive developments in tool design were possible, and tool parts were imbedded in larger and ever more complex tools.

4.7. THE BEHAVIORAL NEUROLOGY OF TOOL USE

As in many other realms of human psychology, some of the earliest insights concerning the mechanisms underlying tool use were obtained by European clinicians working in behavioral neurology. The first neurological model of tool usage was that of Hugo Liepmann from 1908 (Figure 4.12) – a conceptual framework that has proven remarkable robust. Whether due to age, disease or brain trauma, specific impairments in the handling of tools were found in the neurology clinic to occur following specific types of brain damage. Classification of the symptoms of the abnormalities of tool use led Liepmann to develop a theoretical model consisting of three main components, all three of which were located in the left hemisphere.

The three types of behavioral abnormality subsequent to damage to these modules were labeled kinetic apraxia, ideomotor apraxia and ideational apraxia. Although overlap of these syndromes is common, in their pure form kinetic apraxia is the inability to use a tool – a motor failure not involving the peripheral nervous system or the peripheral musculature itself, but an inability to trigger the simple motor routines appropriate for the given tool. Despite retaining normal perception of the tool and an understanding of its proper use, motor behavior itself is abnormal. In contrast, ideomotor apraxia is a deficit corresponding to a lack of understanding of what the tool is to be used for. Combing one's hair with a toothbrush or striking a nail with a screwdriver would be examples of conceptual abnormalities coupled with motor routines that are themselves normal. Patients with ideomotor apraxia typically exhibit misapplications of the right tool in the wrong place or the wrong tool in the right place. Finally, ideational apraxia refers to object "blindness," where the patient fails to understand the nature of familiar tools and their typical uses. Visual perception of a comb or hammer, for example, would not trigger the normal associations of using the tool in dealing with hair or nails. Significantly, modern versions of this early tool-use model have retained the same triadic architecture (Figure 4.12), although the modern understanding of the localization of functions within the brain has become much more complex.

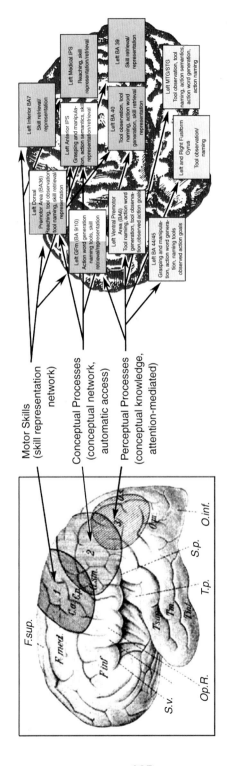

FIGURE 4.12. Liepmann's (1908) triadic model of tool use and the modern version (after Johnson-Frey, 2004, with permission). Johnson-Frey's classification of the subprocesses involved in perceptual (light gray), conceptual (white) and motor processes (dark gray) is more complex but retains Liepmann's triadic functional architecture.

4.8. CONDITIONAL ASSOCIATIONS

For nearly all manual skills, it would be unusual to consciously plan the movement of a particular muscle, but, in principle, such deliberate control is possible, and any complex skill can be seen as a series of such voluntary motor commands undertaken in coordination with countless others. The devil, of course, lies in the details. Simply to push a button to receive a food reward, extension of one finger might suffice. But extending the forearm by unbending the elbow, or leaning forward at the shoulder with the finger out-stretched or any combination of finger/elbow/shoulder movement could accomplish the same task. The weighting of every synapse at time intervals of a few milliseconds for each muscle fiber at multiple sites along the arm/hand/finger trajectory will determine the overall motor activity. As a consequence, even a "simple" manual task will necessarily be a complex symphony of commands to the motor system.

What is known about the brain circuitry involved in tool-usage is that the answers are as complex as are the tool-usage tasks themselves. Fortunately, the cognitive neuroscience of, in particular, the control exerted by the frontal lobes is gradually becoming known (Fuster, 2007; Passingham, 2008). Figure 4.13 shows the generalized mammalian brain circuit for motor activity of all kinds. The basic architecture indicates that the control of striate muscle is under direct neocortical control, but, moment by moment, whether or not the musculature will respond to a cortical command will depend on the relative strength of excitatory impulses from the motor cortex and inhibitory impulses from the brainstem (striatum and pallidum).

So, what does it mean to have "higher-level" control over the commands emanating from motor cortex? Surely, every squirrel collecting acorns has motor cortex that is either engaged or disengaged in motor control of the hands, so how are primate motor activity and human toolmaking different? The principal difference lies in the mechanism for "control of the control" – a switching function known as conditional associations. The sensory input itself will often be a cue in a specific location in visual/auditory/tactile space and the motor response will necessarily be a complex hierarchy of neocortical commands to a variety of striate muscles, but that much can be said for any stimulus–response task performed by a mammal. What distinguishes primate motor control is the still-higher level of control exerted on motor cortex from so-called premotor cortex, which itself is controlled by so-called prefrontal cortex (Figure 4.14).

What is unusual about a motor task undertaken by a primate is that there occurs, in effect, a weighing of alternatives and the selection of an

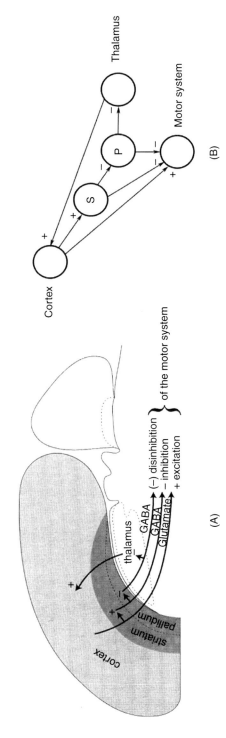

FIGURE 4.13. The organization of the descending projections from the mammalian neocortex to the motor system (after Swanson, 2003, p. 177, with permission). (A) shows the known anatomy with the known neurotransmitters. (B) depicts the same architecture as a neural network that connects to one skeletal muscle. The balance of excitatory and inhibitory impulses will determine the actual motor output.

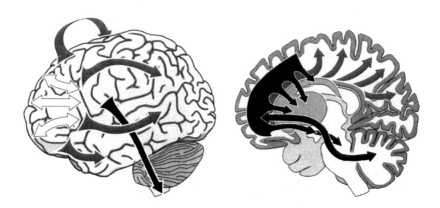

FIGURE 4.14. A summary of the extensive "top-down" control that prefrontal cortex exerts over most other regions of the brain. The white arrows indicate the large expansion of prefrontal neocortex. Gray arrows from prefrontal cortex indicate the projections to many brain regions, and black arrows indicate new (hominid) projections from neocortex to the brainstem, as well as new (hominid) projections to and from the cerebellum (after Deacon, 1997, p. 256, with permission).

appropriate motor response depending on the strength of a third cue. The associative link between sensory stimulus and motor response is "conditional." Instead of a response following directly from a sufficiently strong sensory stimulus, in a conditional association task, a motor response is selected on the basis of conditional cues. When conditional-cue W is present, the response to stimulus X is Y; otherwise, the default response to X is Z. In principle, this type of conditional response holds true for each and every muscle involved in the motor act. In an animal experiment where a go/no-go decision must be made at the highest cognitive level, the "decision" whether or not to extend a finger will depend on the prior "decision" to bend or flex the elbow, which will depend on the prior "decision" concerning the torque of the shoulder, which will in turn be influenced by the prior "decision" about leaning forward of the trunk at the hips. Condition upon condition and all their possible permutations are too much to think about, but, by employing regions of prefrontal cortex to transiently store various patterns of responses, flexible behaviors to changing circumstances are possible. Whether signaled by sensory cues or internal memories, the commands for motor activity can be regulated, delayed or suppressed until alternative scenarios are weighed.

The complexities of the cognitive control exerted over motor cortex is normally summarized as "planning" and can be interpreted as a talent made possible by massive increases in inhibitory control over stimulus–response pairs. From both clinical studies and experimental work on apes, we know that the ability to process so-called conditional associations first evolved in the apes and requires intact regions of prefrontal cortex. The anatomical and behavioral details are daunting – and well beyond the scope of this brief review – but a study by Petrides (2005) can be taken as representative of current brain research on this aspect of higher-order cognition. He states that "Whereas lesions of the mid-dorsolateral frontal cortex cause a severe deficit on tasks that require the monitoring of information in working memory, the caudal dorsolateral lesions do not affect performance on such tasks, but yield a massive impairment on tasks ... that require the selection between alternative competing responses based on conditional operations" (p. 297). Brain-damaged chimpanzees "perform very poorly on conditional associative tasks ... that require that the animal select between different possible responses in a given situation according to conditional rules, such as if A, select X, but if B, select Y.... Patients with lateral frontal lesions ... also perform very poorly on such tasks" (p. 301).

Tasks employing conditional associations are complex and defy a simple, unified description, but the important evolutionary point is that they are generally beyond the capacities of nonprimate brains that lack a significant mass of prefrontal neocortex. Clearly, a huge cognitive leap is made possible by the availability of stored "conditions" and the ability to pick and choose among alternative responses by means of neocortical suppression of the undesirable options. Details of experimental design and the localization of brain mechanisms are the essence of the cognitive neuroscience of "higher cognition," in general, but the more basic argument of the present chapter is simply that for "executive control" typical of human tool use, a flexible switching mechanism is essential. Issues of the functions of specific regions within prefrontal cortex, species differences, individual differences, learning effects and laterality effects (Aron et al., 2004) remain controversial. As Ramachandran (2011, p. 265) has commented, "The dorsolateral frontal area is required for logical reasoning, which involves paying attention to different facets of a problem and juggling abstractions.... How and where the precise rules for this juggling arise is anybody's guess."

4.9. CAUSALITY

Elucidation of the brain mechanisms underlying tool use and toolmaking is in progress, but the big unanswered questions concern the cognition that drives the motor behavior. Given a primate organism with dexterous hands capable of the necessary manipulation of external objects and enough prefrontal cortex to sequence the appropriate motor actions, how could it have imagined that one stone should be struck by another stone to obtain a tool? How did it come to understand cause and effect?

Lewis Wolpert (2003) has been focused on this issue for some time, asking: "What served as the prime mover in the evolution of the human brain? ... A key issue is the relationship between language, tool use and causal beliefs; there may have been a mutual positive synergy between all three." Wolpert explicitly rejects "social innovations" as the prime mover, and, lacking any clear evidence for early language capabilities, he has argued that a belief in causality in the making of tools is the essence of the cognitive revolution of *Homo sapiens*. The crucial issue is the distinction between learned associations – common among all animal species – and an understanding of the *connectedness* of things and events that underlies the concept of causality.

In modern-day animal experiments, it can of course be shown that the mere exposure to two nearly simultaneous events will lead to a sufficiently strong association between them – say, a lever press and the appearance of a food pellet – such that, without any understanding of mechanisms, an animal will learn to press the lever for the reward. An early hominid could equally well learn the same kind of association, but to link the "reward" of a stone with a sharp edge and one's own actions would require the motivation to undertake such action in the first place. The laboratory rat trapped in a cage with nothing to do but press the lever will learn the association after perhaps 30 trials, but the chances of an hominid witnessing multiple associations between hammer stones and the sharp flakes produced from core stones are small, unless it is already exploring the cognitive issues of cause and effect. Associative learning will not suffice.

The central "negative" insight is quite simply that an understanding of cause and effect is not a dyadic association. The psychological theory of behaviorism (supported by the philosophical theory of positivism) that refuses to peak behind the surface cues of input and output is, in this regard, blatantly wrong. Yes, only two, real-world components are involved, but the cognizing animal organism works to link these components in whatever way that will maximize utility. For the pathetic laboratory rat in an artificial world with an omnipotent human experimenter designing the conditions of its life, the only possible link between cause and effect is its own desperate flailings in the cage. But for the cognizing organism in a natural environment, countless factors might serve as the mechanistic glue connecting cause and effect. It is the relevance of "other factors" that our big-brained ancestors eventually discovered in their chance encounters while in search of sharp stones.

Some hominids may have stumbled onto sharp stones in rainy weather and concluded that rain was an essential factor for successful stone hunting. Others may have found sharp stones after the death of an infant and concluded that hominid sacrifice was useful. But eventually the link between the forceful impact of a larger "hammer stone" onto a smaller igneous rock must have been recognized as the crucial third factor – a causal connection that was far more reliable than other factors that were ultimately dismissed as superstitions. The relationship between the hammer stone, A, and the core, B, was rewarding, provided that C – the vector of manual force joining A and B together – was also present. Our ancestors were of course not alone in drawing valid conclusions about cause and effect in the natural world, but a significant mass of prefrontal cortex that allows for the cognitive rehearsal of conditional associations was probably necessary to sort out the many possible connections among events in a complex environment.

Research on the triadic mechanisms that underlie an understanding of cause and effect is already well advanced in both primate studies (e.g., Berthelet & Chavaillon, 1993; Tomasello & Call, 1997; Povinelli, 2000; Premack & Premack, 2002; Gomez, 2004) and developmental studies (e.g., Sperber et al., 1995; Gibson & Ingold, 1993; Tomasello, 1999) – although the "triadic" label is not frequently employed. From the perspective of the hypothesis of triadic cognition, the number of cognitive components involved in understanding cause and effect is the crucial factor. Similar to the perception of the major, minor or chromatic mode of three-tone configurations, where their simultaneous or sequential performance as melodies or chords has no influence on their overall affective sonority, the qualitative leap from dyadic associations to triadic cognition is due to the addition of a third component. When employed in a tool-use task, of course the sequential performance of the act will strongly influence the success of the outcome, but the cognitive insight is the focus on three, not two, components.

4.10. CONCLUSIONS

The reality of certain forms of unimodal triadic *perception* undertaken by the modern human brain was amply illustrated in Chapters 2 and 3, and there is little doubt that, whatever our many psychological shortcomings, as a species we exhibit some unusual three-component perceptual talents in the visual and auditory domains. The question addressed in this chapter has been whether the evolution of tool use might also have a triadic perceptual foundation. The first triad that can be identified already in the tool use of the capuchin monkey concerns the simultaneous perception of three concrete objects. Typically, this entails (i) a tool, (ii) the object to be affected by the tool and (iii) an appropriate material "context" within which the task can be completed. At that level, tool use is a rarity in the biological world but is not uniquely human and not unambiguously triadic. It becomes explicitly triadic, however, when the "context" is a specific platform without which the tool cannot be successfully used to accomplish a specific task (and not an unspecified "environmental setting" or "general circumstances" that might include dozens of relevant factors). The understanding that a specific task can be achieved if and only if three material objects (hammer, target and anvil) are kept in mind is the conceptual insight underlying the most primitive examples of human-like tool use. As a hypothesis concerning the evolution of higher-level cognitive talents, the emergence of tool use at the capuchin level might suffice to make the main point about triadic *perception* – with all further developments seen simply as recursive elaboration. But the inevitable follow-up question is why tool use and toolmaking took hold in the human line, while the motor talents of all other species have not included toolmaking.

Why only *Homo sapiens*? The most convincing answer can be stated in terms of brain anatomy. From the fossil record, we know that simple hand-axes were made by hominids during a time of massive expansion of *posterior* association cortex (Holloway, 1999) where visual, tactile and auditory sensory information first converged at the neocortical level. If we are to rely on *empirical evidence* to answer the question of what new behavior drove that rapid growth of the brain, toolmaking is virtually the only candidate answer. We can of course speculate endlessly about changes in mating behavior, a sense of mortality, intraspecies empathy or religious insight by our early ancestors, but we have *evidence* only with regard to tools and gradual changes in the shape of the human skull. That evidence suggests the emergence of a new form of trimodal association in the inferior parietal lobe as the neural ingredient underlying the new forms of tool use and toolmaking behavior. It is therefore a plausible hypothesis to surmise that the earliest neuronal algorithms for handling three independent streams of information evolved precisely there – and precisely for the purposes of toolmaking. As a consequence of those neurological changes in association cortex early in hominid evolution, a capacity for trimodal perceptual processes developed and has continued to be the driving force in the cognitive evolution of our species.

Of all the forms of triadic perception, cognition and (later) social interaction discussed here, the significance of cognition that employs three modalities is the hardest to appreciate for the simple reason that the integration of sensory modalities is already something that is second-nature to all normal human beings. "Drawing connections" between visual patterns, auditory signals and haptic sensations is so obvious that failing to understand multimodal relationships or overlooking the meaning of simultaneous sights, sounds and touches would be the surprise! But, clearly, inter-modal connections are possible, easy and "obvious" primarily because we have large cortical regions where two or three types of sensation converge on one and the same cortical neurons. Such neurons are anatomically in a position to make connections between sensory signals that impinge on the organism through different sensory organs. Multimodal cortical neurons are there – sitting in association cortex – because, over the course of evolution, those of our early ancestors who, by chance or good fortune, had more such neurons in association cortex more readily "made the connection" between diverse sensations. Today, we are coolly unimpressed with our own abilities here, but comparative neuroanatomy teaches us that specifically trimodal associations evolved first in the primate line – and became serious business among our early ancestors making stone tools.

The first concrete signs of *recursive* cognition are hafted tools, where a second object is tied to or glued onto a first object using a third material (hemp or pine sap) to produce a new tool with new uses. Such imbedding is the beginning of an endless process of fitting pieces together to obtain functionality beyond the sum of the parts. A few animal species may have the cognitive seeds of such recursion in mind, but it appears that only *Homo sapiens* has evolved a capability for the necessary conditional associations that can be used at will and that can allow for behaviors involving the manipulation of objects.

In conclusion, the evolutionary step into the triadic complexities of human cognition appears to have begun with toolmaking. As a species, we began our remarkable cognitive ascent as "Man, the toolmaker," but the next step was to "Man, the communicator."

5

Human Communication: Language

In the preceding chapters I outlined the idea that various three-component *perceptual* processes underlie some of the most interesting behaviors undertaken by human beings. Music and visual art are quintessentially *human* activities, and the triadic nature of the perception of harmony and pictorial depth is relatively straightforward. Although "skilled motor activity" is not a human uniqueness, several aspects of tool use demonstrate the importance of, to begin with, triadic visual perception, as reviewed in Chapter 4. Moreover, the subsequent emergence of handedness, cerebral laterality and toolmaking is intimately connected to the evolution of trimodal sensory neocortex, where the coordinated use of visual, tactile and auditory information underlies the skills of toolmaking.

In comparison with the unimodality of auditory music and the unimodality of visual art, the trimodal nature of the perception necessary for toolmaking makes the entire triadic argument more complex, but it is important to keep in mind that, once sensory phenomena have been converted into nerve impulses, there truly are no more sights, sounds and touches! Whatever the nature of the "higher-level" coordination of such sensations, they end up being digitized into the language of neurons... and the essential question concerns only the number of independent streams of information. So, although language is essentially an auditory skill, it is at the rather abstract level of "independent streams of information" that we must address questions concerning the cognition underlying language capabilities.

In the present chapter I summarize what is known in the field of linguistics concerning the basic cognition of language – and show once again not simply that triadic processes are involved, but that it is the step from dyads to triads that makes syntactic flexibility possible. The main argument concerns the basic grammatical manipulations that we all master as young children – and later struggle with as students of foreign languages. The bottom line is that the syntax of phrases is triadic and has been described explicitly as triadic in the terminology of "transformational grammar" since the 1950s (evolving into "x-bar theory," "head-driven phrase structure grammar" and "principles and parameters" over the years, but maintaining triadic phrase structure at its core).

Fortunately or unfortunately, syntactic triads are combined recursively to produce the extreme complexities of everyday verbal communications that make our species so unusual (Corballis, 2010). It is therefore both a simple and a complex story – simple because of the triadic structure of phrases, but complex because there are no fossilized triadic artifacts left over from earlier times, only the fully recursive, multiple-component complexity of modern languages. Nonetheless, the transition from truly simple, three-component phrases to complex sentences can be coherently explained in the terminology of modern linguistic theory, and that argument leads inexorably to the notion that the "Universal Grammar" underlying the human capability for language is, in essence, the ability to handle cognitive triads.

But before we get to the nuts and bolts of linguistic theory, we need to consider the overall architecture of language and how it evolved among our early ancestors.

5.1. THE TRIPARTITE ARCHITECTURE OF LANGUAGE

The most coherent summary of language as it is currently understood has been provided by Ray Jackendoff (2010). Over the course of more than 30 years, he has addressed primarily questions concerning the architecture of language in a modern setting: "what" cognitive processes underlie language capabilities and how they are related to one another. Jackendoff's work is relevant because he has advocated a model of human language competence that is explicitly "tripartite." Language, he argues, entails three simultaneous, related but rather independent, cognitive processes: phonology, syntax and semantics (Figure 5.1).

Perhaps more so in linguistics than in other disciplines, the study of language is a minefield of knit-picking jargonophilia – so that extra efforts must be made to translate the linguist's important insights about language into a terminology that normal people can understand. Right from the start, Jackendoff's tripartite parallel architecture of phonology, syntax and semantics can be more comfortably restated as the issues of pronunciation, grammar and meaning. These are the three processes that, at a very young age, we learn to handle simultaneously in using language and that appear to

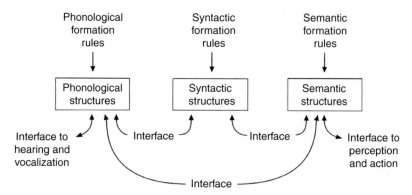

FIGURE 5.1. Jackendoff's "tripartite parallel architecture" of language (2002, with permission).

be already too complex for infrahuman species – often because of phonological problems (the difficulty of enunciating different vowels and consonants) and *always* because of syntactic limitations.

Phonology is the ability to produce the sounds of language with the organs of speech and to perceive those sounds in the auditory modality. Syntax – arguably the main issue discussed in academic linguistics over the last century – is the ability to use rules of grammar to decide on a correct sequence of speech sounds. And semantic processes are the nonverbal, conceptual notions that precede syntactic formulation and speech behavior. Syntax appears to be a distinctly human capability, but it is less clear whether or not there is anything distinctly human about the understanding of meaning. On the one hand, "we now know that animals, especially higher primates, *do* think, [but] the way we differ qualitatively from animals is that we have the ability to convert our thoughts into communicative form, via phonological and syntactic structure" (Jackendoff, 2002, p. 84). Together, these three processes make language possible – and make us the "talking ape."

Each of the realms of phonology, syntax and semantics is a hotbed of academic controversy, but few linguists would deny the importance of specifically those three functions, their relative distinctiveness and the need for an understanding of how each process relates to the others (Figure 5.1). Variations on Jackendoff's tripartite architecture have been proposed. Notable is a model from Willem Levelt (1999), who also argues that language is comprised of three functions: a conceptualizer, a formulator and an articulator. The conceptualizer is equivalent to Jackendoff's semantic module – the process by which intentions, motivations and preverbal meanings are organized. Prior to their formulation as syntactic sequences and delivered as speech, these "meanings" are, by their very nature, difficult to specify, but they are the motivators that drive speech. Indeed, the reality of nonverbal concept is Levelt's starting point for understanding language usage: without "something to say" – without the need or desire to inject a linguistic comment into a social scene – all subsequent syntactic and phonetic manipulations are pointless and, indeed, unmotivated. Once the nonverbal motivation has been isolated and elevated above the background noise of unrelated and less urgent feelings (i.e., conceptualized), then it must be translated into a form that can be expressed verbally. Levelt refers to this process as "formulation." It involves, above all else, the sequentialization of the individual conceptual components following an arbitrary, but locally agreed upon, convention concerning word-ordering (i.e., syntax). The dos and don'ts and internal coherency of syntax are the core issues of traditional linguistics and are arguably the most complex, most daunting aspects of modern linguistic theory. But after a conceptual structure has been (i) compartmentalized and (ii) sequenced by the formulator, it still requires (iii) motor expression (i.e., phoneticization). This is the work of Levelt's "articulator." The utterance of speech sounds is of course not strictly necessary for language capability – both writing and the sign languages of the deaf are indication that the sequentialized meanings of oral language can be communicated in other forms. Nevertheless, the use of spoken language probably preceded other forms of communication and is in any case the motor behavior that most people most frequently use today for communication.

Although some consensus has been obtained concerning the "tripartite architecture" of language (see Section 5.2), linguists vehemently disagree about the evolutionary importance and cognitive complexity of these capabilities (Table 5.1). Despite the acknowledged necessity of all three, many have argued that only one is the "core talent" – the essential qualitative leap that leads directly into human cognition, while the other two pieces of the tripartite model represent merely quantitative improvements on the more primitive capabilities seen in other species.

There is also little consensus on the nature of the interface mechanisms at work among these three main modules. Given that an individual has an explicit "meaning" (feeling, motivation, concept) in mind, how is meaning linked to the arbitrary structures of syntax? And what are the rules for linking an arbitrary syntax to the motor mechanisms of speech? Moreover, the link between semantics and phonology – the realm of emotional

TABLE 5.1. *The emphases placed on different aspects of language by various authors*

Author	Time limits	Key characteristic of language	What was and who had Proto-Language?	Emphasis on
Lieberman	Early hominids	Speech	Proto-speech; Neanderthals	Phonology
Studdert-Kennedy & Goldstein		Discrete infinity	Phonetic capacity	Phonology
Hurford		Subject–predicate dichotomy	Prephonetic, presyntactic, presemantic capacities; *Homo erectus*	Syntax
Bickerton	*Homo sapiens*	Syntax	Symbolic representation, no syntax; *Homo erectus*	Syntax
Szathmary	*Homo sapiens*	Recursive embedding	Chimpanzees	Syntax
Bierwisch	*Homo sapiens*	Universal Grammar	Proto-lexicon	Syntax
Deacon		Universal Grammar		Syntax
Hauser & Fitch		Recursion		Syntax
Corballis		Vocal speech	Gesture	Syntax
Klein		Context integration	Lexical content	Semantics
Comrie		Lexicon		Semantics
Pinker	Gradually over 5 million years		Intelligence	Semantics
Davidson	*Homo sapiens*	Symbolic representation		Semantics
Tomasello	*Homo sapiens*	Capacity for sharing		
Arbib	*Homo sapiens*		Gesture	
Dunbar		Grooming		

After Crow, 2002.

prosody – is an issue of fundamental importance that no one yet knows how to deal with.[1]

While the tripartite architecture of language is widely acknowledged, each of the three pillars of language contains its own mysteries and conundrums. The main puzzle of phonology is that, in order for languages to be mutually understandable, some regularity of the sound structure across speakers must be maintained, but in fact quite different pronunciations of the same words – by men, women and children in various accents and dialects – are rather easily understood by native speakers. If each phoneme were pronounced as distinctly as the spelling in written text, or playable as an isolated tone on a piano, the theory of both speech production and perception might be easy, but the rapid-fire overlap of speech sounds that are acoustically quite different, individual by individual, makes both the physiological production and the acoustical perception of speech highly complex. It remains unclear why language is so "easy" for normal people.

The puzzle of syntax is that it is so extremely complex – and different for every language on Earth – and yet, as native speakers, we can judge the correctness or incorrectness of individual sentences with surprising ease. Whatever the rules of grammar may be, we learn them at a young age and eventually use them "without thinking." Our mastery of the complexities of syntax in our native tongues eventually becomes nearly perfect, but how can that be if we actually experience only a tiny fraction of all possible sentences? And how can mother-tongue syntax be so easy, while picking up a second language is so hard?

---

[1] The problem of emotional prosody concerns the issue of "grounding" the seemingly arbitrary sounds of speech to a nonarbitrary biological reality. The pitch rises and falls that convey some of the most powerful emotional effects in speech must somehow connect with our biological beings in order to have impact. One form of grounding is certainly through the "frequency code" (Chapter 2) – where rising pitch connotes weakness and falling pitch connotes strength. As is the case with music, an analysis of the pitch dyads alone in emotional speech (low-to-high, high-to-low) *does not* and, in principle, *cannot* be used to deduce the subtleties of emotional states. Well-known examples are the difficulty of distinguishing – solely on the basis of properties of the acoustic signal – between the positive affect of joyous utterances and the negative affect of angry utterances (both characteristically employing large and rapid pitch rises) or between the positive affect of serenity and the negative affect of sadness (both characteristically employing slow pitch falls). People are in fact rather good at hearing the emotional tone in the speech of others, but simple measures of pitch range, interval size, velocity or variability (etc.) will not suffice for identifying emotional valence. Similar to the analysis of musical harmony, however, both pitch dyads and pitch triads in the pitch contour of speech prosody can be related to positive and negative affect (Cook et al., 2006). Progress in research on emotional prosody will undoubtedly be possible once the reductionist obsession with pitch dyads has been transcended!

Even if the puzzles of syntax and phonetics were completely solved, we would still need to explain the mechanisms that underlie our understanding of "meaning," the third enduring conundrum of language, semantics (Hurford, 2007). As Jackendoff (2010, p. 81) has commented: "This is the area of linguistics where the professionals, after years of research, still have not settled some of the most basic questions." While the empirical study of differences in syntax among diverse languages has progressed remarkably over the past century, and the acoustics of phonological processes has also attained a scientific foundation as a consequence of computerized methodologies, the philosophical issues of semantics – "the meaning of meaning" – remain to this day a conceptual nightmare. This is not merely the problem of matching numerous word-sounds to specific meanings, but the more difficult problem of (i) perceiving cause and effect in the wider world and (ii) expressing that cause and effect with word-strings. Semantics is not only the amazing verbal "labeling" of objects – that 2 year olds and chimpanzees are capable of – but the transmission of the meaning of causality through sentences – a linguistic talent that takes children a few more years to master but seems to elude the animal brain.

The evolution of these three components of natural languages needs to be considered individually, but the first problem that must be considered concerns the empirical validity of the tripartite architecture of language. Who, in fact, maintains that this kind of tripartite theory is correct? Here, we can rely on the clinical experience of neurologists, neurosurgeons, neuropsychologists and a small army of researchers whose daily work is built on this view.

5.2. BEHAVIORAL NEUROLOGY

The tripartite model of phonetics, syntax and semantics is best understood, not as an abstract theory, but as formalization of the three core cognitive mechanisms underlying language competence that were established in European clinical neurology more than a century ago. Formal "proof" that language is tripartite would be a difficult undertaking, but the importance of a specifically three-component model is demonstrated by the fact that – whatever theorists might say – clinicians have found such a model useful for describing and classifying behavioral syndromes in patients suffering from brain damage. Disturbances of specific brain regions lead to characteristic disorders that can be described as some combination of abnormalities of phonology, syntax and/or semantics. Notably, the language centers known as Broca's area and Wernicke's area play specific roles in language

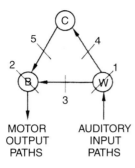

FIGURE 5.2. One of the earliest models of language functions was developed by Lichtheim (1885). As Donald (1991, p. 47) has commented, the early clinical cases "served as the basis of the first modular models of language that postulated the existence of several independently variable linguistic subsystems ... and the [localizationist] approach Wernicke took survives, unmodified in its essential features, to this day." The brain localization of conceptual processes (C) is uncertain/diffuse.

production and comprehension, and their approximate localization in the anterior and posterior neocortex of the left hemisphere provides an important grounding for many neurological syndromes. Debates continue regarding their specific functions (e.g., Grodzinsky & Amunts, 2006), and individual differences and the natural diversity of brain organization mean that there is a wide distribution of normal language functions distributed over the cerebral cortex. For this reason, a neat-and-clean localization of tripartite functions is an overly optimistic view of the modularity of the brain, but the reality of three processes is well established in the clinic (Figure 5.2 and Table 5.2).

Experimental work using noninvasive techniques to study the brain activity of normal subjects generally supports the tripartite model of language in the sense that foci of activity can be detected in relation to semantic, syntactic and phonological errors (e.g., Figure 5.3). Questions concerning the independence of and interconnections among the three main components of human language remain largely unanswered, but there is currently no reason to believe that the empirical facts about language perception will be viewed in a radically different (nontriadic) way in the future. At this rather "macroscopic" level of brain organization, the unusual human capability for coordinating three rather independent cognitive operations suggests that normal language capabilities are built on a triad of processes, and that characteristic abnormalities arise when the triadic balance is lost.

If a triad of independent language functions is an empirical reality, it must be said that difficult questions yet remain about the sequence of the

TABLE 5.2. *The clinical disorders of language*

	Syndrome	Speech	Comprehension	Repetition	Naming	Lesion site
1.	Wernicke's aphasia	Fluent, empty	Poor	Poor	Poor	Posterior
2.	Broca's aphasia	Poor, nonfluent	Good	Poor	Poor	Anterior
3.	Conduction aphasia	Fluent	Good	Poor	Poor	Arcuate fasciculus
4.	Transcortical sensory aphasia	Fluent	Poor	Good	Poor	Parietal lobe
5	Transcortical motor aphasia	Little	Good	Good	Not bad	Frontal lobe
	Anomic aphasia	Fluent with circumlo- cutions	Good	Good	Poor	Anywhere
	Global aphasia	Virtually none	Poor	Poor	Poor	Large

Note the close correspondence between the modern classification and Lichtheim's early model (the numbers in Column 1 and those in Figure 5.2).
After Banich (1997).

evolutionary emergence of those elements and their mutual influences. So, before proceeding to the central argument of this chapter concerning the triads of syntax, let us briefly consider what is known about the evolution of language. To begin with, why and how did the vocalizations of early hominids attain their symbolic meaning?

5.3. THE EVOLUTION OF LANGUAGE

Unlike physics and chemistry, where decisive experimental discoveries and useful technological artifacts have strongly influenced theoretical developments, in the science of linguistics the "plausibility" of arguments has played a dominant role. "Soft science" is the scornful phrase used by those engaged in "hard science," but, to the surprise of the lab technicians, the broad outlines of paleontology and global climate science provide a few unambiguous landmarks that cannot be compromised. As a consequence, the theory of language has proceeded, perhaps slowly, but inexorably toward a broad consensus concerning the implicit rules that govern language and

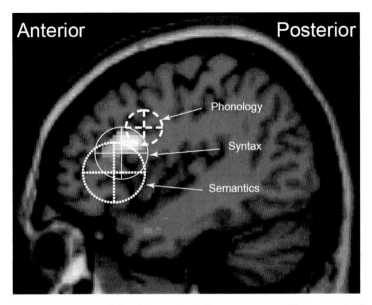

FIGURE 5.3. Locations of the three principal components of language in frontal cortex in an fMRI study where subjects were required to detect anomalies of pronunciation, grammar or meaning (from Petersson et al., 2004, with permission).

how symbolic communications emerged in a biological world. Questions concerning localization of functions within the brain are generally not addressed by linguists, and the error bars in the evolutionary argument are normally many tens of thousands of years in magnitude, but notable progress in the science of linguistics has been made through the efforts of many scholars, outstanding among which are the linguists Noam Chomsky and Derek Bickerton.

Chomsky's main contributions (1975, 1985) have been twofold: (i) in effectively moving linguistics from its comfortable home among the Arts and Letters and establishing it as an analytical, reductionist, empirical science and (ii) in advocating the idea of an underlying, human cognitive skill known as "Universal Grammar." What Chomsky has declared, to the chagrin of many in evolutionary biology, is that language is a specifically human competency that cannot be explained by simplistic appeals to the power of natural selection. While the supremely self-confident behaviorist "recognizes no dividing line between man and brute" (Watson, 1913), linguists find it difficult to justify that ideological stance – and have focused on the science of language to formalize that dividing line. Virtually no one in the sciences denies the reality of evolution, but Chomsky has been forthright in stating that there is a qualitative jump from animal communications to human language. As the most cited scholar of all time and a specialist in language, Chomsky has been in the unique position to throw cold water on easy answers concerning the origins of language that, from a linguist's point of view, do not address the central issues. Language is special, and the incoherency of creationist religion does not remove the huge problem of explaining its evolution in a biological world where symbolic thought is otherwise apparently absent.

Bickerton (1990, 2009, 2010) has studied the so-called pidgin languages that emerge when people with different mother tongues live together and are forced to communicate with a patchwork of mutually unintelligible lexicons and grammars. That research topic has given focus to the problems of communication among *normal* people in *abnormal* situations. His recent contribution has been to provide a link between the realm of animal calls and the realm of symbolic vocalizations – focusing explicitly on the nature of the gap between "man and brute." Although his ideas have developed over several decades, they are delineated most remarkably in a recent book, *Adam's Tongue* (2009), where the birth of language is placed squarely within the context of what is already known from the fossil record about human evolution. Because of the importance of understanding language as a product of the natural world, let us begin with (a condensed version of) the evolutionary scenario outlined by Bickerton.

The setting is the rather arid grasslands in East Africa some two million years ago. For millions of years prior to then, hominids had lived in Africa – originally in dense jungles, later to become fertile woodlands, and gradually changing into dry savannahs. Those ancestors walked on two legs and had brains that were twice as large as chimpanzees – probably a consequence of the need for more cerebral cortex when living in relatively large social groups (as is typical of modern primates, Dunbar, 1996). And primitive stone tools were in use (Figure 4.3) – clearly indicating an intelligence that already surpasses that of modern-day chimpanzees. But, the overwhelming pattern of early human evolution following the divergence of the human and chimpanzee lines 6 million years earlier was the slow pace of change. Even with fancy names like *Australopithecus*, *Homo habilis* and *Homo erectus*, those early ancestors were still clearly apelike – probably naked, homeless and rather loosely organized into clans not too different from modern day chimpanzees and bonobos.

Archeologists have provided a good understanding of the changes in human behavior that occurred over the most recent 100,000 years as we enter a period where fossil relics are relatively numerous. But the traces of our prehistory over the course of the preceding 3 million years in East Africa show surprisingly few changes – precisely that period when our unique humanity is likely to have emerged. The absence of change in the fossil record is strong indication that those hominids had achieved a secure position within a sufficiently stable environment – so stable that the mother of invention, raw necessity, was rarely encountered and unchanging continuity meant survival. Edible vegetation was all around them, there for the picking in the warm jungles of Africa. The need for tools, shelter and clothing was minimal. Life was good – and as noisily nonlinguistic as modern-day chimpanzee societies!

Eventually, however, climatic changes on a global scale did in fact arrive – demanding some behavioral changes. Indeed, Bickerton's insight is not primarily a new idea concerning language itself, but a careful consideration of the biological niche in which our early ancestors found themselves. On the one hand, they were anatomically similar to their chimpanzee ancestors in many respects, but three significant changes had occurred over the course of several million years: (i) a shift to bipedal locomotion, probably a consequence of the gradually disappearing jungle environment, (ii) a modest increase in brain size – including the emergence of trimodal association cortex – and (iii) with the freeing of the hands, the ability to make, carry and use simple stone tools (Figure 5.4).

Each of those changes was revolutionary and each requires an explanation to provide a full understanding of our evolution. But none of those transformations – alone or in combination – indicates why those happy campers would need to sit down and invent language. If their biological niche had not been in the course of changing from bountiful jungle greenery to the severity of arid savannahs, our ancestors might have remained similar to the chimpanzees and gorillas that have continued to survive without inventing language in the deep jungles of Africa. But the environment in specifically East Africa was in fact in transition – making survival on abundant jungle fruits no longer possible.

The broad outlines of the geological and fossil record (as summarized in Figure 5.4) provide the major landmarks within which the evolution of language needs to be understood. Concrete indications of language are completely absent, but the descent of the larynx to a position similar to that of modern *Homo sapiens* began about 1.0 mya – after which the phonetic sounds of language, as distinct from animal calls, would have become possible (Lieberman, 2006). Moreover, the remnants of climatic changes are unambiguous in showing what kind of environments our apelike ancestors

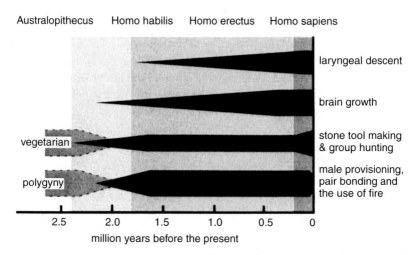

FIGURE 5.4. The approximate timing of four important aspects of human evolution. Note that the shift away from a predominantly vegetarian diet (together with stone-toolmaking, hunting and using fire) preceded the explosion of brain size and the vocal tract changes that made speech possible. As of approximately one million years ago, all of the principal material and biological innovations of our species were in place (after Deacon, 1997, p. 409, with permission).

encountered around them. Most outstanding is the fact that, over the course of relatively brief periods of geological time (a few millennia), the Northern Hemisphere repeatedly underwent massive glacier formation and, as a consequence of the transformation of huge volumes of water into ice over the entirety of Europe, rainfall in Africa was drastically reduced – changing jungles into woodlands and changing abundant vegetation into sparse greenery. Again and again, many species on both continents failed to adapt to the cold or to the dry heat, and became extinct.

Our ancestors were already relatively intelligent, moved about on two legs and had dexterous hands that could fashion crude tools and clothing, so that immediate extinction was not their fate. However, the gradual disappearance of nutritious jungle fruits necessitated an increase in meat eating. In fact, fossilized teeth clearly indicate a trend away from a predominantly vegetarian diet. Fruits, insects and small rodents continued to be sources of nutrition, but the energy requirements of an enlarged brain demanded that these hominids find the kind of protein-rich food that only animal meat could provide. Unfortunately, the physical stature of our ancestors was too modest for them to compete with the fierce carnivores of the savannah – lions, tigers, cougars, leopards, cheetahs and hyenas in far greater variety and numbers than alive today. For millions of years, those carnivores had established themselves as the unchallenged predators – fast, strong and hungry enough to hunt down and kill almost any animal species in central Africa. The primates were in no position to challenge them directly.

The general setting noted above has been described in detail by many others (e.g., Whiten, 1999; Klein & Edgar, 2002; Tattersall, 2008) and is as well established as any aspect of prehistory one million years old can ever be. Our understanding of the evolution of *Homo sapiens* remains unsatisfactory, however, insofar as "random mutation" or a metaphysical bolt-of-lightning must be invoked to insert language into the scene. The rather late descent of the larynx (Figure 5.4) – making speech at least anatomically possible (Lieberman, 2006) – argues strongly *against* the emergence of spoken language early in hominid evolution. But why would language in any form emerge only in the primate line and only in this period less than a million years ago? And why have other primates not independently developed language?

These are questions familiar to scholars of human prehistory – often answered with vague assurances that "big brains" plus various fortuitous changes in body anatomy might suffice. But Bickerton has unraveled the puzzle to put the pieces together in a coherent fashion. In retrospect, it is perhaps obvious that it would have taken a linguist (aware of the fossil record, the constraints of evolutionary theory and an understanding of the importance of biological niches) to come up with the whole story because no change in human behavior compares with the evolution of language. Yes, in addition to language, there remain questions about the circumstances that led to a bipedal stance, the oppositional thumb, the monthly estrus, the taming of fire, the emergence of pair-bonding and so on. Unfortunately, all of the other pieces can be arranged in dozens of inconclusive scenarios, where the unanswered question of the origin of language sticks out like a sore thumb.

The stubborn fact is that language shows no signs of *gradual* evolution from the realm of animal vocalizations. The linguistic point that Bickerton has duly emphasized is that, quite unlike human language, animal calls are inevitably and without exception tied to the here and now. A specific monkey call may "mean" leopard and a different call may "mean" snake, but they have those meanings only as warning signals to scramble away from danger *at the present time*. They do not mean the leopard seen yesterday or the possible snake tomorrow. In contrast, a true language can also use such calls as symbolic *words* (i.e., concepts abstracted from the here and now). As Tomasello (2008) has also remarked:

> For all mammals, including nonhuman primates, vocal displays are mostly unlearned, genetically fixed, emotionally urgent, involuntary, inflexible responses to evolutionarily important events that benefit the vocalizer in some more or less direct way. (p. 53)

How then were the animal calls of our apelike ancestors "displaced" – in space and time – to be used as informational units that do not act as triggers for immediate motor activity?

Paleohistorians have often noted this problem, but under the spell of evolutionary theory most have speculated that, given enough time and enough neurons, the repetition of specific alarm calls would eventually generalize into abstract concepts, and symbolic language would – voila! – become possible. Bickerton rejects that story for the simple reason that the effectiveness of animal calls lies precisely in the fact that they alert listeners to *immediate* dangers. Used in any other way, they lose their biological significance, so that a *gradual transition* from "call" to "word" is, in principle, impossible. The same calls uttered in a context where no leopard or snake is present would eventually negate their usefulness: the "boy crying wolf" syndrome. It is therefore not enough to declare that animal vocalizations are early versions of language, for there is an insurmountable barrier to the seemingly easy transformation of one to the other. We must ask: What are the circumstances that would lead to the invention of abstract thought – to the displacement of calls into words? Bickerton (2009, 2010) provides the following answer.

The starting point is the realization that hominids of one million years ago were no longer lackadaisical apes living off the bounty of the land, but, on the contrary, were periodically in a desperate struggle for survival every time Europe froze and East Africa dried out (Figure 5.5). From fossil bones and the stone cuts on those bones, it is known that, out on the savannahs of East Africa, climatic changes had brought our ancestors into a very different biological niche. Unable to survive on jungle fruits, they had become scavengers in search of the bone-marrow and protein remains of large animal carcasses – so-called megafauna. But the crucial point is that they would have been able to eat such high-protein meals *only once the local carnivores had had their fill*. That was not an easy life. No longer in the lush jungle, our nimble and clever, but not particularly mighty, ancestors were forced to compete with the fearsome, fanged, fighting carnivores. One against one, no primate can compete with a tiger, but Bickerton points out

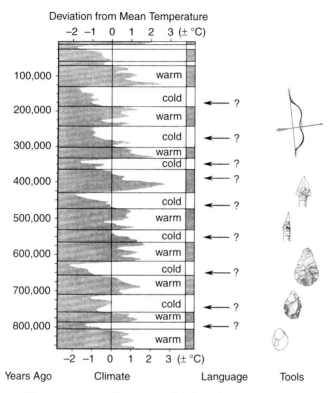

FIGURE 5.5. The innovation of language is likely to have occurred during worsening climatic conditions (transitions from warm to cold) (after Oppenheimer, 2003, with permission), but precisely when remains unclear. Note that there is unambiguous historical evidence on climate change (fossilized plants, isotopes in ice cores), tools (the stone tools themselves) and brain size (fossilized skulls), but no indications of symbolic language until less than 100,000 years ago.

that the prospect of feeding off of megafauna would not be hopeless provided that (i) hominids used their primitive tools – throwing sharpened rocks and hand-axes – and (ii) they confronted the stronger carnivores *in large numbers* (i.e., as gangs of primate "soldiers" able to inflict some pain on lions, tigers and hyenas feasting on animal carcasses). This is the key point: As social groups, they *could* compete – even with the warrior kings of Africa.

The fact that they *did* compete with some success is indicated by the remains of a huge excess of discarded hand-axes at the feeding sites where the bones of megafauna have been found. That historical evidence suggests that large groups of weapon-bearing hominids were actively engaged in fighting for their high-protein meals. Some of the stone tools they created might have had other, more delicate uses, but much of their stone creations were expended as ammunition in trying to keep the carnivores away. That much appears to be certain on the basis of the archeological record, but Bickerton has asked the crucial language question. How could these bipedal, armed, primate scavengers have possibly arrived in sufficient numbers – as a group with one common purpose – at the site of a large carcass to carry off significant quantities of meat? Straightforward calculations on the numbers of carnivores and megafauna per square mile and calculations on the range that our foraging/scavenging ancestors could individually cover in a normal working day indicate a profound problem. That is, primates, such as modern-day chimpanzees and gorillas, typically come together at a home-base in the evening for mutual protection, but diffuse away from home in the daylight in search of food (so-called fusion/fission social organization). Coming upon an edible carcass or fallen animal during daytime foraging would leave an isolated hominid in the hopeless situation of being unable to compete with numerous other species also scavenging for the same food. What to do?

Here the answer is complex – and only a close reading of *Adam's Tongue* (2009) will give the "aha!" insight that Bickerton delivers – but the upshot is that early *Homo sapiens* needed to invent a strategy of "recruitment" – inviting others from the same clan to join in on a scavenging expedition that would be to everyone's nutritional benefit. So, having witnessed a fallen elephant, mammoth, elk or bison, a feast for everyone would be possible, but only if an ape-man together with many, many compatriots could arrive at the animal carcass before scavenging carnivores had reduced it to bones. This is the strategic problem that our prelanguage ancestors faced. Coming across a dead elephant or witnessing the kill of a bison by carnivores, how could a hungry hominid *individual* convince others that there was an unseen treasure just over the hill?

Recruitment of conspecifics to a food source by a lone individual is a rare biological talent, but is in fact known in a few other social species, notably bees and ants. Needless to say, the nonlinguistic mechanisms used by bees (the bee dance) and ants (pheromones) are irrelevant to the human language story, but the communicative need is the same. Just as an ant that discovers a dead caterpillar or a bee that discovers a patch of blooming flowers can exploit that treasure only with the help of friends, an ape-man could enjoy a nutritious meal only with the help of other ape-men who could swarm about the edible carcass in sufficient numbers that even the carnivores could not interfere with their feast. The problem was therefore how an individual could communicate information about a large mass of protein that is spatially and temporally removed from the present situation. The communicative signal must be delivered to many compatriots who themselves cannot see, hear or smell the promised meal. "Alarm calls" will not suffice for recruitment.

The scenario that Bickerton develops concerns the nature of the jump from vocal calls to symbolic communication. Specifically, he argues that simple words – phonetically, much like the dozen or so animal calls of modern-day monkeys – could have had their first *displaced* use specifically in this kind of scavenging scenario. Having witnessed an animal carcass that carnivores would devour if action was not soon taken, our stone-age genius would need to return toward home base to recruit a small gang of hand-axe-carrying allies using a vocalization that meant, say, "elephant." That vocalization would likely have been nothing more than a mimicking of the elephant's bellow and made possible by the evolution of mirror neurons (long employed in the realm of toolmaking) (Rizzolatti & Sinigaglia, 2008). But the "elephant" label would have never been used previously except when an elephant was in direct sight. Now, it was used to indicate a "future elephant meal."

The cognitive connection between the mimicked elephant noise and edible elephant meat would have been a straightforward dyadic association, but the mimicked sound of an elephant was now intended to have displaced meaning: "elephant meat is available for all of us (if you follow me)." From modern-day animal research, we know that understanding spoken labels for objects (typically food that is not in the immediate vicinity) is not beyond the capabilities of chimpanzees – and arguably within the grasp of cats and dogs. But, without a human trainer, the associative link between an arbitrary vocalization and an invisible real-world object is normally absent. Only in the unusual circumstances of desperately hungry social primates, one of whom knows where food lies, would there be a chance that a familiar auditory label, "elephant," could be (i) uttered and (ii) understood to mean the existence of an elephant meal displaced in space and time.

The precise circumstances of course remain uncertain, and it is impossible to know how many false starts and failed communications were attempted before the realization that a spoken word can "mean" something slightly different from what an alarm call signals. But the survival scenario is essential to bring focus to the entire argument. These were (1) bipedal primates, (2) in a deteriorating biological niche, (3) already smart enough to mimic each other in the construction of simple stone tools and, crucially, (4) social animals requiring mutual support for survival. Verbal communication was the revolutionary solution to a dire situation – possible, not because "big brains" were just naturally capable of abstract thought, but because a vocalization that was used to have meaning beyond the here and now would allow both speaker and listener to survive. As described in Chapter 4, auditory information processing had attained a new importance in the development of stone-tools and had now become relevant in nonalarm settings, as symbolic speech.

This story about the evolution of language is only a snapshot of the jump to symbolic thought and could have happened in any of the dozen or so successive eras of worsening climate in East Africa (Figure 5.5), but the important point is that it required the successful communication of but one word – protolanguage in its simplest form. With no grammar and no subtleties of pronunciation, the protolanguage did have a distinct vocalization that, to these first symbolic hominids, had a mutually understood meaning concerning a mutually desirable object that was out of sight, but most definitely not out of mind.

The jump to symbolic protolanguage would have allowed our early ancestors to develop initially a small set of words that were qualitatively distinct from the alarm calls. Unlike alarm calls, symbolic words remained in short-term memory without triggering the amygdala to drive escape behavior. On their own, the words alone would not send our primate ancestors scrambling up trees but would engage a previously unknown association between an auditory vocalization and a visual/olfactory/gustatory "idea" about food for which there was no direct, immediate sensory indication. Undoubtedly, gestures and pantomime would have been necessary to communicate the urgency of the feeding situation, but what was new was the use of speech "displaced" from circumstances where the meaning would have been perceptually evident. This was the beginning of abstract thought.

The displacement of animal vocalizations from the here-and-now so that they could be used as words divorced from immediate sensations was a crucial first step in the evolution of language. This was the jump into "symbolic meaning" that some scholars optimistically see as emerging "gradually" over millions of years, but where Bickerton (2009), Chomsky (1975) and Deacon (1997) argue – to the contrary – that the "big-bang" of human evolution had to occur in one generation among individuals who were social enough to search for cooperative solutions but inventive enough as individuals to find a new use for vocalizations. By its very nature, there could be no "transitional stage." Either the vocalization would have its symbolic effect (and we all feast together) or it would be dismissed as meaningless chatter (and we have no meal again today).

Once primate vocalizations had revealed their possibilities for displaced usage, then language was off and running: Other words could be invented, vocal articulation refined, meanings disputed and the nightmare of word combinations (grammar) addressed. The transition from the big-bang of successful word usage to modern language was undoubtedly lengthy and tortuous, and required countless generations of wordsmiths who wrangled and disputed word usage to establish again and again the "correct" use of words in local linguistic communities. But, the birth of symbolic thought was necessarily an event that occurred within the short period of one primate life span – a notion that made use of conditional associations and has been transmitted culturally ever since. Details and articulatory refinements to spoken language could be negotiated over eons, but a near-instantaneous jump into symbolic vocalizations was the starting point.

Such is the current "best guess" about the first semantic step in the evolution of language – dreamt up by Bickerton out of thin air, but supported by the empirical facts, such as they are. It is admittedly an imaginative reconstruction, not "hard science," but based on what several generations of paleontologists have concluded about human existence many hundred thousand years ago. Given a starting point of this general kind, how could the complexities of grammar have evolved?

5.4. SUBJECTS, OBJECTS, VERBS

The use of language begins with "something to say" – the desire to tell others what to do or to indicate the causality of some aspect of the world to let others make their own decisions on what to do. In a prehistoric setting where all of the players were desperately hungry cavemen, an isolated word, such as "elephant," might have sufficed to convey a rather complex idea about cooperative hunting, but in a modern setting with innumerable, equally probable meanings associated with a single word, there is the need to specify what is relevant about "elephant." In fact, the simplest declarative sentences normally contain not only a noun entity (the subject of the sentence), but also a predicate – that is, a description of some property of the noun entity. Normally, the predicate will contain enough subcomponents that the subject of the sentence will be linguistically connected to a second entity (a direct object) through the action of a verb. Therefore, a typical sentence will contain a subject (S), an object (O) and a verb (V), and the alignment of just three such words can be used to explicate countless real-world situations.

Unfortunately, speech emerges from the mouths of speakers as a temporal sequence and the order in which words are presented to listeners can alter the meaning of the sentence. In fact, the basic SOV word-order varies among the 6000+ languages on Earth and all possible permutations are known. Syntactic tricks other than word order are also employed to indicate which noun is the subject and which the direct object, but word sequencing is an inevitable consequence of the temporal unfolding of oral speech; by adopting a default word order agreed on by speakers and listeners in a local community, the causality expressed in the sentence can be quickly understood. Some languages also use the structural "agreement" between subjects and verbs or between verbs and objects, other languages utilize the "marking" of adjectives, nouns and verbs with prefixes or suffixes to indicate their syntactic roles, still other languages use the systematic accretion of word stems into whole sentences (polysynthesis), and a great many languages use combinations of these syntactic rules. But, whatever other techniques are used, the temporal sequencing of the components of a sentence is essential for motor expression.

The crucial point about the sequencing of subject, object and verb is that the commonly used default sequence in a given language is nothing more than a convention – an arbitrary custom, a habit, a locally agreed upon rule, and *not* a "natural law." Within the conventions of English language usage, "Bill likes Susie" tells us something very different from "Susie likes Bill." In each case, one specific meaning is understood because we know that the active agent, the subject of the sentence, normally lies at the start of an English language sentence. There are of course ways of altering the meaning with the same SVO word order: "Bill is liked by Susie" or ways of maintaining the same original meaning with different word order: "Susie is liked by Bill." But, as English speakers, we are on the lookout for these kinds of syntactic manipulation that will alter the meaning implied by the default word order. And we are on the lookout because we are interested in deciphering the causality being expressed by a speaker in the conventions of our common language.

Within the limitations of short-term memory, the human brain is capable of the manipulation of multiple sets of three-word phrases, such as SVO, and capable of conditionally assigning a unique causal interpretation to any of the possible permutations of the triadic structures. In the simplest case, the convention concerning sentence structure allows us to assign a meaning to the relationships among three consecutive words: subject, verb and direct object (Figure 5.6).

For reasons of historical serendipity that are unlikely ever to be deciphered, the default order in English, Chinese and Russian is SVO, but for equally obscure reasons the default word order in Japanese, Hindi and

FIGURE 5.6. The permutations of subject-object-verb (SOV) wordorder, with the percentages of world languages noted in parentheses. A shift in the position of any word in a three-word sentence will change the meaning or make the sentence meaningless, but it is precisely the changed wordorder that would make sense in a different language.

Turkish is SOV. So, "Bill likes Susie" and "Bill Susie likes" are perfectly unambiguous ways of describing the same relationship between Bill and Susie in the conventions that have come into place in different linguistic cultures. The unnaturalness of SOV word order to SVOers and of SVO to SOVers is undeniable, but it is known that 87% of the world's languages are about equally divided on this fundamental issue of where to place the main verb (Figure 5.6). Amazingly, all of the other permutations of S, V and O are also known and are the "natural" order in distinct cultures for native speakers who have been raised in communities with these conventions. The bottom line is that, provided that both the speaker and the listener know what sequence of words to use to specify causality, it makes precisely no difference what that order is.

The conventionality of word order is thus apparent, but there remains a question concerning the overwhelming predominance of default grammars beginning with the subject. Deutscher (2005) explains this predominance from the simple fact that most spoken sentences – in all cultures and in all eras – are first-and-foremost about ... "me." I want this. I did that. I saw, heard and felt something. And then I did this and that. My comments may eventually come around to you and the unimportant events in your life, but not until I have had my say! For this reason alone, starting the sentence with the subject "me" in order to explain the causality in "my life" is arguably the default sequence in the majority of the world's languages. With a slightly less jaded view of humanity, Jackendoff (2002) has argued that "agents" naturally come first, followed by "patients" and the objects of agent actions – with the placement of the verb itself being a less important issue than the precedence of S over O.

In terms of physical causality, the sequentialization of real-world events is of course not so easily specified. "Susie slapped Bill" does not unfold as three separate events; without Bill's presence and the slapping action, Susie does not precede anything in the real world. There may, however, be a natural sequence to the psychology involved; when Susie slaps Bill, there is only one simultaneous event involving both the agent and the patient, but Susie was clearly the initiator of the action of her own hand, and Bill was the unsuspecting (?) recipient. For this reason, it is perhaps understandable that many languages favor the S_ _ sequence.

But even if there is a psychological rationale for the predominance of SOV and SVO languages, the fact that all other combinations have emerged in different language communities is noteworthy. If SVO and SOV languages do the "natural" thing of putting "me" first, it is even more remarkable that other linguistic cultures put "you" first, or have opted for specification of the action first, only to fill in with the details of who, whom and what later: "Love I you!" Indeed, the default word order in Gaelic and Hawaiian is VSO and VOS. The convention behind such word order implies that "Here is what happened, and, by the way, these are the actors involved." Although linear sequential causality on, for example, the billiards table may be clear enough, language is often employed to describe how people *perceive* actions in a world where causality is anything but obvious and typically open to contrary interpretations. So questions of "naturalness" aside, perhaps the most important discovery of comparative linguistics is the simple fact that word order – the sequence of spoken words indicating agents, actions, patients, objects and players is not all that important in conveying meaning. It is just a local rule – as arbitrary and liable to change as the pairing of word sounds and word meanings in the lexicon.

Arbitrary or not, the word-order convention is a convention with many implications. One of the most interesting was discovered by Joseph Greenberg (1968), who found that languages using the SVO word order have a strong tendency to construct prepositional phrases with the prepositions (*in, on, to, above, within,* etc.) *preceding* their noun phrases. In contrast, SOV languages use "postpositions" instead of prepositions, and construct phrases where the noun is followed by the "preposition." So, in SVO languages such as English, prepositional phrases employ the preposition before the noun (e.g., "on that desk"), but in an SOV language such as Japanese, the phrase becomes postpositional, so one says "that desk on" (*sono tukue no ue*). Similarly, relative clauses in SVO languages typically follow the noun which they describe, but precede the same noun in SOV languages ("the man we met yesterday" in English and "yesterday we met man" in Japanese).

Alone, the convention of SOV word order or the convention of prepositional phrase word order or the convention of relative clause placement might be easy enough to handle, but in multiple, overlapping combinations they lead to monstrous inversions of entire sentences. The English sentence "He put the book I read on the desk before breakfast" becomes in Japanese "Breakfast before he I read book desk on put" (*asa gohan no mae ni kare ha watashi ga yonda hon wo tsukue no ue ni oita*). You may wonder: Why is the friendly Japanese tourist with no real interest in linguistics tongue-tied at the hotel front desk? It's because he would ask: "Why linguistics in real interest no with friendly Japanese tourist hotel of front-desk at tongue-tied is?" In other words, he is understandably busy with, for him, totally unnatural linguistic gymnastics while the receptionist just wants to see his credit card.

Speakers of Mam, a Polynesian language, have found their own default word order in an unusual way. The placement of the verb among various nouns is less important than the placement of the nouns in terms of their volitional control over events. Human beings have volition and therefore come first. Animals seemingly have less volitional control and cannot precede a human agent in a sentence, but will precede plants and minerals. So, it is grammatical to say "The man killed the animal" but not that "The animal was killed by the man." Similarly, "The animal killed the man" is not grammatical, and must be stated as "The man was killed by the animal." In the same way, one cannot say "The [poison] plant killed the animal" or "The sand smothered the plant" because the correct ("natural") word order is determined by the supposed animacy of objects (Deutscher, 2005).

In fact, a great many languages use animacy as a secondary factor for determining syntax. Primary factors are more frequently word order, agreement between nouns and verbs, and case-marking, but often a default word order and case marking are used redundantly to specify the causal relationships among the agents/patients. The predominance of animacy in Mam is extreme, but a cross-cultural comparison by Bates and MacWhinney (1989) indicates that animacy is often relevant (Table 5.3).

Understanding the causality implied by three-word sentences, as in Figure 5.6, is something that a normal 3 year old can handle, but it is a known secret that the "language-using" chimpanzees in psychology laboratories around the world have severe difficulties already at this level of grammatical complexity. The one fact that most clearly indicates their utter inability to understand syntax comes from a statistical study of the kinds of sentences used by Kanzi, the "genius" behind many ape intelligence stories. Forget about the passive voice and relative clauses, the average number of words per "sentence" among 2000 consecutive spontaneous sentences constructed by Kanzi was found to be 1.5 – not quite two words in a row (Savage-Rumbaugh et al., 1993)! Such animal experimentation is important in indicating the

TABLE 5.3. *The relative importance of word-order, animacy, case-marking and agreement in various languages*

English	Word-order > Animacy > Agreement
Italian	Agreement > Animacy > Word-order
French	Agreement > Animacy > Word-order
Dutch	Case-marking > Word-order > Animacy
Serbo-Croatian	Case-marking > Agreement > Animacy > Word-order
Hungarian	Case-marking > Agreement > Animacy
Turkish	Case-marking > Animacy > Word-order
Hebrew	Case-marking > Agreement > Word-order
Warlpiri	Case-marking > Animacy > Word-order

After Bates and MacWhinney (1989).

ability of apes to use a symbolic lexicon (whether spoken words or discrete shapes on a computer screen). Like young children, chimpanzees can be trained into the symbolic realm; and this is the most remarkable result of animal language research (Bickerton, 2009). But clearly sentence construction is not a chimp strength. In contrast, the complexities of syntax are a notable human capability – and allow us to indulge in conceptual juggling to consider tentative scenarios of real-world causality that are indicated by syntactic conventions. That is, once having ascended to the symbolic level to represent ideas about the real world through language (that both we and the chimpanzee can do), human beings can then employ arbitrary rules of syntax to specify causal relations and can easily imagine other types of causality simply through manipulations of word order. (Significantly, in the research study where Kanzi remained "stubbornly stuck" at a "mean length utterance" of 1.5, an 18-month-old human child, whose language development was being compared to that of Kanzi, advanced from an average of 1.9 to 3.2 words. The construction of sentences containing three words was soon possible.)

The basic pattern of word order (or comparable syntactic manipulations) is the level at which most students of foreign languages encounter linguistic theory, but, from the linguist's point of view, the SVO story is just an issue of surface structure, the sequencing of speech acts that does not necessarily touch on the deeper issues of cognition. Although the foreign language student is actively concerned specifically with surface structure and topics such as word order, to the linguist, the dissimilarities of surface structure (where to place the verb, what order to use for nouns and adjectives, whether a preposition precedes or follows a noun phrase, etc.) in expressing one and the same meaning in different languages suggest the existence of a common "deep structure." And herein lies the main argument for the triadic nature of syntax.

5.5. UNIVERSAL GRAMMAR

From the linguist's perspective, the tripartite architecture of language (Figure 5.1) is a well-established model concerning *cognitive* organization, but it is not unique to language: It is the cognitive architecture of goal-directed behavior. In other words, the three modules of language are the linguist's restatement of the nature of actions involving (i) an imagined goal-state, (ii) potential manipulations to arrive at the goal-state and (iii) motor routines to realize them. Similarly, the cultural curiosities concerning SOV word order are interesting but are issues concerning the surface structure of language, whereas the heart of linguistic theory concerns deep structure, and specifically so-called Universal Grammar – the commonalities of language that exist in spite of surface differences.

The essence of Universal Grammar is the claim that there are principles of language use that are innately understood by all normal human beings (but not other species) – and, moreover, understood from a young age due to relatively brief exposure to language use. As a statement of the fact that people can learn languages, Universal Grammar would not be controversial, but the implication that the ability to encode and decode linguistic structures is innate and unique to *Homo sapiens* leads directly to difficult questions concerning cognitive mechanisms. Just what are the principles that we all know (or learn to know by the age of three)? Linguists have an answer ... initially proposed by Chomsky but refined by countless theorists over the past half century into a coherent theory of syntactic regularities.

At the level of describing the syntax of Indo-European languages, it can be said that the principles of Universal Grammar were established by Chomsky already in 1957 with the basic parsing techniques of Transformational Grammar. Sentences can be broken into their constituent phrases, and the individual phrases themselves can be broken into subphrases and ultimately reduced to individual words. Although not often stated as "triadic," the three-component nature of the theory of phrase structure is in fact well known. Bickerton (1990) has discussed specifically this point, asking:

> How do we know how to construct phrases? Because we have – somehow – a kind of template or model of what a phrase must be like. Not just a noun phrase: any kind of phrase. For the remarkable thing is that phrases of all kinds, including whole sentences (for a sentence turns out

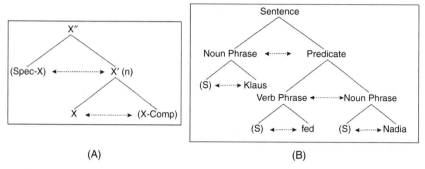

FIGURE 5.7. (A) Universal phrase structure. Every phrase has a "head," which can be linked to optional specifiers (Spec-X) and complements (X-Comp). The structure of all sentences and phrases in Transformational Grammar is therefore fundamentally triadic with, in principle, limitless embedding of phrases within phrases. "The most critical part, the only part that can be there on its own, is the head. The head of a phrase cannot be larger than a single word and that word must belong to the same class as the phrase. Noun phrases have nouns as their heads. Verb phrases have verbs, and other types of phrase (adjective phrases, prepositional phrases, and so on) follow the same pattern" (Bickerton, 1990, p. 63, with permission). (B) The complete phrase structure of the three-word sentence "Klaus fed Nadia."

to be just a big phrase with lesser phrases in it), are constructed in the same way. A phrase consists of three parts. (p. 59)

The triadic structure of phrases is familiar to most high school students from the tedious process of "parsing" a sentence into a noun phrase and a predicate: S → NP + Pr, followed by further subdivisions (NP → article + noun, Pr → verb + direct object, etc.) until we are left with a sequence of the "lexical items" that are actually enunciated in producing speech. Over the course of several decades, the general argument about phrase structure has been refined and formalized as "Head-Driven Phrase Structure Grammar" (HDPSG, Pollard & Sag, 1994), the central idea being that each and every phrase in a sentence contains a "head," below which two components hang, and below each component (then acting as a head) further components may hang (Figure 5.7A).

The head of a sentence will necessarily link a noun phrase and a predicate that, together, give the sentence a meaning beyond the meanings of its individual components. The predicate may be nothing more than a descriptor of the noun phrase: (The stone) [NP] (is round) [Pr]. Or the predicate may specify a verb that describes the action of the noun phrase: (The stone) [NP] (rolled down the hill) [Pr]. Depending on the language under consideration, the temporal sequence of the sentence can be either NP + Pr or Pr + NP, but their linkage within a single phrase is a universal feature, required in order that the sentence has meaning beyond the listing of labels for unrelated objects and events. By definition, phrases link two conceptual units through a head.

Ultimately, the complexity of language derives from (1) the linkage inherent to phrase structures, (2) the fact that the components of phrases may each function as phrase heads, below which lie an indeterminate number of further phrases and components, and (3) the fact that the actual temporal word order of any phrase is arbitrary (but fixed within any linguistic community). In addition to the problems of pronunciation and the memory load of storing the meaning of individual words, it is the arbitrariness of the sequencing that makes learning new languages so complex. From simple inversions – "postage stamp" to "timbre-poste" in translating from English into French, "eine Frau" [a woman] to "emakume baten" [woman a] in translating from German into Basque – to shifting multilevel, hierarchical phrases from one place to another, each language has its own detailed rules for phrase construction. As Cedric Boeckx (2006) has commented, "Although linguists have discovered that there is only a finite way of combining words and phrases, they have not been able to reduce the gap between different word orders yielding the same meaning. This seems to be a fact about how language is built" (p. 40). In other words, the cognitive templates that are used for the construction of phrases in any given language are arbitrary (in the sense that other languages may use different templates), but, once adopted, they cannot be reduced to still-simpler structures. Even phrases containing just two words are joined by an implicit head and are triadic in nature; they cannot be restated in terms of dyads or serial dyads without losing their fundamental meaning.

The one nonobvious "trick" that has been a part of phrase structure theory since the 1950s is the idea that the three components of a phrase (the head and the two items linked beneath it) need not be instantiated as spoken words but may be tacitly "understood." As such, they do not appear in the "surface structure" at all but are an essential part of the deeper cognitive structure nonetheless. That trick means that even an extremely simple, three-word sentence such as "Klaus fed Nadia" has – following the theory of phrase structure – the 11-components shown in Figure 5.7B.

To nonlinguists, HDPSG may appear to be an unnecessarily pedantic way of saying something fairly simple: That the words in a sentence are connected in a (culturally) specific way that indicates normally one unique underlying meaning. But linguists have shown that, by teasing apart the regularities of "phrase structure," the *same* meaning can be expressed in different languages with *different* surface structures (i.e., word orders). The linguist's theoretical argument is that the apparent simplicity of even a three-word sentence hides a deeper complexity that the human mind has an innate capability to understand – simply by thinking in terms of triadic phrases. Although counterintuitive on first consideration, the nearly fourfold increase in the linguistic components needed to go from the surface structure ("Klaus fed Nadia") to the deep structure of a sentence with a specific meaning (Figure 5.7B) is typical of the explication of causal mechanisms in reductionist science. To take this apparent complexification to be an example of unnecessary academic obfuscation of something fairly simple is to misunderstand what linguistic theory has achieved. In expanding the sentence into a multilevel structure consisting of triads, it becomes general – and expresses the hidden "universality" of countless comparable sentences in thousands of different languages.

It is important to understand that the presence of the invisible "head" structure is at the core of linguistic analysis. The linkage of the two elements beneath each head in, for example, a noun phrase is essential (i) to distinguish between the noun and its description and (ii) to join them together into a semantic unit. As a unit structure, the phrase can then be inserted into sentences of varying complexity. Note that the order of the two elements hanging below each head is entirely a matter of local linguistic conventions and can be reversed: The deep structure (the internal connectedness of the head and the two elements hanging beneath it) is important, but the surface structure (the actual sequencing of phrase components) is not. This general principle gives the structure of any phrase or hierarchy of phrases the three-dimensionality of a "mobile" sculpture (Baker, 2001) (Figure 5.8). Pass a gentle breeze through any linguistic culture and word order will change, while meaning remains constant.

The dynamics inherent to the conventions of word order is of course trivial in a simple three-word sentence (e.g., translating between SVO and SOV) (Figure 5.6), but the innate genius of all possessors of Universal Grammar is that they are able to rotate multiple phrases around their "heads" and thereby shift the entire appended substructure – without losing the meaning (Figure 5.8). Of course, the ability to render such transformations in "real time" is nontrivial – and normally requires the "intense training" inherent to learning a mother tongue in childhood – but the transformations of the wind-blown mobile are themselves mechanical: simple inversions of hanging components connected through a head.

The conventions of word-ordering must be understood for paraphrasing within a given language – for switching between active and passive voice,

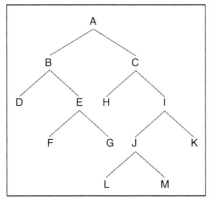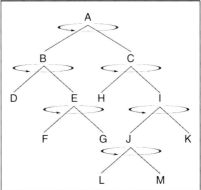

FIGURE 5.8. The 2D parsing structures illustrated in grammar books (left) assume the conventions of a particular language. But in the different languages of the world, the same sentence can be displayed as a 3D mobile (right), where each head can swivel on itself, thereby reversing the order of its hanging elements.

for changing the emphasis from one topic to another and for distinguishing between statements and questions – but the rules are fixed within a given linguistic community, and not beyond the learning capabilities of all normal people. The complexity of mobiles becomes apparent, however, when translation between languages is undertaken. Using the arbitrary conventions about what word-order is "correct" in two distinct languages, we learn two sets of rules about sequencing that transform the chaos of all possible configurations of the mobile into a finite set of transformations that allow for two sentences with radically different structure to have the same, unique meaning (Figure 5.9).

So, is a competence in using Universal Grammar essentially a linguistic talent? Do we have phrase-structure mobiles in our heads, or does phrase structure reflect something deeper about the ways in which human beings think? Guy Deutscher has asked the same question (2005, p. 225): "Is the aptitude of the brain to learn and handle hierarchical linguistic structures just a consequence of a more general cognitive capacity, manifested also in the way we process visual information, for instance? The answer is that no one really knows."

The triadic hypothesis, however, suggests that there may indeed be a "more general cognitive capacity": human beings understand conceptual units to consist of three elements. In language, the conceptual unit of a "phrase" always contains two elements hanging from a head and, as such, the elements cannot be disconnected from one another regardless of their

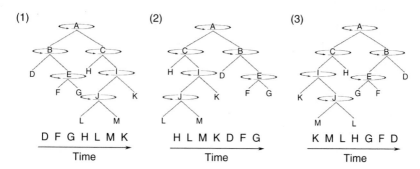

FIGURE 5.9. Rotations around selected phrase-heads lead to speech utternces with dramatically different "surface structures," but identical dependencies (semantic relations). Translation between Ex. 1 and Ex. 2 is easy because it entails only the one rotation at Head A, whereas six rotations are required to translate between Ex. 1 and Ex. 3 to obtain a complete reversal of word order of the sequentialized speech. The importance of the triadic architecture of head-driven phrase structure is apparent if we compare the number of dyadic operations (deletions and insertions of word units) that would be required to translate between Ex. 1 and Ex. 2 with the *one rotation* around Head A that achieves the same result.

placement in the surface structure of a sentence. That means that, similar to the artistic mobiles of Calder, the phrases in sentences never get tangled up with other phrases and can always be moved about as phrasal units. Regardless of changes in emphasis within a sentence, the phrases maintain their unique meanings and can be joined with other phrases within higher-level head structures. Making use of the triadic phrases of Universal Grammar, normal language speakers are capable of the conditional assignment of meaning to any such structural triad – and to bundle them together recursively up to the limitations of short-term memory.

Chomsky and colleagues (2007) and Corballis (2010) have recently emphasized specifically recursion as the key development of the human mind. While their argument is coherent, they begin their discussion at the level of the recursive imbedding of phrases. To be sure, the everyday complexities of human mental processes derive from multiple imbedding. So, when I say "I know you think he is wrong," we are already deep into the multi-imbedding issues of "theory of mind." But an understanding of the conceptual issues of language begins at a much simpler level. The "simple" phrases contained in the sentence "I know you think he is wrong" are already beyond ape cognition: "I know X." "You think Y." and "He is Z." are triadic linguistic phrases that delineate a specific relationship between two components in a manner characteristic of the phrase structure of human languages. Each such phrase is already beyond chimpanzee talents – and, when used recursively together, they produce a conceptual complexity that young children would not grasp, but a 7-year-old would.

The bottom line is that the syntax described initially in Transformational Grammar and later formalized as Head-Driven Phrase Structure Grammar is cognitively triadic at its core: each phrase is a triad and the combination of two phrases through a head will always form a higher-level triad. That means that the speech of people wanting to explain ideas about causality in the real world through words requires the transformation of a jumble of simultaneous semantic notions into triadic clumps, in which the sequence of spoken words is determined by adopting conventional phrase-structure rules of grammar and is dependent on the rules adopted by the local language community, but the triadic chunking is universal. What Deutscher refers to as "somewhere deep in our cognition" is nothing more, but nothing less, than an innate capacity for triadic manipulations.

To linguists, the assertion that "phrase structure is Universal Grammar" may seem anticlimactic. Universal Grammar is a bold theoretical claim asserting the uniqueness of the human mind, while phrase structure is the bane of schoolboy Latin lessons – too familiar to be interesting, too simple to be profound! But the basic argument underlying phrase structure is as simple and deep as the mystery of hearing emotions in musical triads or seeing illusory depth in the alignment of three visual cues. A linguistic phrase achieves the linkage between two independent concepts, implying a causal relationship between notions that need not be connected but that – within a phrasal unit – are indeed connected. As Piattelli-Palmarini (2010) has also noted, "The founding intuition [of traditional grammars] is that certain substrings of words within sentences strictly belong together and constitute a relevant subunit. Those grammars had ascertained that the inventory of such units is extremely reduced [noun phrases, verb phrases, adjectival phrases and prepositional phrases], and that they are, at a deep level of analysis, the same in all languages, the world over" (p. 150).

In spite of the profound success of phrase-structure grammar, recent developments in linguistic theory (Linguistic Minimalism) have been criticized by both Bickerton (2009) and Tomasello (2008) for having lost the strengths of traditional Transformational Grammar. While implicitly wanting to retain the successes of the traditional theory, the new "minimalist" ideology has abandoned the triadic structures of Transformational Grammar with a dubious reformulation in terms of "merging." The basic idea is that the grammatical or ungrammatical nature of linguistic structures can be stated exclusively in terms of the properties inherent to lexical items. A given noun, for example, can or cannot be merged with a neighboring lexical item, say, a verb, depending on the specific properties of the noun and the verb. The fundamental structure is therefore no longer the phrase, but the word. As such, the minimalist program attempts to pack all of the structural complexity of phrases, dealt with traditionally in terms of syntactic structures (and their triadic transformations), into each and every lexical item, and then deals with grammaticality in terms of the "mergability" of contiguous lexical items. It should be noted that the reduction of complex structures to more fundamental "local interactions" is a common aspect of scientific analysis, but Linguistic Minimalism has succeeded only in shifting the puzzle of triadic phrase structure to another location. Instead of resolving issues of, for example, the regularities of word order in terms of allowed and disallowed phrase transformations, the *same* problem is stated as an issue of the properties inherent to the merging of lexical items with their own unique set of "parameters" that allow certain types of merging, but not other types. In Minimalism, the individual words carry hidden within themselves all of the triadic regularities of phrases, but nothing has been gained in such a reformulation of the rules of syntax, while the triadic nature of phrase structure becomes hidden. For the computer programmer working on the problems of machine translation, it may be a simplification to reduce all of syntax to a list of lexical items – each with a few dozen "parameters" indicating what kinds of merging it can participate in – but it is highly unlikely that people cognize in that manner! The coherency of individual phrases and phrase combinations is what we consciously understand.

It is curious that the "merging" principle in Linguistic Minimalism reflects the same kind of "desperate reductionism" that is found in the failed program of explaining musical harmony in terms of tone intervals: A highly complex dyadic mechanism is hypothesized to work between adjacent lexical items (tones), but is unable to explain the meaning inherent to the simplest triadic phrase patterns (harmonies). It remains to be seen whether or not the minimalist research program could actually be made to work in parsing and translating, but discarding a half-century of analytic successes from phrase structure theory appears to be hard road.

Within the framework of phrase structure grammar, there are yet many unsolved problems that, theoretically, might lie outside of the "principles and parameters" of a final theory of Universal Grammar. Some have commented that the validity of Universal Grammar remains uncertain because, if true, it should have long ago allowed for flawless, fully automatic ("algorithmic") translation among foreign languages – and that has not in fact been realized. Despite decades of promissory notes, ever-faster computers and significant research funding, so-called machine translation remains the "fusion energy" of cognitive science – year after year, always just a mere three decades away from practical application – telephones interpreting, robots chatting and language barriers collapsing all around us!

A less pessimistic view is that, despite the continuing struggles of artificial intelligence (AI) research, phrase-structure already explains a sufficient diversity of the structural regularities of language that its fundamental importance can no longer be doubted. In this regard, an emphasis on technological applications, such as machine translation, misses the main point. Although there remains a gap between the successes of Universal Grammar and the complexities of real language, the problem does not lie so much with Universal Grammar, as with the overly ambitious claims appended to the fundamentals of linguistic theory. In a word, no scientific theory can deal with the sloppiness of real-world events!

The obvious comparison is with quantum mechanics (QM) – widely heralded as the greatest, most powerful and most precise theory in all of natural science. To be sure, some of the computational successes of QM on small physical systems are remarkable, are truly unrivaled and will never be quantitatively challenged by the results in biology, brain research or linguistics. Nevertheless, the severe limitations on the application of "algorithmic" QM to real-world physical systems should be fully appreciated before dismissing Universal Grammar as "soft science." In QM, exact results are obtained for systems involving one-nucleus and one-electron – *a two-body interaction!* Approximation techniques must be introduced already for the one-nucleus/two-electron case – not to mention the 13-body problem that is the water molecule. Nobody believes that QM is wrong or needs to be replaced with something entirely different, but – in its basic "algorithmic" form – it is nearly useless until various, rather ad hoc parameter settings are decided on. It is brilliant when applied to extremely simple systems, but it requires a host of additional assumptions, approximation techniques and adjustable parameters that are determined in the light of empirical reality for application to real-world problems.

The same might be said for Universal Grammar. As of today, the rules of HDPSG can be applied flawlessly to only a relatively small number of carefully selected examples, while exceptions and linguistic oddities lie around everywhere. I maintain that the exceptions do not disprove the theory, but rather simply indicate the complexities of real-world communications and the fact that, when people talk with one another, there are likely to be various, often contradictory undercurrents of human psychology in play – influencing the choice of words and syntactic structures to a degree not fully appreciated by AI researchers from the 1950s.

Precisely how do we manage the triadic juggling that underlies the competency for Universal Grammar? And where does that happen in the brain? Answers are not yet available, but the switching mechanism involved in conditional associations is undoubtedly involved and is known to occur in the lateral prefrontal cortex. Georg Striedter (2005, p. 344) has speculated as follows: "The disproportionate enlargement of the lateral prefrontal cortex ... probably enhanced the ability of humans to suppress stimulus-bound behaviors and to replace them with more arbitrary, voluntary acts.... Without the large lateral prefrontal cortices and the massive neocortical projections to the medulla and spinal cord, humans probably could not have learned to speak." In other words, the capacity for tentatively linking two strictly speaking independent notions through the head of a phrase, we form a semantic unit that itself can be linked with other notions. Real language is normally a complex hierarchy of such phrases, but the core trick is nothing more than the conditional associations formed by triadic phrases, piled one on top of another.

5.6. CONCLUSIONS

Let us conclude the discussion of language with a glance back at the traditional explanation of the cognitive discontinuity between *Homo sapiens* and all other animal species – the idea stated (but not explained) in the Bible: Man alone has "The Word" and has thereby transcended the realm of beastly behavior. As Merlin Donald (1991, p. 24) says, "God gave Adam a primordial tongue, which most Western writers assume to have been Hebrew, while Arabs assume it to have been Classical Arabic. These theories of divine intervention could be considered the original form of discontinuity theory." But it is specifically the discontinuity that modern evolutionary theory will not let stand. Insofar as "discontinuities" imply metaphysical miracles, scientists have sought mechanisms that will bridge the gaps.

The Bickerton scenario (Section 5.3) might suffice to explain the step into symbolic thought, but many others have emphasized the stark difference between the "isolation" of nearly all animal species and the community of the social lives of *Homo sapiens* ... even the early hominids prior to the blossoming of language abilities. Most noteworthy in this regard is the work of Michael Tomasello (1999, 2003, 2008). Without discussing the specifically triadic nature of syntax, noted earlier, he has emphasized the triadic basis for human social interactions (see Chapter 7) and demonstrated experimentally how triadic dynamics differ from the dyadic dynamics that utterly dominate the lives of apes. He has also reported on the implications of triadic cognition for the emerging social behavior of young children and focused on the theoretical issue of the transition – both developmentally and evolutionarily – from dyads to triads. He maintains that, failing to achieve triadic interactions, apes (in the wild) are incapable of symbolic communication and remain at the level of dyadic signaling. Conversely, somehow being instinctively capable of social triads, the social dynamics that human infants participate in from an early age lead into simple triadic and eventually recursively complex, triadic language behaviors. Aside from arguments concerning syntactic triads, Tomasello suggests that prior to the revolution brought about through language, there had already occurred a mental change that catapulted our early ancestors beyond the rather unremarkable social lives of other primates.

The developmental picture that Tomasello paints is strongly supported by experimental work and has been illustrated in terms of attentional triads that have an admirably concrete plausibility. For all the importance of that empirical work, the evolutionary scenario that he outlines implies that the crucial *first step* into the unique mentality of *Homo sapiens* was the realization by our early ancestors that they were all intentional beings – members of the same "human family" with needs, desires and thoughts that were common among themselves, but not shared with other species. Again, there is a plausibility to this hypothesis, but questions must nonetheless be asked concerning where this intraspecies empathy might have originated.

Unlike the strong experimental foundations underlying most of Tomasello's theorizing, he attributes the key to our humanity to a subjective realization of intraspecies similarity that, for reasons unknown, our ancestors achieved, but other primates did not. This new level of subjective consciousness, in turn, became the basis for all forms of social *cooperation* upon which human culture is built – leading in turn to the emergence of human language, human families and human civilization in a synergistic blossoming of higher cognitive talents. The simplest version of such empathy manifests itself as the triadic social interactions that underlie "joint attention": you, me and our common task. In a developmental context, it is most commonly: mother, child and focus of their common concern (Chapter 7). Tomasello's emphasis on this evolutionary step is understandable: The uniquely human ability to set aside competitive, dyadic "dominance issues" in pursuit of a common goal allows for social cooperation and, once we have cooperating primates, subsequent cultural developments would not be a surprise. In terms of explicit, measurable behavior, a natural tendency to cooperate is where, generally speaking, we part company with the apes. But the evolutionary starting point, "intraspecies empathy," still requires a causal explanation, if we are not to retreat to a "hand of god" miracle to get the ball rolling.

The general plausibility of Tomasello's argument notwithstanding, it is essential to address the problem of having the entire story of humanity hanging on a rather vague assumption concerning the emergence of a new form of intraspecies, subjective feeling – totally unconnected to the historical record on fossils and brain anatomy. That assumption remains undocumented, in principle cannot be proven and – if not a fatal flaw in the theory – is, at best, an obscurity in need of attention. In order to avoid a reversion to premodern speculations that revert to arguments about a "subjectively-knowable, but empirically-unverifiable soul," let us consider a reversal of the Tomasello hypothesis in which triadic cognitive processes came first, and eventually led to joint attention and cooperative social interactions. The hypothetical evolutionary process then becomes: With expansion of trimodal association cortex, tool use and toolmaking became possible among our prelinguistic ancestors – and the first concrete evidence of unusual human talents is indeed stone tools (Chapter 4). The social cohesiveness, joint attention and mutual empathy that are also characteristics of *Homo sapiens* would therefore have been a *consequence* of the recognition by our stone-age ancestors that their compatriots were engaged in unusual tool-related behaviors that were instrumental for survival on a daily basis. As Lewis Wolpert (2003) has commented, "It has been technology that resulted from causal beliefs, not social interaction, that has driven human evolution."

Only subsequent to the emergence of such technological behavior did there occur what might be referred to as a change in consciousness. Here, mirror-neurons would have played a crucial role (Arbib, 2002; Rizzolatti & Sinigaglia, 2008; Ramachandran, 2011) in alerting hominid toolmakers to the related toolmaking activities of other hominids. Seeing one another handling stones, making tools and using them to obtain food, our ancestors of 2 million years ago would have known that their compatriots were engaged in, to say the least, atypical animal behaviors. At the same time, those early tool users did *not* perceive comparable behaviors in other animal species, and, as a consequence, they would have known that, whatever the internal thought processes of other animals might be, the thought processes that had led hominids into the world of tool artifacts were absent in the animal world. Of course, apes, gorillas and a host of now-extinct hominids behaved much like our ancestors for most of the waking day – foraging, grooming, eating, defecating and copulating in familiar ways – but they did not use tools. Those animals were (and still are) cognitively dyadic – intelligent and highly competent at that level – but generally incapable of the triadic processes underlying tool usage and toolmaking, in particular, and almost never exhibiting joint attention on a topic of common concern.

In *not* detecting triadic thought processes in animal behavior – neither tool use, gaze following, pointing nor the triadic processes of infant socialization – our ancestors would have instinctively concluded that the members of other species were not like themselves. Empathizing only with fellow hominids who handled tools, *Australopithecines*, *Homo habilis* and *Homo erectus* engaged *with* one another and did not lead lives of isolation typical of animal species. On the contrary, they began to cooperate with compatriots in triads of joint attention, ultimately leading to group activities and the full complexities of human civilization. In other words, the temporal sequence of the differentiation of man from ape began with (1) behavior based on triadic cognition (toolmaking), (2) mirror-neuron-based recognition of species similarities and, only subsequently, (3) a subjective feeling of companionship, empathy and human brotherhood.

The time required for full socialization and the development of a truly human consciousness was undoubtedly long, but it is more parsimonious to think that the triadic cognition of tool use and learning through imitation came *before* the empathetic feeling of species solidarity than vice versa. To state this argument negatively, there is no evidence for intraspecies empathy prior to toolmaking, and no indication of qualitatively unusual developments in social cooperation prior to tool use. For this reason, on the basis of the available empirical evidence, the culture of toolmaking behavior is likely to have *preceded* the subjective consciousness of a "society of empathetic toolmakers."

Once a cooperative, social consciousness had become established, then it was just a matter of appropriate circumstances to achieve the many other hallmarks of humanity. In this respect, Tomasello is correct in emphasizing the importance of a change in consciousness that leads to human socialization – followed by cooking, enjoying dance and music, burying the dead, wearing jewelry and putting on clothing and makeup. But to put the change in subjective consciousness first in the evolutionary story is to evade the question of causal mechanisms. So, in a slight reworking of the Tomasello hypothesis to bring primitive stone tools to the forefront, I maintain that there is no need to postulate an *inexplicable* change in consciousness – provided that a prior *explicable* change in cognition can be demonstrated (i.e., the creation of stone artifacts and their use primarily in the social activities of hunting and scavenging megafauna) (Chapter 4). In summary, following the million-year evolution of the triadic brain for toolmaking, a gradual change in intraspecies empathy would have become possible and culminated in the evolution of the ultimate form of intraspecies, empathetic social activity – linguistic communication.

Replacing "subjective consciousness" in the Tomasello scenario with cognition as the prime mover in human evolution essentially eliminates the need to postulate "miraculous" evolutionary changes to initiate the emergence of *Homo sapiens*. But, if triadic cognition plays the dominant causal role in producing characteristically human behavior, what, we must ask, can be said about subjective consciousness itself?

6

Consciousness

The main concern of the previous chapters has been how human beings *cognize* in characteristically human ways, whereas the present chapter raises the thorny problems of consciousness. Why discuss consciousness in a book on cognition? If answers can be found concerning the different ways that different species think, what is the point of bringing up the more difficult topics of feeling, subjective awareness, qualia and the "internal perspective?" If we process information in ways unfamiliar to the apes, what more needs to be said?

My answer is that, in order to maintain that the unusualness of the human mind is a consequence of triadic *cognition*, it is essential to show that the mechanisms underlying subjective awareness are *not* where the human mind is special. To reiterate the conclusions of previous chapters, if an animal species is incapable of hearing the harmony contained in a three-tone chord, incapable of seeing the illusory depth in a 2D picture, incapable of the syntactic manipulations needed for understanding speech and so on, then clearly it cannot experience the meaning that human beings find to be intrinsic to those stimuli. Of course, the same sensory *stimuli* will enter the animal brain as sensations, but will not have the symbolic significance that they have for us if the processing ends at the level of dyadic associations. In other words, the triadic cognition itself is unusual – "different," "special," "complex" and "higher" – but it is unusual because of the triadic manipulations and not because we "feel" our cognition in ways that are fundamentally different from the ways that other animals feel the neuronal processes running through their heads. We cognize in ways that animals do not, but from a purely physiological perspective it is, in principle, unlikely that we, and only we, sense our neuronal activity in a way that brings us into some sort of transcendent, spiritual dimension that the lowly beasts have not found.

The argument to be developed in this chapter is that modern neurophysiology can explain the fundamental nature of both cognition ("thinking") and consciousness ("feeling"). And these neuron-level considerations indicate that the dimensions that make *Homo sapiens* qualitatively unusual are cognitive – how we think. As will become clear in the following sections, I maintain that the unusual character of the neuron *as a living cell* provides coherent answers to questions concerning both cognition and subjective consciousness, but the answers concerning consciousness have been largely overlooked by scholars who are more comfortable dealing with the questions of consciousness at a rather abstract, philosophical level than in considering the issues of cell biology.

Unlike the previous chapters where the psychological arguments could be demonstrated with familiar examples from everyday life, this chapter focuses initially on what neurophysiology has taught us about neuronal functions. For this reason alone, the discussion may strike some readers as "academic" and less "real" than the down-to-earth psychology of triadic cognition. Be that as it may, a proper understanding of the neuron leads to certain conclusions about consciousness that simply could not have been drawn prior to the many cellular and subcellular discoveries in the fields of electrophysiology and molecular biology over the past 60 years. In other words, I argue that the scientific *foundations* of both cognition and consciousness can be rigorously understood at the cellular level, where actual mechanisms are known. Undoubtedly, the basics concerning neuron structure and function will be familiar to most readers, but the argument concerning the neurophysiology of consciousness is relatively new (Cook, 2002, 2008) – with all of the relevant experimental findings being a product of late twentieth- and early twenty-first-century neuroscience.

In essence, my argument about consciousness can be summarized in two words: excitable cells. Neurons are excitable, whereas most other cell types in animal organisms – and virtually all of the cells in plant organisms – are nonexcitable. As will become apparent, an understanding of the physiology of cellular "excitability" is considerably less sexy than many seductive, rather abstract ideas about the human mind, but it is at the mundane level of cell biology where a huge insight awaits concerning how different kinds of living cells interact with their environments. Excitable cells, and particularly neurons, *are unusual as cells* – and from their collective activity there emerges both of the phenomena that are the essence of, broadly speaking, animal psychology: The information processing known as *cognition* and the subjective feeling known as *consciousness*. In brief, I will show that these two topics, cognition and consciousness, can be meaningfully distinguished and correspond to different aspects of neuronal functioning.

We start with a fact stated in all textbooks on neurophysiology, but rarely examined with regard to the puzzles of consciousness. That is, there are *two* unusual functions of the neuron: (i) synaptic secretion of neurotransmitters leading to intercellular communications and (ii) the intra-cellular action potential that sends the nerve impulse from the cell body to the synaptic terminals. In most discussions of the mind, only synaptic functions are considered to be of relevance (with all other physiological processes being relevant only insofar as they support synaptic communications). Indeed, in the *Synaptic Self* (2002), Joseph LeDoux has been admirably explicit in stating his focus on a synaptic explanation of mind:

> Because psychological and behavioral functions are mediated by aggregates of cells joined by synapses and working together rather than by individual neurons in isolation, the contribution of [other] properties of a cell to mental life or behavior occurs only by way of the role of that cell in [neuronal] circuits.... Synapses are ultimately the key to the brain's many functions, and thus to the self. (p. 64)

In contrast, while it is universally acknowledged (in the world of cognitive neuroscience) that *cognition* can be explained on the basis of synaptic activity, the argument of the present chapter is that – at the cellular level – *sentience* is a consequence of the momentary "openness" of the neuronal membrane to the external world during the action potential. In other words, the fact that complex neuronal systems "feel" is due to the unusual, neuronal bursts of *transparency* to the in-flow and out-flow of ionic charges as the action potential impulse is passed down the neuronal axon. This is not "feeling" in the sense of "a human being having emotional experiences" (and a further argument is required to get from the neuronal level to the whole brain level), but it is "feeling" at the cellular level – quite literally, the neuron *in isolation* sensing the condition of its extracellular fluid – making contact with its surrounding world (by absorbing a small portion of the local ionic charge) as it goes about its cognitive business of synaptic communications.

The chief merit of a cell-physiology approach to the problem of subjective consciousness is that it avoids the necessity of bringing philosophical dualism into the scientific discussion of the mind but, at the same time, acknowledges the reality of the subjective dimension of neuronal experience. Of course, LeDoux is correct in emphasizing the importance of synapses for cognition and behavior, but, as most commentators in the modern consciousness literature acknowledge, the psychology of the human "mind" is that *plus subjective feeling*. So, in addition to LeDoux's orthodox emphasis on the synapse, the argument to be developed in this chapter is that neuronal sentience is the starting point for discussions of subjective feeling. The still unsolved problem of *self*-consciousness (awareness of the trajectory of one's own personal history through a lifetime) is an area where further insights are yet required, but I maintain that the more fundamental problem of subjective awareness ("consciousness") has been essentially solved at the single-neuron level. The animal feeling of "being alive" – awareness of one's current state in relation to a larger surrounding reality – is a topic with straightforward scientific answers, provided only that we make efforts at building our understanding from the cellular level.

6.1. THE MAIN QUESTION

During the 1990s, there was a resurgence of interest in the problems of consciousness. While never an unpopular topic – and, according to some, the *only* topic in all of psychology that is truly important! – functional brain-imaging made it possible to study subtle questions concerning states of consciousness in normal human subjects in a harmless, noninvasive manner. Together with modern experimental techniques for presenting sensory stimuli and measuring behavioral responses with millisecond accuracy, questions about conscious and unconscious processes have been addressed – and conclusions drawn concerning where and when information is processed in the brain.

Some of the insights of such brain-imaging studies have been mentioned in the preceding chapters, but it is still widely believed that certain aspects of consciousness itself remain unsolved, and possibly unsolvable, puzzles. Specifically, the majority of commentators on the core issue of "subjective awareness" maintain that we are far from having any coherent answers at all. The question why there is such a thing as subjective feeling ("qualia," "awareness," "feeling," "soulfulness," "an internal perspective," etc.) in a material world seems to go beyond the realm of empirical science. Even if we had clear answers about brain localization, the question of "why" leads into murky philosophical realms where answers can no longer be framed in terms of brain anatomy, synaptic transmission, cerebral blood flow and the other bits and pieces of quantitative neuroscience. According to some, the enigma of consciousness demands answers that simply cannot be delivered by natural science, but rather must be based on verbal gymnastics and clever abstract notions without correlates, much less causal mechanisms, in neuroscience.

In the consciousness literature, a surprisingly large number of commentators have expressed utter pessimism about ever finding a solution to the problem of subjectivity and have declared that, in asking questions about consciousness, we have run into a brick wall that science cannot penetrate. Not everyone is convinced that answers are impossible, but even the optimists maintain that a widely accepted scientific consensus – comparable to those already achieved in atomic physics and cell biology – is at least decades, if not centuries, away. In other words, despite the remarkable progress in brain science throughout the twentieth century and the modern capability to observe the functioning of the living brain objectively through brain imaging, many commentators believe that we have fundamentally no clue how to explain the internal subjectivity of our minds. While not abandoning empirical science entirely, kicking the ball down the road to future generations is a remarkable declaration of impotence. Yes, they seem to be saying, the physical apparatus is all about neurotransmitters and neuronal connectivity, but the "stuff of the mind" that each of us knows subjectively is somehow different – and certainly not visible as electrical brain waves or cerebral blood flow! Such deep pessimism is in effect a confession that hardcore, empirical science really has no idea how to proceed in the elucidation of the material basis of the mind. So, is subjectivity an eternal paradox that will never be answered? Or, if such pessimism is not warranted, what is yet missing that will allow us to find coherent answers?

In *The Quest for Consciousness* (2004), Christof Koch has summarized most of the relevant facts pertaining to the consciousness debate, but, regarding subjectivity, he admits that: "Why qualia have their subjective feeling remains an enigma" (p. 310). Steven Pinker has also expressed frustration on the dilemma of subjective consciousness in *How the Mind Works* (1997). With characteristic clarity, he writes that

> The entities now commonly evoked to explain the mind – such as general intelligence, a capacity to form culture, and multipurpose learning strategies – will surely go the way of protoplasm in biology.... [Those presumed] entities are so formless, compared to the exacting phenomena they are meant to explain, that they must be granted near-magical powers. (p. 27)

Later, having raised a dozen or so of the seemingly unanswerable questions in the consciousness literature – Can consciousness exist in computers? In a Petri dish? In a zombie? – Pinker then uncharacteristically admits defeat:

> Beats the heck out of me! I have some prejudices, but no idea of how to begin to look for defensible answers. And neither does anyone else. The computational theory of mind offers no insight; neither does any finding in neuroscience. (p. 147)

An even clearer statement of the perplexity of subjective consciousness has been written by John Duncan (2010). He notes that calling it a "hard problem" implies a disingenuous optimism that we think the mechanism underlying qualia is just another of a long line of scientific challenges, when in fact it seems to lie outside our normal framework of reasoning.

> I would prefer to hear qualia described as 'the problem that has us so befuddled that we don't even know what sort of problem it is'... We have not failed to solve this problem; we have failed even to engage on it... At least for now, I think we have to accept that what I would call not the hard but the impossible problem is just that – not only outside science but beyond all useful conception. (pp. 147–148)

V. S. Ramachandran (2011) has made similar comments:

> Qualia are vexing to philosophers and scientists alike because even though they are palpably real and seem to lie at the very core of mental experience, physical and computational theories about brain functions are utterly silent on the question of how they might arise or why they might exist. (p. 248)

But let us examine the findings from neuron science before we raise the white flag.

In contrast to the previous chapters concerned exclusively with questions about cognition, the present chapter presents the view that scholars in consciousness studies are looking in the wrong direction and have paid far too little attention to basic neuron physiology. This view is seemingly paradoxical, since there is an abundance of facts from brain science that are regularly marshaled to bolster hypotheses about consciousness. Moreover, for virtually everyone concerned with the scientific study of consciousness, it is axiomatic that the starting point is the neuron: *All* of the systems that unambiguously display signs of consciousness are, without exception, built from *neurons*. That much is known and agreed upon!

But a paradox remains. The remarkable discoveries about the structure and function of neurons are today so "obvious" and so widely known that they have invited profuse speculation on the possibility of how *nonneuronal* information processing systems (principally, computers) might someday display analogous properties of mind, rather than inquiry into what is unusual about neurons themselves. The present chapter returns to basic neurophysiology to ask the obvious question: What is so special about nerve cells? Before we entertain science-fiction ideas about inorganic forms of consciousness, and before we declare that understanding consciousness is an impossibility, let us consider the phenomena that we know are made possible by the neuron, a most unusual cell type found only in animal species.

As was the case in previous chapters, we do not want to get lost in specialist details that only a laboratory neurophysiologist would understand. To be sure, a few technical details will be discussed, and some speculation about future research will be made, but the aim here is to shed light on the consciousness debate by focusing on what is known for a certainty about neuronal functions. For that purpose, controversial aspects of neuron physiology will be avoided. Dubious interpretations of, for example, the role played by so-called microtubules (á la Hameroff & Penrose, 1996) will not be entertained. The pharmacologist's long-standing, but undelivered promise that the mysteries of the mind might be resolved by future discoveries concerning neurotransmitter biochemistry (á la LeDoux, 2002) will also not be pursued. And the possible relevance of the physics of the electromagnetic field surrounding biological systems (á la McFadden, 2002) will not be explored. Similarly, although there is already a huge literature in which the likelihood or unlikelihood of silicon "minds" is discussed, the focus of the present chapter is on the undisputed facts concerning the involvement of neurons in all currently known manifestations of mind. At some point – truly generations down the road – meaningful discussions concerning artificial systems might be possible, but only once the fundamentals of neuron physiology are nailed down.

These other topics – the physicist's fields, the chemist's molecules, the biologist's organelles and the engineer's circuits – are all worthwhile research topics and the source of fascinating speculation and late-night bull sessions. So, let us not place a ban on even wild-eyed fantasies, but let us be frank in stating that these are among the many topics where coherent, explicit hypotheses are generally absent and where empirical tests are difficult or impossible. The extravagant claims, unrealized dreams, false predictions and endless balderdash are too extensive even to list, but it seems fair to ask what in the world ever happened to those self-aware robots promised since the 1950s? Where is the "chemical theory of mind" and the little pills to micromanage our moods? And if mind really does have an electromagnetic penumbra that can be measured outside the skull, just where are the experimental demonstrations of the paranormal? Advocates will continue to hyperventilate over future possibilities, but it seems that the reality of subjectivity has proven to be far more complex than many once imagined. The promissory notes will continue to fill the popular press, fuel Hollywood blockbusters and probably do no real harm, but speculation just ain't science. In fact, a level-headed assessment of brain research, in general, leads to some rather modest conclusions.

The purpose of the present chapter is therefore to demonstrate how far the firmly established facts of neuron physiology can already go in explaining the properties of mind. These are the scientific facts that everyone agrees on, but (with some notable exceptions) no one talks about. My basic contention is that – just as most modern commentators on consciousness routinely discuss robot technology, zombie mentality, recursive programming and the possible implications of quantum mechanics – the known and not overwhelmingly complex facts about neuron physiology should also enter the discussion on consciousness. At present, most commentators touch on many issues of questionable relevance to the topic of subjectivity, while steadfastly ignoring the elephant in the room, *the neuron*. In contrast, the question raised here is: What is unusual about the neuron itself that makes it the central player in the consciousness story?

6.2. THREE LEVELS OF DISCUSSION

Whatever the ultimate solutions may be to the problems of consciousness, it is certain that at least three distinct levels of inquiry will be required. The first can be broadly described as philosophical – and entails precise definition of the problems and the scope of topics that must be included. To a large extent, these are issues of conceptual coherency and appropriate terminology that have been addressed by Bernard Baars (1988, 1997), David Chalmers (1996), Daniel Dennett (1991, 1996) and John Searle (1992, 1997, 2004), among many others. Modern consciousness researchers are in debt to these scholars for their efforts at delineating the core issues that must be clarified in a scientific account of consciousness – and, simultaneously, for their careful dismissal of the peripheral issues that simply cannot be addressed in a scientific manner.

The second level is psychological – and concerns the mental states that, from the perspective of brain science, need to be accounted for. The distinctions between conscious, subconscious, unconscious and preconscious states, awake and asleep, and the subtle issues of attention (focus and fringe) are important here. These topics overlap with the conceptual topics of the philosophers, but the emphasis in the psychology of consciousness is not on abstract concepts, but rather on phenomena that can ultimately be described in terms of brain processes and studied experimentally. The psychological issues have been debated by many who are concerned with relatively macroscopic issues of brain circuitry. Clearly, the bone of contention, in general, is the mapping between the complexities of brain anatomy and the varieties of psychological experience – and progress has indeed been made. The different roles of the brainstem, neocortex and limbic system are real – and form the basis of clinical behavioral neurology. But few scholars would even suggest that MacLean's "triune brain" has solved the problems of consciousness. Somehow, a deeper understanding of brain functions must be attained! Eventually, topics concerning the computational architecture of robot "awareness" will also undoubtedly be addressed, but "silicon psychology" is, by most accounts, a secondary issue that might someday provide independent and perhaps definitive evidence for or against any hypothesis concerning consciousness in living brains, but is today largely a matter of speculation.

In any event, the broad strokes of philosophical and psychological work – Levels One and Two – are essential groundwork that must precede and guide subsequent empirical studies, but it is unlikely that any philosophical or psychological conclusion will be considered firm until support from Level Three, neuron physiology, is also in hand. Physiological research is essential because it allows for the quantitative predictions and double dissociations that are normally crucial for the acceptance of scientific hypotheses. And, most importantly, reductionist neuron science brings with it the realistic promise of elucidating causal mechanisms. This is not to say that the philosophical and psychological discussions that currently make up the bulk of work in consciousness studies are superfluous, but that undeniable, irrevocable progress will probably require a neurophysiological underpinning – an explicit, empirical grounding that Levels One and Two simply cannot provide.

Stated historically, philosophers have identified and debated the essential phenomena for millennia; neurologists and psychologists over the past century or so have added the important brain structures that house the phenomena; and over the past several decades neurophysiologists have added the microscopic and indeed submicroscopic neuronal detail that can potentially provide a thorough understanding of causal mechanisms. All three realms are part of the consciousness story and clarity at all three levels will be needed, but it is worth emphasizing that neuron science is

TABLE 6.1. *Five approaches to subjectivity, their "solutions" and proponents*

Field	Solution	Modern proponents
A. Philosophy	No solution; it's an eternal paradox!	Chalmers, Hardcastle, McGinn
B. Neuropsychology	Build a better model of the brain!	Baars, Crick, Damasio, Edelman
C. Quantum Mechanics	Employ the uncertainty principle!	Hameroff, Penrose, Stapp, Walker
D. Robotics	No problem, just give us the algorithm!	Aleksander, Dennett, Taylor
E. Neurophysiology	Examine the neuron more closely!	Fuster, Koch, Llinas, MacLennan, Singer

"new": electrophysiology and molecular biology from the latter half of the twentieth century have informed us of a cellular world that was simply unimagined to previous generations of scholars. That is where new insights have been gained.

So, let us begin by reviewing current ideas on consciousness that attempt to connect these three levels.

6.3. FIVE APPROACHES TO SUBJECTIVITY

Scientific inquiry into the material basis of consciousness is no longer considered to be a topic of impossible complexity, but it is arguably the case that the difficult issue of *subjective awareness* had already been debated to a complete stalemate by the mid 1990s. While there are as many hypotheses as commentators, several discernible schools of thought have emerged – corresponding to the different levels at which mechanisms have been proposed and future progress is anticipated. For the sake of explication, the labels "philosophy," "neuropsychology," "quantum mechanics" and "robotics" will be used here to denote the four *dominant schools of thought*. *Paradoxically, all four views are demonstrably weak on the one topic* that all agree must be included in a final theory, that is, the fifth pillar in consciousness studies: "neurophysiology" (Table 6.1).

6.3.1. Philosophy: Subjectivity as an Eternal Paradox

Virtually all conceivable perspectives on the problem of subjectivity have been defended by philosophers, but the one view that has found no support in the scientific world is the declaration that a solution is impossible.

McGinn (2004) has staked out precisely that philosophical position, maintaining that the human mind is capable of understanding the *problem* of subjectivity, but not capable of devising a *solution*. Pinker (1997, p. 561) summarizes this position as follows: "Maybe philosophical problems are hard not because they are divine or irreducible or meaningless..., but because the mind of *Homo sapiens* lacks the cognitive equipment to solve them. Perhaps we cannot solve conundrums like ... sentience." It is, to be sure, an interesting conceptual argument: We, as a species, are clever enough to see that a problem lurks, but, by the very nature of the brain apparatus that we have to work with, we are unable to transcend the limitations of our imperfect brains and look at ourselves objectively. In this view, we have identified the appropriate door labeled "subjectivity," but now realize that it will remain forever locked!

Although not embracing despair as explicitly, other philosophers (Chalmers, 1996; Levine, 2001; Nagel, 1986) have defended dualistic approaches that also deny the possibility of a *scientific* explanation of subjective awareness. These attempts to explain the nature of subjectivity without reference to material mechanisms have the undeniable merit of *not evading* the central issue: *Subjective consciousness is the one real, unquestionable fact of human existence!* Unfortunately, there is no apparent compromise between a scientific, materialistic view of the universe and a dualistic view that postulates the independent existence of a subjective realm with properties somehow disconnected from those of the material world. As Dennett (1991, 1996), Searle (1997, 2004) and other defenders of, broadly speaking, realist, materialist philosophy have noted, the dualist world-view leads to an impasse, where the only "resolution" is declaration that subjective consciousness is a permanent mystery. Perhaps the acceptance of dualism is a satisfactory answer for a small minority of philosophers, but it strikes most others as antithetical to the entire scientific approach and is an obvious dead-end. So, just as the mysterian dualists categorically deny that the materialist scientist has anything to contribute to the core debate on consciousness, the scientist in turn rejects the suggestion that the game can be played on a field where material mechanisms have no role, creating a stalemate. There must be a better way than dualism.

6.3.2. Neuropsychology: Subjectivity Through Sufficiently Complex Neuronal Circuitry

With a focus somewhat different from that of the philosophers, psychologists concerned with consciousness have advocated ideas about possible brain mechanisms. Two recurring themes are concerned with the neuronal circuitry required to sustain consciousness. The first is a quantitative argument that a "critical mass" of sufficiently interconnected neurons must be present. The second is a qualitative argument about the complexity of the circuitry itself.

The quantitative argument turns out to be anything but quantitative: No one has offered numerical estimates concerning the number of neurons required to sustain subjectivity. Is it a 1000-neuron minicircuit that provides a spark of subjectivity, or do we need 10 billion to get started? Or, is a three-neuron feedback loop the crucial subjectivity circuit? Nobody knows and nobody guesses, but there is occasional mention of the need for a minimal brain size to foster consciousness. Here, the general argument is that an overabundance of neuronal connectivity is probably required to calculate the implications of sensory input and to allow reverberation along countless pathways – and thereby give brain activity the overwhelming "richness of endless possibilities" that we experience as subjective awareness. So, instead of a critical threshold of complexity, the argument boils down to a "more the merrier" view. This type of "big brain" hypothesis has several well-known difficulties: The Neanderthals had bigger brains than modern *Homo sapiens* and became extinct. The dolphin, whale and elephant have much bigger brains than we have, but they don't exhibit signs of comparable, much less more advanced, consciousness. And some human patients with hydrocephalus have lost more than 90 percent of cortical neurons, but graduate from college, live normal lives and are clearly not "unconscious." The difficulty in explanations based on neuron numbers arises in trying to specify more precisely why a small neuronal circuit does not sustain subjective awareness, but a multineuron circuit does. Making an appeal to "complexity" or "hierarchical structure" might be pointing in the right direction, but, without a mechanism, the argument lacks persuasive power and cannot be tested.

The qualitative neuronal argument is normally stated in terms of the need for feedback circuitry, recursive algorithms or topological mapping between brain regions. Edelman (1989, 2004) believes strongly in this approach – and attempts to distinguish the complexities of brain processes from simple thermostat feedback mechanisms using the term "reentry." Again, the direction in which this hypothesis points is intuitively appealing, but the mechanism that allows for the emergence of subjectivity from any type of feedback circuitry remains unclear. Only Chalmers (1996) has been brave enough or foolhardy enough to bite the bullet here, and essentially declares that the engineer's feedback thermostat sitting on the wall simply must have a primitive consciousness that is multiplied some trillion-fold in the human brain. Fair enough, as far as it goes, but where do we find indication of an increase in *subjective awareness* in a modern computer chip in comparison with the thermostat on the wall? Increased computational complexity is certainly there in the chip, but is that really the same dimension as subjective awareness?

Both the qualitative and the quantitative arguments may be correct in suggesting that there are minimal conditions necessary for "higher" cognitive functions; however, neither argument gives any indication of where precisely *subjectivity* arises. Even if a "threshold of complexity" could be explicitly defined, the essence of subjectivity seems to be its qualitative, "internal" character, as distinct from the well-understood, logico-deductive processes of cognition. Of course, we can explain how the thermostat works in an entirely mechanistic manner, but asking how the thermostat might feel about its own mechanistic activities seems to be an absurdity. As a consequence, while hypotheses about large-scale brain circuitry are of course needed to explain the full spectrum of mental states, the jump from simple feedback mechanisms to the issue of subjective awareness remains essentially unexplained. Few commentators on the problem of consciousness would object in principle to ideas about complexity, feedback or the involvement of many neurons, but true explanatory power will require facing the issue of subjectivity, and not simply declaring it as an unexplained consequence of large numbers.

The apparent gap between the straightforward mechanisms of circuits (whether electronic or neuronal) and the subjective feelings (presumably experienced by those circuits) has been labeled "the explanatory gap" by Levine (2001). He maintains that, even if we had the complete circuit diagram for the human brain – all 10 billion neurons and their countless connections, the complete map! – we would not learn from the circuit diagram how it feels to run cognitive processes through it. Of course, the circuit diagram for the human brain remains an utter fantasy, but is an accomplished reality for the microscopic bug, *Caenorhabitis elegans*. Its entire behavioral repertoire – from sensory input to motor output – is known and predictable. But do we have any insight into what it feels like to be this bug? Levine's argument is that there is a conceptual "gap" that is not – and, in principle, cannot be – bridged by a circuitry diagram: Subjective awareness is something more than or something different from logic gates. The "explanatory gap" argument is therefore often stated as the reason why "consciousness studies" are up against an insurmountable barrier. Progress comes to a halt right here! Circuit diagram in hand, we have no idea how to proceed! Many scholars in consciousness studies would of course debate that point, but the core argument that a flowchart depicting a series of logical operations does not equate with the subjective feeling of that process seems strong. There are two different types of psychological phenomena involved here – cognition and consciousness – and neither is an obvious consequence of the other.

6.3.3. Quantum Mechanics: Subjectivity Through the Uncertainty Principle

To most people who have pondered the problem of consciousness, the idea that answers might be found in atomic physics seems an absurdity – a non sequitur, a nonstarter! But, in fact, various, alluring quantum mechanical arguments have been advocated to explain the mysteries of the mind (Eccles, 1994; Green & Triffet, 1997; Hameroff, 1994; Satinover, 2001; Scott, 1995; Stapp, 1993; Walker, 2000). Examination of those arguments shows that they all rely on inherently controversial *interpretations* of Heisenberg's uncertainty principle, so it is essential to understand what that "principle" is – and, indeed, what it is not.

In 1927, Werner Heisenberg noted that there is an absolute limit to the precision with which we can measure a physical phenomenon. The uncertainty can be overcome in measuring one property only at the expense of increasing the uncertainty in experimental precision concerning a second, related property. For this reason, the uncertainty relations are expressed as an inequality of so-called conjugate variables: $h/2\pi < \Delta m \cdot \Delta p$ (where h is Planck's constant – a constant of nature with a known numerical value – and the uncertainty in the mass, Δm, and the uncertainty in the momentum, Δp, of a particle can be measured with a total uncertainty that will always exceed $h/2\pi$).

That formula was a great insight, but there has been continuing controversy concerning its interpretation. From the outset, two schools of thought were apparent, and their differences have *not* been resolved over the ensuing decades. On the one hand, Heisenberg himself saw this uncertainty as indication of a "fuzziness" inherent to physical reality. Since it is experimentally not possible to attain greater precision than $h/2\pi$ in measuring m and p, it makes no sense, according to Heisenberg, to pretend that a particle nonetheless has a mass and a momentum that together are more precise than $h/2\pi$. The other school of thought, however, maintained precisely the opposite. The uncertainty relations are accurate indication of the limits of experimental precision when measuring physical systems, but they do not tell us anything about the inherent properties of the physical systems themselves.

The difference between these two interpretations is between a "realist" school, advocated by Einstein (Schrödinger, de Broglie and Planck himself), and a more revolutionary "positivist" view, advocated by Bohr, Heisenberg and Pauli (see Penrose, 1989, for a coherent review). The latter view (the "indeterministic" or Copenhagen interpretation) has in fact captured the popular imagination and is often stated as the "accepted" interpretation in popular science books, but it is noteworthy that Einstein and the two other originators of the quantum revolution, Planck and Schrödinger, never accepted the indeterministic view. (Planck, whose constant is an essential part of the uncertainty argument, is reported to have regretted his unwilling contribution to the indeterminist philosophy advocated by Heisenberg. Schrödinger favored a statistical interpretation of his wave equation, with no implications about individual particles. And Einstein ended up with a reputation for being an old reactionary unable to accept the sexy revolution of an indeterministic reality!) Even among the next generation of atomic physicists who contributed to the consolidation of quantum mechanics, Bohm, Lande, Bell and others argued *against* the dominant Copenhagen School (while the vast majority of practicing physicists are clearly agnostic, while going about the experimental business of microphysics). In a word, the meaning of the uncertainty relations remains an unresolved issue in fundamental theoretical physics (Cushing, 1994).

In consciousness studies, the idea that there might be an absolute limit to precision in subatomic experiments is of no intrinsic interest, but if there is an inherent "wiggle room" in physical reality – a fuzziness in the nature of reality itself – then implications might be drawn about consciousness: Maybe the clockwork mechanisms of a cold and mindless universe have an inherent vagueness into which phenomena from other dimensions might sneak! Although speculative, the strength of this type of suggestion is that it is based on a well-established mathematical relation in atomic physics. Its chief weakness, however, is that the most illustrious figures in twentieth-century physics failed to agree on an interpretation of the uncertainty relations in its application to physical systems (in a brief overview, Wikipedia summarizes 16 "common" interpretations!). It is therefore rather presumptuous of later academics to declare an unambiguous "winner" in order to bolster a particular view of consciousness. Simply stated, no application of the uncertainty principle to the problems of consciousness will be uncontroversial.

As prominent as this unresolved debate was in the first half of the twentieth-century during the quantum revolution, it is important to note that the philosophical argument concerning uncertainty has had no concrete implications concerning the measurable properties of physical systems. There is agreement concerning an absolute limit to measurable precision in microphysics and ballpark estimates of particle sizes and velocities are often made using the uncertainty relations, but the main debate has been about the logical consistency of the two opposing viewpoints. Both sides insist that the implications of the contrary view lead to *logical* absurdities, but neither side has provided evidence of physical phenomena that the other camp cannot explain.

With regard to consciousness studies, the only certainty here is that any hypothesis about subjectivity based on a given interpretation of quantal uncertainty leaves open the possibility of a contrary view. That in itself does not present a problem, but it is not logically correct or historically accurate to maintain that "quantum physics demands" a certain view of determinist/indeterminist philosophy, much less that "modern physics implies" a certain view of the nature of consciousness.

Most participants in the consciousness debate acknowledge the theoretical possibility that a lower-level phenomenon might provide answers to questions at the brain level. In general, that is the established approach of reductionist science, but the mechanism by which quantal indeterminacy (in external reality or in human understanding) might give rise to *subjectivity* remains a difficult challenge for theorists. Let us assume that the uncertainty is in reality itself. How does that fact allow "subjectivity" to enter into physical systems – as distinct from little green men or a picosecond of time reversal? Maybe there is some wiggle room here, but how does it imply the possibility, much less the inevitability of subjective consciousness in an otherwise mechanical universe?

However such questions might be answered, the most outstanding difficulty in developing an "uncertainty principle theory of consciousness" is the apparent reliance of all known forms of consciousness on specifically *neuronal* systems. If subjectivity somehow has its origins at the level of Planck's constant, then all matter consisting of atoms must contain the same seeds of subjectivity through the uncertainty principle. It therefore follows that related manifestations of the subjectivity that we experience and know to be real at the level of human consciousness should arise in other kinds of complex material systems too. That is an interesting possibility perhaps – and a clear invitation to explore quantum mechanical panpsychism, but the quantal theory of consciousness must therefore not only explain the unusual status of animal and/or human consciousness, but also find empirical support for nonneuronal subjectivity in other physical systems. That is a tall order, to say the least.

6.3.4. Robotics: Silicon Subjectivity

The engineering approach to the problem of consciousness begins with the assumption that all manifestations of mind are "algorithmic" (Aleksander, 1996; Taylor, 1999). In other words, whatever the detailed circuit diagram might be, the phenomena described as "cognition," "consciousness," "psyche" or "mind" arise from synaptic signaling among neurons. If this starting assumption is correct, then there is, in principle, no obstacle to recreating those algorithms in artificial (silicon) systems. As complex as the algorithms may be, it is, from an engineering perspective, simply a matter of discovering the computational code by studying neuronal systems, and then implementing silicon versions to give robotic systems identical capabilities – not only identical cognitions but also identical subjective feelings. The engineer's argument is bold and refreshingly direct: "Give us the algorithm and we will give you a conscious robot!" But deep problems remain.

From the point of view of the engineer's algorithmic argument, the belief that "there may be a mysterious internal perspective that is the essence of subjective awareness" is simply mistaken – a premodern myth without foundation. Therefore, the idea that "consciousness is *not* a computation, but rather something unspecifiably different," is, to the engineer, a red herring in need of extirpation from the entire discussion, rather than the central issue in need of explanation (Chalmers, 1996). The philosopher, Daniel Dennett (1991), has been one of the most vociferous advocates of this engineering view. His arguments have the clear virtue of solving an age-old problem (subjectivity) by discarding it, but – the allure of Dennett's prose notwithstanding – most scholars in consciousness studies remain unconvinced that the puzzle of subjectivity has such an easy solution. "Don't think about it!" is not advice that philosopher's are likely to follow. And throwing the baby out while eloquently talking about the bathwater is not progress.

On the contrary, interest in consciousness studies is most frequently *motivated* by the "baby": The intuition that the circuit logic that is our own thinking is somehow (!) qualitatively different from the subjective feeling that accompanies such thinking. Dennett would tell us to be brave enough to chuck our naïve intuitions into the garbage bin, and admit that we really don't know what the difference is between a thought and a feeling simply on the basis of introspection on what is passing through our minds! To be sure, the history of science is littered with discarded ideas about ephemeral phenomena that simply cannot be defined. But, at the same time, just how many of our intuitions can we do without? Most people would probably admit that intuitions, in general, are not final answers, but to dismiss an intuition as entirely groundless – a "funny feeling" that just should not have happened! – is a radical rejection of one's own thought processes.

Without going to the counter-intuitive extreme that Dennett advocates, even many psychologists are convinced that the problems of subjectivity will eventually find a solution in neuronal circuitry: "Subjective feeling may be different from thinking, but, in the end, there really is nothing but neuronal circuitry that might explain them both." Given what is known about the neuronal contribution to *cognition*, the engineer's hypothesis that a slightly different form of circuitry might suffice to explain consciousness is taken to be the "default" position: "There simply *must* be a clever piece of neuronal feedback that produces subjectivity!" The "cognitive theory of consciousness" advocated by Baars (1988) is perhaps the clearest statement of this view and implies an algorithmic (brain circuitry) solution even to the problem of subjective awareness. Clearly, both the strengths and weaknesses of this type of argument lie in the starting premise. We are left with the unanswered question: Is subjective awareness algorithmic?

6.3.5. Neurophysiology

The neurophysiological approach accepts that subjectivity is indeed one of the central problems in the study of the human mind (as emphasized by Chalmers, 1996, in particular) but assumes that the entire issue can be addressed on the basis of realist, materialist philosophy (as defended by Dennett, 1991) and has a specifically biological solution (a crucial argument repeatedly made by Searle, 1997, 1998; Llinas, 2001; Revonsuo, 2005; and others):

> Consciousness is, above all, a biological phenomenon. Conscious processes are biological processes ... Conscious processes are caused by lower-level neuronal processes in the brain. (Searle, 1998, p. 53)

Why this even needs to be stated is a bit of a mystery, but it is of course a matter of common sense among neurophysiologists. Llinas (2001) notes that: "[The subjective property of] qualia must arise from, fundamentally, properties of single cells ... amplified by the organization of circuits.... That which is summed (single cell primitive qualia-like properties) is what must be understood" (p. 212). Later, he notes that "nerve cells must be capable of [producing] 'protoqualia'" (p. 218). Surprisingly, the idea that experimentally measurable properties of the neuron must underlie a scientific theory of consciousness has not often been defended, but it is in essence the "default" view in the framework of reductionist, empirical neuron research. If neurophysiology fails to provide indication of the biological basis for qualia, we may be obliged to look elsewhere – to information theory, "higher-level" neuronal circuitry, quantum uncertainties or dualistic metaphysics – but it is arguably too early in the scientific research effort to declare categorically that neuronal properties cannot explain the nature of subjectivity.

The main strength of arguments based on neuron physiology is their reliance on known cellular and biochemical mechanisms. Moreover, unlike the quantum mechanical attempts at reductionist explanations, the physiological argument does not exchange the conundrum of consciousness for perplexities at other levels. On the contrary, a physiological approach offers the possibility of (i) understanding the connection between cellular mechanisms and whole-brain phenomena, and (ii) experimentally testing that connection using the established methods of brain research. In a word, the idea that there may be a neuron-level explanation of the phenomenon underlying subjectivity is, in principle, not outlandish – but here the devil is certainly in the details. What neuron-level mechanism is responsible, and how does it work?

Note that a cellular view of mind is entirely *orthodox* with regard to the underlying mechanisms of *cognition* (see Section 6.4.1), but an analogous approach to *consciousness* has not generally been considered appropriate. The rationale for rejecting a reductionist, cellular approach lies in the fact that consciousness is a "global," "organism-level" experience (Velmans, 2000) – quite aside from whatever detailed biological mechanisms may be involved. That is to say, subjective consciousness is, by its very nature, the overall feeling of being alive, awake and aware – the feeling of being me! Even though countless metabolic processes of the internal organs make this possible, we remain blissfully unaware of any and all biological subprocesses, and normally focus our entire conscious attention on the relatively macroscopic psychological processes at hand – me in interaction with my world, regardless of what my cells are up to!

In other words, the reason why the neurophysiological approach to consciousness remains a minority view is that it implies that the key insight to an understanding of subjectivity lies at the single-neuron level (Llinas, 2001; Cook, 2002, 2008; MacLennan, 1996; Singer, 1998). It is entirely appropriate to ask: Is that not the wrong level to address the problem? Shouldn't we be asking questions about the "whole-organism" nature of subjective feeling as we typically experience it in real-world situations? If there is any single feature that characterizes the feeling of what it means to be "conscious," it is that consciousness is "unified": me, being here, now (Bayne, 2010). There is not normally a rainbow of different consciousnesses, much less 100 billion! There is only one, and that singularity does not seem to be consistent with the plurality of neuronal processes that are continually at work in the brain. The neurophysiologist's response is that a rigorous understanding of both cognition and consciousness will require both levels of inquiry. Starting with a core argument at the neuronal level, we eventually must proceed from the microscopic to the macroscopic, ultimately to explain the nature of "whole brain" subjective awareness in the familiar language of folk psychology. But the macroscopic theory will necessarily be built from a microscopic mechanism that will necessarily have the neuron at its core.

So, despite such initial skepticism about neuron physiology, the philosophical strength of a neurophysiological approach is that it promises real answers. It explicitly rejects a return to dualist philosophy and all mysterian, rabbit-out-of-the-hat arguments that do not offer causal mechanisms that are open to experimental demonstration. To date, the primary weakness of a neurophysiological theory of consciousness is that, despite general "in principle" arguments such as suggested earlier, there have in fact been few hypotheses that even claim to explain the emergence of consciousness directly from neuron physiology. The following section is an attempt to fill that void.

6.4. THE NEUROPHYSIOLOGICAL SOLUTION

In contrast to the inconclusive conclusions of the four dominant schools of thought outlined earlier, I maintain that plain-vanilla, conventional, laboratory neurophysiology has already provided the basic empirical facts needed to explain both cognition and consciousness, but that, as of today, the mechanisms underlying consciousness are not widely recognized as such. The solution is, in essence, a cellular-level argument with a solid foundation, but one that – for various reasons noted later – has *not* received significant attention in the modern "synaptocentric" view of brain functions.

As will become evident, the neurophysiological view directly addresses the question of why the phenomenon of "subjectivity" exists in a material world, but notably does not have any of the bizarre metaphysical implications that dualism or panpsychism have. In the end, it is nothing more than an interpretation concerning biological mechanisms – and indicates the material processes that are responsible for *sentience* at the cellular level. That reductionist understanding of neuron functions does not itself influence the experience of subjective awareness and will not give us glimpses into a "deeper" reality, but it may allow the field of consciousness studies to move beyond the current stalemate over subjectivity to address the many outstanding, unresolved issues of the mind (specifically, the many problems concerning the brain circuitry involved in conscious vs. unconscious, attended vs. unattended, etc., processing).

The neurophysiological solution to the problem of *subjective awareness* is readily understood if we first consider the related, but much easier topic of *cognition* in neural systems. By distinguishing between (i) the elementary facts about neuronal information processing (known for a certainty and universally accepted in neuroscience) and (ii) inferences about whole-brain cognition (not known for a certainty, but widely considered to be a plausible, working hypothesis), it is straightforward to show how a related two-level approach to the problems of consciousness provides insights into subjectivity. Therefore, I will begin with a review of the current, relatively uncontroversial understanding of cognition (in animal brains) and later turn to the seemingly more difficult problem of what subjective awareness could be in light of the known physiology of the neuron.

6.4.1. Cognition

The many topics in human *cognition* are often classified among the "easy" problems of psychology. Chapters 1–5 have already indicated that there is nothing trivial about cognition, but the problems of cognition are in principle "easy" – or, at least, soluble – because we have a solid understanding of how information processing can take place via synaptic communications among neurons. From a professional neurophysiologist's perspective, the layman's understanding of neuronal communications may be rather simplistic, but I would argue that it is not substantially in error. That understanding is essentially as follows:

> Multiple excitatory and inhibitory inputs to any given neuron are received principally on the neuron's dendritic spines and are summated at the neuron cell body. When a threshold of activation has been exceeded, the neuron initiates a so-called action potential that travels away from the cell body down the axon, resulting in the release of neurotransmitters at the synapses of the axonal terminals. Depending on complex details concerning cell types and the molecular structure of neurotransmitters and receptors, the signal propagated across the synapse will have

either excitatory or inhibitory effects on the receiving (post-synaptic) membrane. In essence, physically-, biologically- and informationally-discrete neuronal units participate in communications throughout the brain, ultimately to drive behavior by synaptic activation of the somatic musculature.

The physiologist would want to add much meat to those bones, but it is likely that this skeletal description would remain intact in a more rigorous version of the neuronal mechanisms of cognition (Figure 6.1).

The point of this undergraduate-level synopsis of cognition is simply to illustrate the fact that the hardcore physiological science of neuronal communications remains quite far from the "whole-brain" psychological phenomena that we experience as cognition. When we think about the conclusions that might be drawn from a specific configuration of sensory stimuli, we normally have little understanding of the step-by-step, neuron-by-neuron deductive processes that unfold in our heads. Those processes might be explicable in terms of small-scale neuronal circuits, but "thinking" is quite something else: It is a "whole-brain" phenomenon. And yet, in spite of that yawning gap between what we *know* about neurons and what we *experience* in the act of thinking, there is little skepticism about whether or not "real" cognition is a complex summation of synapse-level information processing. Of course it is! The scientific establishment is, for good reason, fully convinced that there is no alternative model that can be used to account for any aspect of human or animal cognition other than neuronal information processing. Given what is known about the on/off and excitatory/inhibitory characteristics of neurons *and* the mechanics of logical inference, there is simply no doubt that, in principle, all forms of information processing (i.e., *cognition*) can be performed by neuronal circuitry. This is the essence of Crick's (1994) "astonishing hypothesis" – an idea that was perhaps once "astonishing" to the man in the street, but is today entirely orthodox, cognitive neuroscience. Today, we are all "synaptocentrists!" Given this strong neuronal foundation, no one in the modern era has argued that there is a conceptual crisis in cognitive psychology simply because we do not yet have any examples of the detailed circuitry of acts of human cognition and do not know the mechanisms of neuronal coordination that allow one set of neurons to drive behavior, whereas other neurons fire without behavioral consequences. On the contrary, those puzzles are seen as part of the technical challenge that will eventually succumb to advances in recording techniques, brain-mapping, etc. As discussed later, the leap from synaptic "information processing" to whole-brain cognition

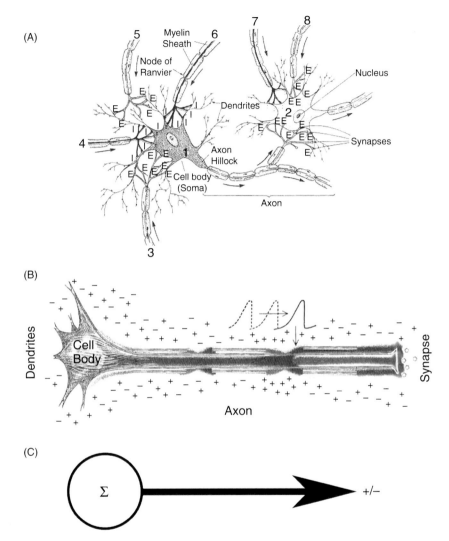

FIGURE 6.1. Progressively simplified cartoons of the neuron. (A) is a semirealistic depiction of the interactions among eight neurons. The direction of information flow is indicated by the arrows. (B) shows one idealized neuron. Dendritic input is summated at the cell body, where the action potential is initiated. The impulse is propagated along the axon to the terminal synapse, where it triggers the release of neurotransmitter molecules. (C) shows the unit structure in artificial neural networks: a cell body where input stimuli are summated and an output axon having excitatory (+) or inhibitory (−) effects on downstream neurons.

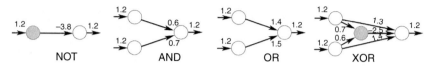

FIGURE 6.2. The core logic functions underlying all forms of cognition (Penrose, 1989). Assuming that all neurons have activation thresholds of 1.0 and each synapse has some degree of excitatory or inhibitory effect on its postsynaptic membrane, these kinds of circuits can realize all of the functions necessary for any definable logical operation. Real cognition may entail countless redundancies and structural complications, but the underlying logical mechanisms are likely to be this simple.

remains to be explained in a fully rigorous, scientific way, but it is not a conceptual mystery. Although still debated, the most likely mechanism involves the temporal coordination of neuronal firing: synchronization (e.g., Crick & Koch, 2003; Singer, 1998).

So, despite the fact that many questions about *whole-brain* cognition remain unanswered, it is known that neurons can achieve the necessary and sufficient operations for logical deduction. In fact, artificial systems with simulated neurons can be constructed to perform *any cognitive task that can be adequately defined*. As a consequence, neuroscientists are confident that *cognition* occurs predominantly via the *synaptic* interactions of neurons. Glial cells, liver cells, cerebral blood flow, electromagnetic waves and other biological, chemical and physical phenomena are certainly involved, but they have their effects on cognition *by exerting influences on the synaptic activity of neurons* (LeDoux, 2002).

The essential point that requires emphasis, however, is that we have a detailed scientific understanding of *not* the staggeringly complex problem of "whole-brain cognition," but rather of the basic information-processing units of cognition rooted in the cellular-level science of neurophysiology. That is to say, we are quite certain about what might be called the "protophenomenon" of cognition: synaptic activity in small neuronal circuits that realize basic logical functions, such as NOT, AND, OR, XOR and so on (Figure 6.2).

Such circuits can be implemented with as few as 2~4 simulated neurons, or in much larger neural networks whose effects are summed probabilistically to achieve the same computational result. It is generally presumed that millions of these small circuits run in parallel and act together to have effects that we refer to as "cognition," but it is important to point out that empirical demonstration of the actual circuitry responsible for "real cognition" – from sensory input to motor output – has been achieved in only the simplest of invertebrate systems. Specifically, the research of

Brenner (1974), Kandel and colleagues (1967) and Walhout et al. (1998) has shown that the behavior of *C. elegans* can be fully explained on the basis of the activity of its 302 neurons and approximately 7500 synapses. If, moment by moment, we know the sensory input to this small organism, we can predict its behavioral output. It is that simple! In larger brains, everything becomes more complex. Not only is the sensory input already an intractably huge problem, but the internal state of the nervous system itself will make mechanistic predictions of output impossible. Moreover, in complex nervous systems, the mechanisms by which neuronal cooperation occurs (fuzzy logic, stochastic summation, synchronization, and/or competition among rival subnetworks?) are unknown, so that many questions remain unanswered. Nonetheless, the simplicity of neuronal logic gates means that there is no doubt in the scientific world that the basic unit of cognition is due to neuronal synaptic activity. No astonishment there.

6.4.2. Consciousness

What about the hard problem of subjective awareness – the qualitative "feelingness" or "qualia" of experience? Clearly, we are no closer to a solution of the *whole-brain* phenomenon of subjective feeling than we are to a solution of *whole-brain* cognition, but again basic neuron physiology provides some indication of what the underlying "protophenomenon" must be. So, before we consider the relative merits of hypotheses about brain-level circuitry and the large-scale coordination of neuronal networks, we should clarify what kinds of processes at the microscopic level might underlie the macroscopic phenomena.

6.4.2.1. *The Sentience of the Living Cell*

Let us first take one step backward and consider the basic characteristics of all living cells. We know that the cell is a collection of organic and inorganic molecules, contained within the protective covering of a plasma membrane (Figure 6.3). Without exception, the cell is the unit structure for all living organisms, and, without exception, each cell is hermetically "sealed" by an enveloping membrane in order to keep the potentially dangerous biochemical effects of the external environment away from the biochemical processes needed to maintain life.

As a rule, a cell's contact with the external world is made via slow, transmembrane import–export mechanisms that are under tight biological control. Needed nutrients are imported molecule by molecule, and biochemical detritus is discharged in bulk through vacuole discharge, while the

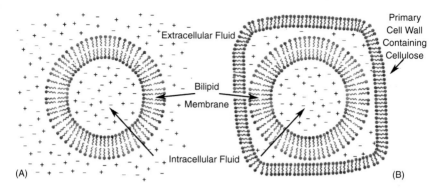

FIGURE 6.3. (A) The basic structure of the animal cell. The solutions both inside and outside of the cell are predominantly water, but differences in ion content produce a charge gradient with the intracellular solution being slightly alkaline. All living cells must maintain that gradient such that the delicate biochemistry of the cell's metabolic processes can continue. (B) The basic structure of the plant cell. The primary cell wall in plant organisms contains cellulose, microfibrils and hemicellulose that provide added protection, but also reduces plant cell sensitivity to local chemical changes.

coherency of the cell membrane is maintained, thereby keeping the intracellular environment at so-called "physiological conditions." For the cells of mammalian organisms, physiological conditions mean a lukewarm temperature and, importantly, a rather neutral (~7.6) biochemical pH. Notable is the fact that an excess of positive or negative charge (i.e., a strongly acidic or alkaline solutions) will lead to the breakdown of the plasma membrane, destruction of functional proteins, eventual dissolution of the cellular "self" and ultimately cell death. For this reason, the *primary* function of the cell is to maintain a barrier between the intracellular "self" and the extracellular environment, between the internal world of controlled biology and the external world of haphazard physicochemical reactions. Homeostatic maintenance of a biochemically viable intracellular state is the essence of cellular life. Given a closed membrane separating the intracellular milieu from the random fluctuations of the external world, the cell can actively adjust the internal concentration of ionic charges to a state that allows for the continuing processes of life. In this respect, the neuron is a typical cell with the need to maintain the "self–other" distinction for its own survival.

6.4.2.2. *Membrane Permeability in Single-Cell Organisms*
A membrane that is closed to the free flow of most molecules is an essential first line of defense for all living cells. Given the known fragility of

macromolecules to acidic or alkaline solutions, the barrier provided by a membrane allows the cell's genetic material to function without external interference. The cell does, however, need to make contact with its external world both to maintain the pH of physiological conditions and to obtain nutrients. "Active transport" mechanisms have thus evolved to carry ions and needed molecules across the cell membrane. Characteristic of active transport is its relative slowness (on the cellular timescale). Individual molecules are first bound to receptors on the external surface of the cell membrane the receptor is then pulled through the membrane, and the bound molecule is finally released. From the perspective of the individual cell, this process of slow and selective import of external materials is the prudent course and allows for biological control of the intracellular state. There are circumstances, however, in which a rapid response to external threats is required. One such response is well documented for the unicellular organism, the paramecium.

As illustrated in Figure 6.4, when a mechanical probe is used to stimulate the paramecium, it responds with movement of its cilia to swim *away from* the potentially threatening stimulus. That which is relevant to the issue of consciousness is how the cell coordinates the cilia to swim *backward* in response to a poke on the "head" region and *forward* in response to a "tail" poke. The key is the cell membrane. Upon stimulation, the membrane in the anterior region opens channels that allow for the influx of sodium ions, leading to the rapid flooding of the cell with an excess of *positive* charge. A similar poke to the posterior region opens chlorine ion channels, leading to an excess of *negative* charge. The mean concentration of intracellular charge in turn affects the orientation of the cilia, so that the sudden inflow of either positive or negative ions will determine the direction of cell movement. The paramecium is thus in a continual state of escape from mechanical stimulation – a survival behavior that produces appropriate directional motility away from "aggressors" using a uniform cilia structure with two functional states. In brief, what the paramecium nicely illustrates is the fact that the electrostatic state within a cell can drive its behavior.

6.4.2.3. *Membrane Permeability in Neurons*

In the case of multicellular animal organisms, behavior is not determined by any individual cell, but *every neuron* continuously responds to the charge gradient of its intra- and extracellular worlds. Unlike the free motility of the paramecium, the anatomically immobilized neuron responds not with "escape" or "approach" behavior, but rather with *two* (causally related, but mechanistically distinct) actions: The first is the *action potential* – induced

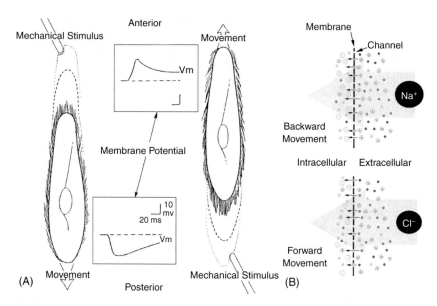

FIGURE 6.4. The paramecium will swim either forward or backward depending on the state of intracellular charge (redrawn after Ogura & Machemer, 1980, with permission). (A) An excess of positive ions orients the cilia toward the anterior region, propelling the cell backwards, and vice versa for negative ions. (B) The appropriateness of the cell's response to external stimulation is due to the membrane's selective openness to the influx of positive ions (above) or negative ions (below).

when the difference in the charge concentration between the internal and external ion solutions exceeds a certain threshold (Figure 6.5A). The second is *neurotransmitter release* – which determines the excitatory/inhibitory polarity of the response. The release of neurotransmitters at the synaptic terminal is a specialized form of the secretion of biochemical signals typical of many other cell types, whereas the action potential that initiates neurotransmitter release is more akin to paramecium behavior, insofar as it involves changes in the intra- and extracellular charge gradient.

What is most remarkable about the action potential is that impulse transmission does *not* entail the flow of (large amounts of) materials in the direction of the synapse. Unlike the signals in electrical engineering, where electrons flow unidirectionally down a metal wire, signal transmission in the neuron entails a momentary opening of the axonal membrane to the flow of ions between the intra- and extracellular fluids – a material exchange that is *orthogonal* to the direction of synaptic transmission (Figure 6.5B). As Graham Cairns-Smith (1996, p. 108) notes: "The action potential is really a pulse of relaxation (not a letting out of air of course, but

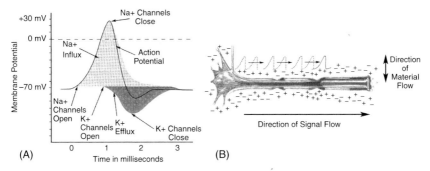

FIGURE 6.5. The action potential. (A) The rapid changes in the membrane potential are a consequence of the opening of ion channels. (B) Signal transmission is the essential "business" of the neuron, but what flows down the axon is not material substances, but rather a sudden process of adjustment of the membrane potential. When this electrostatic "jolt" hits the presynaptic region, neurotransmitters are released. Despite the potentially lengthy propagation of the signal, material flow of ions is *orthogonal* to the direction of signal transmission.

a letting in of sodium ions). Here, contrary to our intuitions perhaps, no energy has to pass along the axon with the signal. What energy is expended flows at right angles to the signal propagation, and recharging is for the benefit of the next signal."

Action potentials thus allow a *cell* to send a signal over distances that are *organismic* in scale (centimeters or even meters), on a timescale that is essentially cellular (milliseconds). Without the action potential (and, instead, reliant on the intracellular flow of molecular signals such as hormones), animal species would respond to external stimuli like plants – slowly bending toward or away from nutrients and predators on a timescale of minutes, hours or days.

On the contrary, animal nervous systems have evolved a communication system that relies on a slightly counterintuitive mechanism of "immaterial" signal transduction. In effect, the action potential is an efficient solution to the problem of long-distance impulse transmission. If neuronal signaling required the directed flow of a large volume of materials from the cell body down the axon, it would rapidly lead to an overwhelming accumulation of molecules at the presynaptic terminals. Meanwhile, replenishment at the neuronal cell body would require slow and energetically expensive transport mechanisms that would stifle neuronal activity. Nature's solution – a solution that allows the neuron to fire rapidly and repeatedly in response to external stimuli – is to replace the *global* process of directed intracellular transport (dendrite-to-soma-to-axon-to-synapse) with many *local* ion exchanges across the cell membrane. By allowing transmembrane ion fluxes that, in turn, trigger similar fluxes at neighboring patches of membrane, the neuron can achieve rapid transmission of the impulse without significant synapse-directed flow of molecular materials. There is, in fact, some transport of neurotransmitter precursors from the cell body to the synapse, but – together with the recycling of released neurotransmitters – there is no need to produce a new batch of transmitters for every neuronal response to stimulation.

Because the bulk of molecular movement during the action potential is the influx and efflux of specific ion types through dedicated channels, material exchange is essentially local and short-range. This leads to a transient intracellular accumulation of *ions* at countless locations along the neuronal membrane (Figure 6.5A), but these "imbalances" can be repaired rapidly and *locally* by means of biological transport mechanisms acting over the entire surface of the axon. In effect, the neuron has solved the problem of rapid signal transmission by allowing a controlled flirtation with the dangers of external electrostatic charge (i.e., moments of permeability of the cellular membrane that hasten the synaptic response). Ultimately, what is achieved through the action potential is long-distance impulse transmission (up to 2 meters) employing primarily short-distance (transmembrane) movement of molecular material (over distances of 10^{-5} meters).

The mechanisms underlying impulse transmission via the action potential are well understood at the single-cell level, but the relevance of the action potential for the phenomenon of subjectivity has been entirely overlooked by a research community focused on "whole-brain" consciousness. When the role of neurons is discussed, it is inevitably in relation to *synaptic information processing*. The action potential is seen merely as a mechanism that supports impulse transmission, but a subcellular process of no intrinsic interest. To be sure, for the issues of *cognition*, this "synaptocentric bias" is fully justified and the action potential is correctly viewed as a slightly complicated mechanism that facilitates the main business of rapid neuronal communications. However, for the issues of *sentience* and the reactivity of the neuron to its external world, the rapid flow of charge across the cell membrane should be the main focus of attention. Specifically, during the action potential, the neuron undergoes the biologically *unusual* process of opening up the normally closed cell membrane to, initially, the influx of positive charge. Such membrane transparency is a "violation of the first law" of cellular life: maintenance of a barrier between the internal and external worlds. But, by allowing for a brief thermodynamic free-for-all during which large numbers of charged ions flow into and out of the cell, the neuron

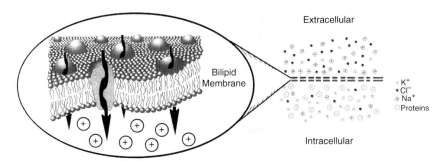

FIGURE 6.6. A section of neuronal membrane showing multiple channels open to the flow of sodium ions. During the momentary permeability of the membrane, ~100,000 ions *per channel* enter the cell, thereby transforming the cellular interior from a negative to a positive charge state. The transmembrane concentration of sodium ions changes from a ratio of 50:1 to parity in the course of a few milliseconds. It is this electrostatic shock that underlies the unusual sentience of animal nerve cells.

participates in behavior that can be important for survival of the organism as a whole. For each firing neuron, hundreds of thousands of sodium channels imbedded in the axonal membrane near the cell body are thrown open, and 800~1200 million ions per second per channel (Levitan & Kaczmarek, 1997, p. 69) rush into the cell under the influence of the potential gradient (Figure 6.6). The significance of these large numbers lies specifically in the fact that the ion flux momentarily transforms the cellular interior from a negative pH to a positive pH (Figure 6.5A) – and that change ultimately leads to neurotransmitter release. Since chemical reactions of all kinds are driven primarily by the interactions among electrical charges, the negative-to-positive switch in internal charge over the course of 2–3 milliseconds is a truly massive overhaul of the intracellular state.

Although the (ion-specific) channels of the neuronal membrane are open during the action potential, there is a moment of free exchange of ions driven by the net electrostatic field surrounding the neuron (McFadden, 2002). It is important to note in this context that the potential gradient itself is not ion-specific. Although the intracellular state can be described in terms of pH (i.e., technically, the relative abundance of positive hydrogen ions), the electrostatic field is a consequence of the net effect of all positive and negative charges – whatever their ionic or molecular origins. When the ion channel "windows" are wide open, the neuron experiences a few milliseconds of thermodynamic *physicochemistry* during which the electrostatic field pulls ions into and then out of the neuron. The membrane then returns to its "sealed state" – preventing the dominance of *chemical* mechanisms and allowing for the reestablishment of the slow and deliberate *biological* processes of adjusting the resting potential through active transport mechanisms. Meanwhile, the action potential proceeds down the axon as neighboring patches of membrane open up – eventually to trigger the release of neurotransmitters at the synapse.

The significance of the action potential is that, in order for the neuron to realize its *all-important cognitive* functions of neuron-to-neuron signaling (essential for controlling the behavior of the animal organism), it experiences not only (1) neurotransmitter release-and-uptake but also (2) the repeated shock of having the neuronal interior switched from the "physiological condition" of mild alkalinity to one of mild acidity. *It is arguably this shock that underlies the unusual sentience of animal organisms.* In the process of cognition, we – animal organisms with brains containing neurons – are continually reminded at the cellular level of the dangers of letting down our guard to the inflow of external molecules! By not having such hyper-excitable neurons in their midst, plant organisms placidly (obstinately, inflexibly, hyper-conservatively!) maintain their own physiological conditions without exposing themselves to the dangers of interactions with the external world. The price they pay is in a lack of sensitivity to that world, an incapacity to cognize about it and an inability for rapid response. What plants gain from their refusal to interact vigorously with the external world through excitable membranes is extreme longevity, but animal organisms have chosen an alternative lifestyle that emphasizes rapid-fire interaction.

Although synaptic activity is what makes *cognition* possible, it is the action potential that is the means by which the neuron directly senses the electrostatic state of its external world – essentially, opening the cellular membrane and letting a small part of the external charged-ion solution rush in. *Brief descent into the world of chemistry makes the biological neuron "sentient" of the current state of its environment in a way that most other living cells – with only their carefully controlled transport mechanisms for making external contact – are not.*

The main point of this chapter is that a proper understanding of the physical chemistry of the action potential suggests that *subjective feeling* has its *cellular* origins in the permeability of the neuronal membrane. Although there is a related membrane mechanism involved in the contraction of animal muscle cells and a weak action potential in certain plant cells (e.g., the Venus flytrap), the action potential in animal neurons is a strong and repeated "electrostatic shock" delivered to *all* neurons throughout the brain during their normal activity. It hardly needs to be stated, but the occurrence of action potentials in primarily *neurons* indicates that the entire range of sentience/qualia/awareness phenomena is necessarily limited to animal organisms that have an abundance of excitable nerve cells. While plant cells enjoy all of the other metabolic properties of cellular life, their cell membranes lack the capacity for "excitability."

Sentience, I maintain, is a consequence of living neurons experiencing the sudden onslaught of electrostatic charge into their interiors. Although Merlin Donald (2001, p. 178) has made similar comments, he does not draw the obvious conclusion: "Asking why these strange ionic ripples can make us aware of everything but themselves is pushing the question too far. There are limits to science." On the contrary, I would push one step further. The lone neuron "feels" something because the ionic dance, from the perspective of the living cell, occurs right at the brink of the living/nonliving worlds: allowing external charge to rush in – in moments of uncontrolled thermodynamics – is as close to cellular death as any cell will experience, before it closes its membrane and spends energy on the reestablishment of physiological conditions.

It is perhaps "astonishing" that sentience would have essentially a cellular explanation, but it is astonishing in exactly the same way that an explanation of cognition on the basis of synaptic activity was once considered to be strikingly counterintuitive (the "wrong level" to address questions about human thought processes) but is today the widely acknowledged scientific orthodoxy (Crick, 1994). Undoubtedly, some philosophers will be reluctant to relinquish center-stage in discussions of consciousness – just as the discovery of DNA moved them to the wings in discussions of the meaning of biological life. But, similar to the story of the progressive understanding of cognition, the relevant mechanisms underlying subjective feeling appear to be cellular. By taking the discussion of consciousness down to the level of the sentience of cellular systems, it is apparent that a mechanism that is already well-understood (but virtually never raised in relation to consciousness studies) might be generalized to the whole-brain level. The next question is therefore how to connect the *cellular* mechanism of sentience to the *organism-level* phenomena of subjective awareness.

6.4.3. Synchronization

The momentary transparency of the neuronal membrane during the action potential allows for a direct, physical "impression" of the extracellular world on the neuronal interior – an electrostatic shock that most other cells in multicellular organisms never experience. The sensing of the neuron's local environs by means of membrane permeability is arguably a type of "sentience" in the sense of being a nearly instantaneous "sampling" of a small part of the external world that dramatically changes the charge characteristics of the cell. That having been said, it is also true that the action potential is only a local *cellular* effect. It remains to be shown how the individual "sentiences" of isolated neurons might work together to produce whole-brain-level, psychological phenomena. Here, the strongest candidate for the coordination of action potentials is again temporal synchronization. As discussed earlier in the case of *cognition*, two distinct levels of biological mechanism were required to explain, respectively, the microscopic protophenomenon (neuronal information processing) and the macroscopic phenomenon (whole-brain "thinking," as we experience it as animal organisms). A similar duality may apply to the problems of (1) neuron-level "sentience" and (2) brain-level "subjective awareness."

As mentioned briefly in Section 6.4.1, temporal synchronization of neuronal firing is the leading candidate for explaining the *cognitive* binding of perceptual features and the unity of mind (Crick & Koch, 2003; Engel et al., 1999; Marlsburg, 1997; Marlsburg et al., 2010; McFadden, 2002; Singer, 1998). While controversy continues about the detailed mechanism (species differences and the relevant synchronization frequencies), the basic idea is that both cognitive and behavioral unity might be achieved by the temporal coordination of otherwise independent and spatially distinct neural networks (Figure 6.7). Experimentally, what has been found is that neurons widely separated in the brain fire synchronously when they are processing related information. The classic example is the synchronous firing of neurons in color-processing cortical regions and shape-processing cortical regions in response to the visual stimulus of an apple. We never experience disembodied "redness" or a colorless "sphere" when looking at an apple. We see a "red sphere," despite the fact that the color and shape processing occur at different locations from one another in the brain. This experiential "binding" of related properties needs to be explained on a neuronal basis. The fact that initially unsynchronized neurons that individually process color or shape spontaneously synchronize in response to, for example, a red apple, is therefore taken to be evidence that the coordination of different brain regions occurs in the time domain. Neurons that fire together over significant time intervals, say, several tens of milliseconds, are subjectively experienced as "belonging to" the same phenomenon.

Similar to the two levels of mechanism needed to explain cognition (synaptic transmission at the cellular level and synchronization at the brain level), the issues of consciousness may also require an understanding of

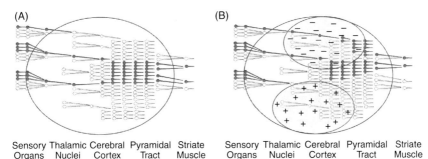

FIGURE 6.7. The temporal synchronization of neurons. (A) Among many firing neurons, the activity of a subset that fires in synchrony (shaded) rises above the level of noise and drives motor behavior. (B) As a consequence of changes in the ion concentrations in cerebrospinal fluid (large oval), regions of relative alkaline or acidic extracellular fluid (shaded ovals) increase or decrease neuronal firing, leading to a different pattern of synchronization and different behavior.

both the cellular mechanism of membrane transparency (mainly during the action potential) and a large-scale, multineuron coordinating mechanism. If indeed unification of *cognitive* activity is achieved by means of synchronization, it would be parsimonious to think that a similar unification of *subjective feeling* could also occur in the temporal domain through synchronized action potentials.

This view of subjective awareness is substantially reiteration of the conclusions already drawn by Crick, Engel, Gray, McFadden, Singer and others, who have argued that "synchronization is subjectivity" (Engel & Singer, 2001). What has been missing from their hypotheses, however, is any indication of how the synchronized summation of many neurons firing together would produce specifically *subjective feeling* (as distinct from cognitive binding). Why would the temporal binding of color- and shape-processing neurons lead to the *subjective* impressions of redness and roundness? Their cognitive togetherness may well be a consequence of their temporal synchronization, but the old problem of why such information processing should also be accompanied by subjective feelings remains unanswered. Unanswered, that is, unless we consider the core "psychology" of the action potential itself.

Clearly, if it makes sense to think of ion-influx as the essence of cellular "sentience," then the temporal coordination of "sentiences" could lead to a higher-level, whole-brain phenomenon. The coordination of "unrelated" cells produces an organismic effect in the same way that the individual muscle fiber cells can influence organism-level behavior only through synchronization with many other muscle cells. The difference between muscle and nerve cells is that the "behavior" of neurons results only in small biochemical changes within the skull. The biochemical effects will be behaviorally negligible until they result in changes in the biochemistry of motor cortex and the promulgation of motor commands that move the striate musculature of the body.

In this view, the key to understanding the importance of synchronization for the subjectivity problem is to understand the action potential as a mechanism involving the flow of electrical charge into the neuron from its environment (and not solely as a mechanism of synaptic activation). It is then conceivable that simultaneous action potentials imply that diverse neurons contribute their individual "proto-qualia" (Llinas, 2001, p. 218) to the organism at the same time to produce a brain-level phenomenon of "macro-qualia." Each and every neuron contributes its own miniscule proto-qualia at the time of firing, but the chance simultaneity of its own activity with that of other neurons will not persist unless there is a third, external driving force. Sensory stimuli provide such a force – driving, for example, redness- and roundness-processing neurons simultaneously for as long as an apple is in view. Synchronized neuronal activity achieves not only the binding of cognitive features by means of the temporal coordination of *synapses*, but it also produces a unity of subjective consciousness, as those same synchronized neurons feel their electrostatic environments in unison.

To reiterate, a scientific understanding of the unusual character of the human mind can be obtained provided that we take *both* of the well-established, fundamental properties of the neuron into consideration. From the perspective of the mechanisms of *cognition*, the action potential is simply a means of long-distance transmission of the neuronal signal from the cell body to the axon terminal, but what is truly unusual about the action potential is that it involves the momentary disintegration of the boundary between the cellular "self" and the extracellular cerebrospinal fluid. This mechanism violates the first rule of cellular existence – the necessity of maintaining cellular integrity in the face of a potentially life-threatening, biochemically active environment. Nevertheless, by allowing the momentary free flow of ionic charges across the cell membrane during the action potential, the neuron can transmit the neuronal impulse rapidly over large distances. In doing so, however, it necessarily experiences a brief rush of positive charges coming into the cell. This is the neuron-level "protophenomenon" of awareness of the external (extracellular) world – the unit of "feeling" – fully analogous to the protophenomenon of neuron-level "thinking" (synaptic transmission) that the lone neuron experiences as a part of a huge network capable of cognition.

6.4.4. Microscopic Protophenomena and Macroscopic Phenomena

When discussing the role of the neuron in brain functions, it is relevant to ask: What precisely does the individual neuron contribute? What can any single neuron on its own "think?" The answer is truly "Not much!" Whatever the actual numbers of neurons involved in sensory, cognitive or motor activities, any given neuron acts solely as a signal-transducing mechanism that connects the activities of a small handful of neurons, and has very little effect on overall brain activity or motor behavior. Similarly with regard to subjective awareness, we must ask: What does each neuron "feel?" Again, the answer is "Not much!" Clearly, it is in direct contact with only a tiny volume of extracellular fluid, so its contribution to the organism's overall subjective state must be extremely small. Nevertheless, the fixed location of any given neuron in the central nervous system means that it is in touch with a very specific portion of the intracranial, but *extraneuronal* "world" – perhaps imbedded in brain regions involved in the processing of visual information, in higher-level cognition or in the motor response of the organism, but always with a definite location somewhere within the overall biochemistry of the living brain. Since that lone neuron's extracellular solution is a small part of the fully connected cerebrospinal fluid that bathes the entire brain, the charge state (pH) of the solution surrounding that neuron will be affected by the "volume transmission" of ions in the extracellular fluid (Agnati et al., 1995). Individually, each neuron will feel only its local state, but it will contribute to the overall feeling of the brain more strongly when the charge gradient is more extreme, because it will experience more action potentials, and therefore more material exchanges with its extracellular world. The anatomical locations of these firing neurons will determine which of many neuronal networks – olfactory, visual, motor and so on – will then contribute more strongly to the current whole-brain subjective feeling of the individual.

While the strength of the subjective feeling of the "external world" made possible by membrane permeability varies depending on the numbers, locations and connectivity of synchronized neurons in the central nervous system, it is important to emphasize that the entire process of neuronal sentience is necessarily neuronal – intracorporeal and predominantly intracranial. Depending on what neurons are active and synchronized, the contents of the individual's subjective feeling might be the feeling of "oneness with the cosmos" or heart-felt empathy with another person or a far more common, mundane sense of seeing the "round redness" of an apple, but the scope of awareness will depend on (1) the subset of neurons that is synchronized and (2) the informational content of the synchronized cognitive processes. Unlike cognition, which can be totally removed from the sensory here and now, whatever the content of the awareness may be, the feeling of awareness is a feeling of "direct contact" with the external world. That characteristic feeling arises simply because neurons individually are indeed in "direct contact" with their extracellular fluids. In this view, the very fact that we have these two classes of words – cognition and consciousness, thought and feeling, heart and mind – simply reflects the duality of neuronal physiological processes. Yes, they are causally connected and influence each other in complex ways, but the neuron itself has two biological modes of interaction with its external "environment."

Fortunately or otherwise, our minds do not make direct contact with any wider "external world" that lies outside the skull other than through the known neuronal mechanisms of sensation, perception and cognition. In other words, our hard-earned knowledge of the external world is a consequence of slow-and-laborious, synapse-mediated information processing, not a consequence of directly "grasping," "intuiting" or supernaturally "being given truths about" the external reality. The fact that we nonetheless sometimes have the subjective feeling of "noncognitive," direct insight into the ways of the world is arguably a consequence of the fact that neurons do indeed open up to their (local) extracellular environs and do indeed let external (ionic) charges come rushing in. They do "grasp" small samples of the extracellular world of electrostatic charge, while they undergo the business of synaptic information processing. And for each and every firing neuron, this sentience of the charged environment is a "feeling" as real as anything a cell might undergo. Such subjective feeling will perhaps be dismissed by some as a meaningless "illusion" or "penumbra" that accompanies cognition and is irrelevant to the well-focused logical processes that are mediated by synapses. But an alternative interpretation is worth considering. The subjective experience of ion exchange can be understood as a second perspective on the neuronal activity going on in our heads – a global feeling that itself lacks explicit informational content or deductive consequences, but is no less real than cognition.

6.4.5. Empirical Tests: Separating Thought and Feeling

Just as the content of cognition depends on temporally coordinated synaptic activity, the content of subjective feeling depends on the coordinated firing of action potentials. As a rule, action potentials and synaptic activity are tightly linked, so that, at any given moment, thought and feeling are focused on the same topic. There exist, however, both normal and pathological states in which the link between action potentials and synaptic activity is slightly weakened. Not only are there neurons that fire without synaptic input (pacemaker-like cells), but changes in the ion content of the cerebrospinal fluid can accelerate or depress neuronal firing, simply as a consequence of changes in the intra-/extracellular charge gradient. The wiring circuitry of course remains unchanged, but the likelihood of neuronal firing changes with changes in the pH of the cerebrospinal fluid. Controlled uncoupling of synaptic activity and action potentials through manipulation of the pH of the cerebrospinal fluid may therefore allow for empirical demonstration of a dissociation of cognition and subjective awareness.

The two areas in psychopharmacology where empirical tests of this hypothesis should be possible concern the mechanisms of the general anesthetics and those of the psychotropic drugs. The first prediction is that a global change in the pH of the cerebrospinal fluid should lead to nonlethal (i.e., not producing direct injury to neurons) alterations of consciousness without direct effects on cognition. It is in fact known that artificial manipulation of the pH of cerebrospinal fluid can induce coma (Kauppinen & Williams, 1998) and, conversely, electrostatic "shock therapy" can bring patients out of a comatose state or out of a pathologically depressed state (Fink, 1999). Both of these "predictions" are of course made in light of known effects, and do not constitute real tests of the hypothesis. Nonetheless, both realities – cerebrospinal pH effects on consciousness and shock therapy – are significant clinical phenomena that are not yet well understood. In both cases, the conventional view is that, somehow, disruption of the ionic balance of the nervous system has transient effects on the associations inherent to neuronal circuitry – with significant clinical side-effects. But, clearly, neuronal associations are fundamentally synaptic phenomena, so it remains uncertain why changes in the ionic state of the cerebrospinal fluid would be relevant. However, if awareness itself relies on charge influx/efflux, then it would be parsimonious to argue that disruptions of charge flow, not neurotransmitter effects, have widespread effects on subjective feeling, and only secondarily on neuronal associations.

A second prediction concerns psychotropic drugs that either have their effects on the permeability of the neuronal membrane or act directly on synapses. Psychedelic drugs that are popularly known to "raise" or alter consciousness (LSD, mescaline, psilocybin, marijuana) are predicted to have their primary effects on membrane permeability and therefore on action potentials (i.e., synaptic effects will be secondary). This is indeed known to

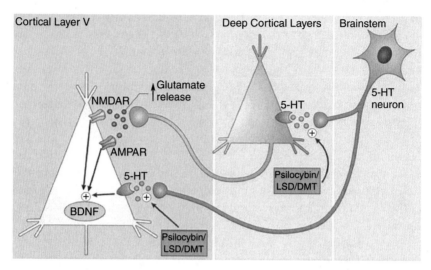

FIGURE 6.8. The action of the psychedelics can be both direct mimicking of neurotransmitter effects, such as 5-HT, or indirectly on the biochemistry of the extracellular space (after Vollenweider & Kometer, 2010, with permission).

be the case for marijuana, since the active component in marijuana, THC, is extremely lipophilic, binds diffusely to the neuronal membrane, but is *not* found in the synaptic cleft (Martin, 1986). Clearly, for any hypothesis concerning the psychological effects of the psychotropic drugs, the absence of those drugs in the synaptic region presents a major problem if effects on neurotransmission are the postulated effect. Either the drugs have their effects through undetected metabolites or other indirect mechanisms, or, as suggested here, they have their effects on the neurophysiology of the action potential itself. The neuronal location of THC suggests an effect on membrane permeability – rather than synaptic activity – and thus an effect on awareness rather than cognition.

The neuronal localization and the biochemical effects of the other psychedelic drugs, however, remain controversial (Snyder, 1996). As illustrated in Figure 6.8, the dual effects of LSD and related drugs have been suggested by Vollenweider and Kometer (2010), who have pointed out two separate neuronal actions in producing (1) hallucinogenic effects and (2) mood-altering effects. LSD, in particular, mimics the actions of the neurotransmitter serotonin in cortical layer V, but it has separate effects on deep cortical layers, ultimately leading to an increase in the excitatory neurotransmitter glutamate in the extracellular space of layer V in prefrontal cortex. Future research may indicate if such effects can be accurately described in terms of distinct cognitive (synaptic) and consciousness (action potential) actions.

Definitive answers are not yet available and many avenues of research remain to be explored, so it is worth mentioning the disastrous effects of the political "war on drugs" on cognitive neuroscience, in general. Insofar as there may be real-world benefits in obtaining a true understanding of the philosopher's favorite topics of cognition and consciousness, it is specifically the drugs that have psychotropic effects that need to be studied. Of course, drugs that have such effects are likely to be abused by some people, but the overpowering interest in mind-altering effects is itself clear indication of its relevance to our understanding of the mind. Legislated prohibition is known for a certainty to be ineffective in preventing use by the public at large, but regrettably it has prevented a broad range of research. The present hypothesis concerning excitable cells, in general, and the neuronal membrane, in particular, suggests that the primary effects of the psychotropic drugs will fall into one of two categories – those affecting initially synaptic activity and those affecting initially the action potential. Answers can be obtained only through experimental research.

6.5. IMPLICATIONS

6.5.1. Qualia

If the logic of the present chapter is correct, the answer to the question why "qualia" exist in a material universe is that living neurons – unlike electrical circuits or other artifacts of engineering science – experience material interactions between the cellular "self" and the extracellular environment. It is a phenomenon of animal, specifically neuronal, *life*. Although the neuronal *information processing* of synapses can be mimicked remarkably well in inorganic systems and thereby reproduce many of the *behaviors* driven by cognition, the action potential entails an exchange of ionic charges between the living cell and its surroundings – an exchange that does *not* occur in nonliving systems. Stated more poetically, the activity of neurons is *not* "cold information-processing," but rather "warm biological give-and-take" – necessarily involving the living processes of biological homeostasis needed to guide animal behavior. Note that this emphasis on organic life is not an argument for vitalism nor a mystical hypothesis about the "organic" properties of carbon versus the "inorganic" properties of silicon. It is, more plainly, recognition that there are principles of material organization at the cellular level that give rise to the very real phenomena that we traditionally label as "life." In the animal kingdom, all of the phenomena of, broadly speaking, "mind" are built upon the phenomena of life and, as far as we can determine, do not exist in nonliving systems. The inevitable conclusion from cell science is therefore that the action potential itself has a rudimentary "psychology" (sentience) that is completely absent in artificial, non-biological systems. The engineer's circuit boards do not recreate and indeed have no need to recreate any kind of self-environment material exchange in undertaking their cognitive functions.

For the limited purposes of information processing, that which occurs in silicon circuits and that which occurs in neuronal circuits may be effectively identical – showing the same pattern of outputs to the same pattern of inputs. But the structural components that undertake such processing are, in the case of the brain, individual living cells that are poised at the interface between life and death – between the self-sustaining metabolism of cellular existence and the dangerous dissolution of cellular integrity. As whole organisms, animals are continually aware of the fine line between life and death and organize their entire lives to maintain life. Moment by moment, maintenance of the physiological conditions conducive to cellular life entails finding enough oxygen in each breath; hour by hour, finding enough water to lubricate our metabolic processes; and day-by-day, securing a sufficient supply of high-energy nutrients to keep the whole system working. But, already at the cellular level, there is the biological reality of maintaining cellular coherency – the protophenomenon of the struggle of biological life that takes its most basic form in the maintenance of a closed cellular membrane – with "self" on the inside and an uncaring universe on the outside. In contrast, inorganic systems have no sense of "self" nor a feeling for the trials and tribulations of biological life. Engineered circuits therefore necessarily lack the subjective experience, the qualia, that are known and felt already at the cellular level by living systems.

For the ambitious engineer wanting to show how a silicon robot can achieve anything a living brain can, it may be worth conceding that someday it might be possible to simulate the homeostasis of each and every simulated neuron in a simulated brain (requiring the equivalent of a modern supercomputer to simulate each neuronal membrane!). The transient loss of homeostasis during each action potential would thereby give the robotic brain an overall simulated "feeling" corresponding to the exercise of different regions of the simulated brain. And, in that respect, a simulated consciousness in an inorganic system would become possible. Far less likely would be a comparable "sentience" in silicon chips that have no material exchange with the local environment lying peripheral to their circuits. Again, the

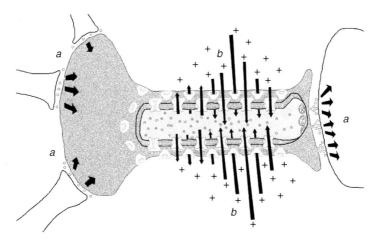

FIGURE 6.9. A cartoon of the two modes of interaction between the neuron and its extracellular world: a, neurotransmitter release-and-uptake (the protophenomenon of cognition), and b ion-flow during the action potential (the protophenomenon of consciousness).

input/output signal transduction may be identical to that of the cognizing neuron, but the unit of information processing in homeostatic neuronal systems and in quantum mechanical atomic systems is so different, that it is difficult to see a parallel between the two realms. Neurons, as living systems, do in fact engage in controlled ion exchanges with their environments, but an analogy with the electromagnetic interactions of electrons moving in relation to the silicon nuclei in a semiconductor seems far-fetched.

Insofar as the action potential and neurotransmitter release are two tightly-linked aspects of neuronal activity, the protophenomenon of cognition will normally occur together with the protophenomenon of consciousness in each and every living neuron (Figure 6.9). Dennett (1991) is therefore correct in maintaining that the philosopher's zombie argument (cognition without consciousness) is a meaningless fiction: neuronal information processing and neuronal sentience happen together, and total suppression of either function suppresses both. In general, a neuron that has been rendered pharmacologically incapable of generating action potentials will not participate in synaptic transmission: similarly, a neuron whose synaptic functions have been chemically suppressed will not contribute to the generation of the action potentials of other neurons. Cognition and consciousness are deeply linked – and "disembodied consciousness" and "feelingless cognition" will remain phenomena found only in the fever swamps of religion and the studios of Hollywood.

Nevertheless, the philosopher's point about zombies seems to apply quite nicely to all nonneuronal (i.e., digital electronic) systems. Any information-processing system that carries out "synapse-like" cognition *without* an exchange of material substances with its environment will necessarily lack the low-level, biological "feeling" of interaction with its external world – a feeling that neuron-based *living* systems clearly do have. It seems likely, therefore, that, as more complex robotic systems capable of evermore human-like cognitive capabilities are constructed, their overt actions – including the simulation of emotional behavior – will of course become more human-like, but they will cognize *without* the qualia inherent to living systems. They are zombies. (Arguments about *other kinds* of sentience based upon *other kinds* of "self/other" interactions [e.g., electromagnetic] in silicon systems might well be developed, but any noncellular form of "sentience" will differ from biological sentience and should therefore go under a different label.)

Although there may be no intrinsic limitations on the information-processing, logicodeductive capabilities of machine algorithms (Aleksander, 1996) and while there may be other types of yet-unknown emergent properties from inorganic systems, *biological sentience is fundamentally not an information-processing algorithm*, but rather a property of living cellular systems engaged in material exchanges with the external world in trying to maintain homeostasis. This cellular account of sentience brings considerable "bad news" for engineers hoping for a world of conscious robots or of plug-in silicon chips to enhance human consciousness. If sentience is due to the fact that each individual neuron senses and reacts to a local electrotonic solution, then inorganic systems lacking the relevant biological exchange mechanisms do not and in principle cannot "feel" anything. They may someday "think" at a level comparable to the best of living brains – and any plug-in chip that enhances human behavior will provide the benefits typical of all external tools and machines, but silicon supplements will not have the neuron's sense of the "danger" of being alive – of taking biological risks in the act of signal propagation. Even if programmed to think about "survival," robots will lack the cellular-level tension of maintaining a balance between the living "self" and the nonliving, external "other."

This biological view of sentience (and therefore subjectivity) is at the same time essentially "good news" for the engineers because disconnecting a robot from its electrical source means only interrupting an ongoing "cognitive" algorithm and does not have implications about biological feeling, much less biological death – which ultimately is the motivation underlying all actions of living systems. In other words, qualia are inherently biological. This is a conclusion that is not born of a failure of imagination or a refusal to transcend our anthropocentric narrow-mindedness and is most definitely not derived from premodern, mystical mumbo-jumbo about the sanctity of life, but it is a direct consequence of the fact that living cellular systems do ionic and molecular things that logic circuits – lacking the burden of self-sustaining homeostasis – do not.

6.5.2. Intuition

Penrose (1989) has argued that mathematical intuition is inherently noncomputable, and, to the mathematician, such intuition is clearly different from run-of-the-mill mathematical cogitation. Although that view is reminiscent of the ineffability of many types of aesthetic insights, in general, the example of mathematical intuition is particularly apt because the "normal" steps involved in mathematical proofs are so close to the basic logic functions of Figure 6.2. Starting with certain definable axioms, it is the process of adding or subtracting corollary arguments that leads to new conclusions. Mathematical intuitions, on the other hand, have a psychological character that is distinct from the deliberate, symbolic manipulations of normal logic – despite the fact that, if successful, the intuition will also be post hoc explicable in a logical, symbolic form.

Although Penrose has gone in search of the noncomputable at the quantal level, the dual functions of the neuron imply that there is already a qualitative difference between "digital" (and inherently computable) synaptic events and the "analog" changes in intra-/extracellular ion concentrations. In principle, of course, ions could be counted and the effects of extracellular charge fluctuations would be fully "computable," but the realm of ionic charges and their electromagnetic fields is many orders of magnitude more fine-grained than synapse-based neuronal computations. At the single-cell level, the synapse is a simple on/off computational mechanism, whereas transmembrane ionic events constitute a continuous gradient of charges with unquantifiable neuron-to-neuron effects due to changes in extracellular ion concentrations indirectly affecting the firing probabilities of neighboring neurons (Figure 6.7). Such effects can in fact be measured experimentally in in vitro preparations (e.g., Agnati et al., 1995), but there is virtually no possibility of a computational theory of cognition that takes gradients of ionic charge as the fundamental unit of information, despite the fact that "the brain's electromagnetic field holds precisely the same information as neuron firing patterns" (McFadden, 2002, p. 31).

I suggest that the noncomputable "feelings" experienced during mathematical intuition – and indeed other subjective feelings, aesthetic insights and gut-level hunches that we all subjectively know to be of a character quite distinct from the plodding, deductive, logical, information-processing operations of "normal" cognition – are consequences of the nonsynaptic (i.e., charge gradient) functions of the neuron (i.e., the effects of "volume transmission" and cognitive leaps made possible by changes in extracellular charge gradients). In summary, intuitions may be unusual and rather unlike normal cognition, but they are explicable on a neuronal basis, provided only that we can think beyond the boundaries of synaptocentric cognition. The idea that even intuitions, hunches and gut feelings have a neurophysiological basis is not as alluringly other-worldly as invoking quantum mechanics or postulating invisible metaphysical causes, but, in general, this is the eye-opening, disillusioning, prosaic way of empirical science. Sorry, Virginia.

6.5.3. The Mystery of Glia

The "synaptocentric bias" of virtually all psychologists interested in brain functions is fully understandable insofar as the psychological topic is cognition. But, as outlined earlier, the topics of sentience, feeling and consciousness require that we consider the neuron's other mode of interaction with its environment – the cross-membrane flow of charged ions. That, in any event, is the main argument of the present chapter. But, if we have reason to include consideration of changes in the ion content of the brain, there is a second cell-type that simply must be brought into the discussion: the glia. As mentioned ever so briefly in most textbooks on neurophysiology, the human brain consists mostly of glial cells – 90 percent by cell number, 50 percent by brain volume. Species with smaller brains have reduced glial content: 80 percent by number in the chimpanzee, 60 percent in the rat, 20 percent in the fruit fly, 16 percent in *C. elegans*, and 3 percent in the lowly leech (Koob, 2009). The approximate correlation with brain-size and/or intelligence is obvious, but just what are the glia doing (Mitterauer & Kopp, 2003).

Glial cells are known to be active "housekeepers" of the cerebrospinal fluid – absorbing neurotransmitters and importing and exporting the ions that play important roles in the action potential. Most textbook discussions leave the glia story at that level, in a hurried return to the topic of the information-processing neurons – about which so much is known! Fair enough, but if indeed the glia somehow "manage" ion concentrations in the cerebrospinal fluid, it follows that they are deeply involved in the "management" of consciousness. This point is indeed the central argument of an extended diatribe by Koob (2009) – an argument against the hegemony of the neuron and all neuron researchers, and a plea for paying more attention to glia. Final answers about glial functions are not yet available (despite some excellent research: Hatton & Parpura, 2004; Verkhratsky & Butt, 2007; Volterra, et al., 2002), but the heavy involvement of glial cells in ion metabolism is clear indication of their role in brain functions. Indeed, the wave of ion-density changes occurring in glia simultaneously with neuronal fluctuations suggest that cellular sentience in the brain is both neuronal and glial. So, without diminishing the importance of neuron functions, particularly in cognition, we can conclude that – pace Koob! – the other major cellular component of the brain is also involved in consciousness. In line with orthodox thinking, it is likely that the glia play a secondary role in supporting neuron functions, simply because the cognitive workhorse, the neuron, drives animal behavior – and it is behavior that determines the evolutionary fate of animal species.

6.6. CONSCIOUSNESS IS UNDERSTOOD, SELF-CONSCIOUSNESS IS NOT

The upshot of the five previous sections is that modern neurophysiology has provided the foundation for a rigorous, materialist, scientific understanding of neuron-level "sentience." That foundation is concerned primarily with the mechanisms underlying the excitable membrane of the neuron and how it allows for the exchange of charged-ions with the external environment through momentarily open ion-channels in the cell membrane. *It is precisely this cellular property that most clearly distinguishes between plant and animal organisms.* Only certain types of animal cells are excitable, the vast majority of which are neurons – and consequently only animal organisms are able to respond rapidly to external stimulation. Of course, if excitability were simply a matter of speed, then there would be few implications regarding consciousness: Animals would be quickly aware, plants slowly aware. But the rapid-fire, "excitable" nerve cell is also involved in the behavioral response to changes in the external world, actively participating in the interaction between itself and its immediate, extracellular environment in a way that "nonexcitable" cells do not experience.

In summary, it can be said that the neuron has two unusual properties. One is its capability for cell-to-cell communication via synapses – and this is known to be the "protophenomenon" underlying the "real phenomenon" of cognition. The neuron's other unusual property involves its excitable membrane. Long overlooked in the consciousness debate and usually dismissed as a minor physiological detail of neuronal functions, the action potential is, if anything, even more unusual than synaptic communication since it is the means by which the neuron maintains a homeostatic state, while allowing material exchanges with its external world. Although rarely mentioned by psychologists, the action potential entails the momentary *permeability* of the neuronal membrane – throwing open the "windows" in the cell membrane and allowing a massive influx of charged ions into the cellular interior in the act of signal propagation. Such openness to the external world is arguably the protophenomenon of cellular-level "sentience" – literally, feeling the charge-state of the electrostatic environment. A general sensitivity to the external pH is of course a common feature of all living cells, but it is greatly amplified and accelerated during the neuron's action potential. Synchronization of the action potentials of the same neurons that are involved in cognitive binding is the likely mechanism by which the sentience of individual neurons is coordinated into the brain-level phenomenon of subjective awareness. The obvious conclusion is that the "feelingness" of being awake, alive and aware – the so-called qualia of modern consciousness studies – has a straightforward solution at the cellular level.

What remains less clear, however, is where specifically *self*-consciousness fits in. Here, it is essential to distinguish between consciousness – which, on the basis of this neuronal definition of sentience, is a property of all animal nervous systems, whereas self-consciousness is a much less common psychological phenomenon that is undoubtedly greatly enhanced through language. Both "consciousness" and "self-consciousness" are frequently and confusingly used to mean awareness of oneself (Jaynes, 1976), but, more precisely, self-consciousness implies awareness of one's own historical path, one's awareness of having a continuous personal history in relation to a specific set of people in a specific cultural setting. In that sense, self-consciousness is clearly more complex than basic animal awareness and may be uniquely human – requiring a symbolic understanding of the continuity of self over time: cognizance of the process of one's own development, maturation, aging and eventual death. Insofar as animals do not have autobiographical memories of where they have been and where they intend to be in the future, they are likely to be self-aware only in the present, but to lack a self-conscious (necessarily language-mediated?) story of where they stand over the course of a lifetime. They are without history, or, as others have noted, "locked into the present." Still, even without a verbal history of self and one's own role in a larger society, it is evident that many animals are conscious, and generally aware of the opportunities and dangers of their current situation. It is that level of conscious awareness that appears to have a rather straightforward physiological explanation, as addressed earlier. Self-consciousness of a personal history remains a largely unsolved problem for future research (see Damasio, 2010).

6.7. CONCLUSIONS

Clearly, if our motivation in addressing the issues of "consciousness studies" were simply to assert the impossibility of a materialist explanation of mind, then a discussion of neuron physiology would be little more than an annoying distraction from our metaphysical preconceptions. And if our underlying desire in talking about consciousness were simply to transplant ideas from the world of magic, the occult or parapsychology into the realm of science, then an understanding of the dual functions of the neuron would be an insignificant addition to current ideas about material reality. And, if – together with a surprisingly large number of consciousness commentators – we are interested solely in proclaiming that quantum mechanics is the only meaningful level of physical reality, then the discussion of details from cell biology would be an irrelevance. But if our goal is to explain why our subjective, mental lives seem to contain more than logicodeductive computations (cognition) and why most commentators in consciousness studies (Table 6.2) have maintained that there is indeed a "hard" problem (i.e., the hard problem of subjective feeling) that cannot be ignored by the science of psychology, then the *dual* nature of neuronal interactions with the extracellullar environment may prove to be a useful insight (Figure 6.10).

Synaptic activity is known to be the *cellular-level* mechanism that makes cognition possible, but that universally acknowledged fact has been demonstrated only in extremely simple, invertebrate nervous systems and simulated in artificial neural nets. In both cases, the cognition is rudimentary, but the causal mechanisms underlying the neuronal control of behavior are clear. In this regard, it can be said that the jump from the "protophenomena" of neuronal communications to the "real phenomena" of whole-brain cognition is, in principle, understood. A similar distance remains between what is known and inferred about consciousness. Although the whole-brain circuitry of consciousness remains to be clarified, neurophysiology has already provided the basic facts concerning the *cellular-level* mechanism that makes neuronal "sentience" possible. I therefore conclude that an understanding of neuronal membrane permeability during the action potential is as important for consciousness studies as is an understanding of the mechanism of

TABLE 6.2. *Modern contributors to the consciousness debate*

Author	Positive contribution	False lead/obscurity/fatal flaw
Nagel Levine	Emphasis on the reality of the feeling of subjective consciousness.	Descent into "dualism," and therefore the impossibility of scientific progress.
Chalmers	Identification of the subjectivity of consciousness as a crucial problem that the science of psychology must explain.	Flirtation with a new brand of "informational dualism."
Jaynes	Emphasis on differential roles of the cerebral hemispheres in normal and abnormal consciousness.	Use of the word "consciousness" to mean "self-consciousness."
MacLennan	Need for a neuron-level explanation of the "proto-phenomena" of mind.	Promissory notes on robotic (silicon) consciousness.
Edelman	Emphasis on neurobiology and commitment to demonstration of cognitive mechanisms in working robotic systems.	Labeling of feedback connectivity as "re-entry connections." Despite denials, main contribution concerns cognition.
Llinas	Emphasis on neuronal foundations of all aspects of consciousness.	Brevity of the discussion the mechanism of amplification of neuron-level "protoqualia" to brain-level qualia.
LeDoux	Emphasis on emotion and feeling. Scientific explanations of cognition alone will not suffice.	Confusions between the issues of subjective feeling and the issues of autonomic system emotional responses.
Searle	Emphasis on the need for a biological solution to the problem of subjectivity.	Failure to pursue the biological argument. Who is this guy, a philosopher?
Pinker Block	Distinguishing between phenomena that can be described as "cognition" versus those that entail "sentience."	Complete neglect of the action potential as one of the unusual characteristics of living neurons.
Crick Koch	"Astonishing hypothesis": cognition is due to synaptic activity. Emphasis on neurobiology.	Implied equivalence between cognition and consciousness.
Dennett	Rejection of dualism. Willingness to dismiss certain kinds of philosophical speculation as pointless.	Failure to acknowledge that qualia and the problem of subjectivity are a central problem in psychology.

Author	Positive contribution	False lead/obscurity/fatal flaw
Aleksander	Emphasis on materialism. The engineer's commitment to building artifacts to demonstrate principles.	Failure to acknowledge the reality of subjective consciousness. Fantasy of robot consciousness.
Engel Singer	Emphasis on the importance of neuronal synchronization for explaining cognitive binding.	Failure to explain how synchronization itself can lead to subjectivity.
Baars	Insistence that consciousness is a tractable problem for experimental psychology.	Explicitly equating cognition and consciousness: "A cognitive theory of consciousness."
Damasio	Emphasis on the importance of a neuron-level "proto-feeling" explanation underlying the issue of subjective feeling.	Difficulties in integrating the issues of "feeling" with the cognition of "self."

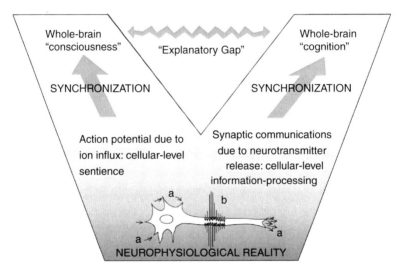

FIGURE 6.10. Synopsis. The perennial enigma of the human mind is the gap between thought and feeling. This idea has been formalized by philosophers as the "explanatory gap" (Levine, 2001) and remains the joy and terror of psychopathology. The gap is unbridgeable (except at the cellular level) but understandable.

synaptic transmission for cognition. For both cognition and consciousness, the crucial neuronal processes are known, and temporal synchronization of many neurons working together is the likely whole-brain mechanism that transforms "protophenomena" into "real phenomena."

Antonio Damasio (2010) has expressed some support for this neurophysiological view of qualia. He notes that the neuron-level argument will not serve as an explanation of the whole-brain "feelings" that we typically describe with words such as "happy" and "sad," but fundamental neuronal functions "suggest the presence of forerunners of a 'feeling' function":

> There are intriguing nuances to this idea. The specialization of neurons relative to other body cells comes, in good part, from the fact that neurons, along with muscle cells, are excitable. Excitability is a property that derives from a cell membrane in which local permeability for charged ions is allowed to travel from region to region over the distance of an axon. ... The temporary but repeated opening up of the cell membrane is a violation of the nearly hermetic seal that protects life in the neuron's interior ... [S]uch vulnerability would be a good candidate for the creation of a moment of protofeeling.... I regard this line of inquiry as worth pursuing, [but emphasize] that these ideas should *not* be confused with the well-known effort of locating the origins of consciousness at the level of ... quantum effects. (Damasio, 2010, p. 197)

The theoretical importance of a neuronal view of subjective awareness is that it averts a possible line of argument that "human beings are special because they have subjective awareness unrelated to the material universe." I maintain that, in order to explain our "specialness," a focus on cognition will suffice – and, within the realm of cognition, a focus on specifically three-body cognition indicates how we differ from other cognizing systems (Chapters 1–5).

7

Other Human Talents

The grounds for thinking that certain of the highest-level cognitive capacities of humankind involve three simultaneous processes have been outlined in Chapters 2 through 5. Tool use, language, visual art and music have identifiable triadic aspects that appear to be essential in elevating those talents to some of the most unusual forms of behavior on Earth – behaviors that other species rarely, if ever, exhibit. Despite denials by certain animal researchers ("Humanity's special place in the cosmos is one of abandoned claims and moving goalposts," DeWaal, 2005, p. 188), our special cleverness is there to be seen by anyone willing to look. But that much is obvious. The main argument of these chapters has been that this cleverness is not something magical, mystical or metaphysical, but rather an explicable form of cognition undertaken by the human brain. The human brain is unusual in identifiable ways.

In fact, various authors have defended related ideas about "triadic perception" and "triadic cognition" in diverse fields over many years. The most important of those insights have already been discussed, but there are still other unusual human talents that appear to have a triadic basis, a few of which are mentioned briefly later.

7.1. RHYTHM PERCEPTION

The perception of harmony (and therefore the major/minor modality of most popular and classical music) is explicitly a three-tone pitch phenomenon, but there is much more to real music than harmonic triads. A proper understanding also requires consideration of the other dimensions of music, most importantly, the temporal aspects known as rhythm.

Interestingly, musical rhythm is also known to *begin* at the level of triads: three-pulse phenomena. Similar to the phenomena of pitch, pulse triads

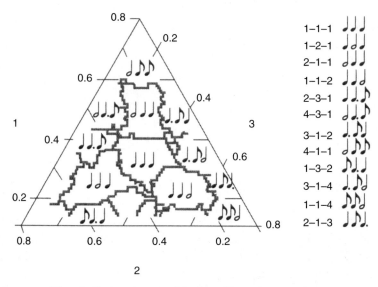

FIGURE 7.1. The triadic description of rhythms (from Desain & Honing, 2003, with permission). The duration of the first three intervals are indicated on the triangular sides labeled 1, 2 and 3. Connoisseurs of rhythm would undoubtedly maintain that the interesting rhythms are those arranged along the edges of the triangle, but most music employs only the four simpler rhythms in the core of the triangle.

only scratch the surface of the complexities of the temporal unfolding of music, but the crucial point is that, in contrast to the fascination of musical rhythms, "beat" (pulse dyads) is as interesting as a toothache. Elephants, apes and parrots can maintain a steady beat, but the addition of but one more time interval catapults the monotony of pulse into the complexities of rhythm. For the professional drummer, the rhythmic complexity of the two different time intervals in a three-pulse pattern is trivial, but it is relevant to note that, already at that level, we are the only species "keeping time."

Leonard Meyer discussed the basics of rhythm decades ago within the framework of Gestalt psychology (Cooper & Meyer, 1960), but the most thorough modern examination of rhythm has been done by Peter Desain and Henkjan Honing (2003). One of their core ideas can be summarized as in Figure 7.1, where the rhythmic intervals used in various time structures is represented as a "rhythm triangle."

Maintaining a steady beat clearly indicates an auditory awareness and the motor entrainment that underlies musical rhythm. But the jump from beat to rhythm is as huge as the jump from hearing the consonance/dissonance of pitch intervals to hearing the modality of harmony. In the present context, it is of interest to note that the vast majority of musical pieces can be described in terms of the time durations separating three beats (two time intervals) (Figure 7.1). The importance of that classificatory scheme has been stated most clearly by David Huron (2006, p. 195): From a collection of 8356 rhythms noted in the Barlow and Morgenstern *Dictionary of Musical Themes* (1948), he found that fully 86.6 percent were the simple meters that lie at the heart of the rhythm triangle (denoted as 1–1–1, 1–2–1, 2–1–1 and 1–1–2). Drummers may consider this thin fare, but in fact most music gets by with these basic tempos.

When pitch triads are combined with the dimension of rhythm, we suddenly encounter auditory phenomena that most people would recognize as small fragments of real music: melodies. The cognitive complexity of combinations of simple rhythms with simple harmonies is already a topic not for the faint-hearted, but the fundamentals have been addressed by Eugene Narmour (1990) and David Temperley (2001), and the future of the cognitive psychology of music undoubtedly lies in that direction.

7.2. FACE PERCEPTION

The basics of face perception have been studied for many decades within clinical neurology, where a selective loss of face recognition capabilities has often been noted (Gruesser & Landis, 1991). Deficits are typically found with bilateral damage to the fusiform gyrus (fusiform face area, FFA), sometimes with unilateral right FFA damage, but rarely with unilateral left FFA damage (Figure 7.2). Functional MRI studies have revealed greater activation on the left with the processing of facial features but greater activation on the right with configurational processing (Maurer et al., 2007), suggesting that both hemispheres actively participate in different aspects of face perception. As with language deficits following brain damage to language centers, damage to the FFA is best known for the selective impairment of specifically face recognition, but it is unlikely that the FFA is devoted solely to the processing of face information. Abnormalities of the recognition of body form and other spatial configurations have been reported, suggesting that the FFA processes various geometrical configurations – one of the most important of which for the social lives of human beings is the triad of facial features.

Interestingly, there is a simple and well-known triadic aspect to face perception that has still not received the attention it warrants. As demonstrated by virtually any eye-tracking experiment that uses human faces as visual stimuli, it is known that, when we look at a face, we look predominantly at the eyes, the nose and the mouth (Figure 7.3). For checking the bilateral

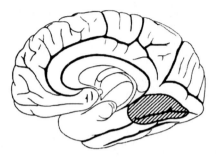

FIGURE 7.2. The fusiform face area is heavily involved in face perception (from Gruesser & Landis, 1991, p. 263, with permission).

FIGURE 7.3. The triad of facial features – eyes, nose and mouth – that are the main focus of attention in face recognition (data from Yarbus, 1959).

symmetry of the face, we normally look at both eyes – making it a four-feature scan, but half a face will suffice for recognizing an individual or evaluating an emotional state. There is of course a hairline, jaw structure, ears, cheeks and other contextual cues within which the eyes/nose/mouth are viewed, but the eyes, nose and mouth are the most prominent facial features and the most reliable cues for the global task of pattern recognition.

What is known from behavioral studies is that focus on isolated features or even features in combinations of two are not the determining factors in the perception of faces. There is a higher-level structure – sometimes referred to as "geometric cues," "configurational cues" or "second-order relations" (Maurer et al., 2007) – that is at least as important as lower-level features. It is specifically the triangular geometry of the relative positions of eyes, nose and mouth – a second-order structure that helps us to distinguish one particular individual from many others with similar first-order characteristics. Could the FFA be a region of neocortex devoted to the perception of triangular geometry?

7.3. JOINT ATTENTION

Michael Tomasello (1999, 2003, 2008) has defended the idea that the social interactions of apes are fundamentally dyadic, whereas normal human adult social interactions are triadic. As it turns out, there are many caveats (concerning the stages of childhood development, intentionality, the effects of captivity and training, species differences, etc.) that researchers in this field would add to such a simple statement of human uniqueness, but the fundamental point is both important and robust. In the wild, chimpanzees, in particular, are incessantly competitive with one another and rarely, if ever, cooperative. Their lives are dominated by direct competition for food or mates ("Who is dominant here?") and grooming behavior that helps to maintain established dominance hierarchies. Human interactions of course also include a dyadic component – and competition for resources and mates is a recurring theme in any social situation – but there are typically also periods of triadic interaction – a joint focus of attention on a common goal. When dyadic power struggles or mating behavior continues unabated, a cooperative venture is doomed. But once the dyadic dynamics are settled or put on hold, triadic dynamics become possible and cooperation becomes paramount as the focus shifts to: "What can we do together?"

Triadic interactions come naturally to human beings and begin at a surprisingly young age:

> At around 9–12 months of age, human infants begin to engage in a host of new behaviors that would seem to indicate something of a revolution in the way they understand their social worlds.... [They] begin to flexibly and reliably look where adults are looking (gaze following), to use adults as social reference points (social referencing), and to act on objects in the way adults are acting on them (imitative learning). These behaviors are not dyadic – between child and adult (or child and object) – but rather they are triadic in the sense that they involve infants coordinating their interactions with both objects and people, resulting in a referential triangle of child, adult, and the object or event to which they share their attention. (Tomasello, 2003, p. 21)

The contrast with ape behavior is striking. Even as adults, apes do not point in the wild and rarely point in captivity. In other words, simply indicating an object or event external from a dyadic power struggle is *not* a normal part of the repertoire of ape social behavior. By its very nature, pointing is triadic, and both an understanding of adult pointing behavior and the use of manual pointing develops at a young age in humans. Indeed, the ability to establish joint attention with others is so important that an absence of declarative pointing in the second year of life is diagnostic of autism (Baron-Cohen, 1995). Meanwhile, apes are absorbed in power and sex; cognitive engagement on a third topic outside of the dyadic relationship is, for them, unthinkable.

The triadic nature of human social interactions has already been outlined in convincing detail by Tomasello (1999, 2003, 2008), but the evolutionary emergence of this behavior remains uncertain. In line with the arguments in Chapter 6, I would argue that the uniqueness of *Homo sapiens* cannot be delineated (in any meaningful scientific way) in terms of our subjective feeling (of empathy, qualia, spiritual connectedness, etc.) with others of our species. Undoubtedly, we do empathize with other human beings, but it is uncertain how we might distinguish our "intraspecies empathy" from the intraspecies recognition that, for example, dogs of varying size, shape and color feel for one another. Yes, there is a sense in which we intuitively know that we can engage with other human beings, regardless of superficial differences, but to argue that we alone empathize in some sort of unquantifiable subjective manner is effectively to leave the realm of empirical science and to retreat to the position of Biblical scholars in asserting that the jump from the ape mind to the human mind is inexplicably metaphysical.

An alternative hypothesis is that it is specifically our facility with "threeness" in various realms that is our unusual cognitive talent – one manifestation of which is joint attention. Gaze following and rudimentary socialization begin in the first year of infancy and eventually bloom into social cooperation (Figure 7.4) and, ultimately, into the multifarious aspects of human civility. We immediately perceive this triadic facility in the behavior of all normal human beings and fail to detect anything similar in other species – animal species that we *correctly* perceive as locked into their dyadic associations and instinctual trains of isolated, asocial cognition. Conversely, human relations characteristically leave open the possibility of joint attention and sharing a topic of common interest – and do not require, first and foremost, resolution of the dyadic dominance and sexual issues that also may be in play. That common topic may well be something as uninteresting as today's rainy weather, but it rains on us both and may have implications for our cooperative activity.

In contrast, even among human beings the stare of hatred or the look of fear is explicitly dyadic and does not allow for an external third topic for our joint consideration until the dyadic dominance issue has been settled. Similarly, the look of love – whether an amorous invitation or a lustful leer – leads in the direction of a dyadic relationship, not toward cooperation on a third issue – an issue that is not about "us" to the exclusion of all else. When

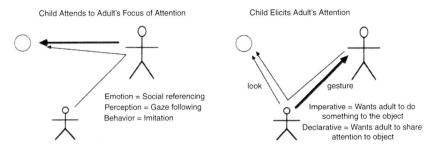

FIGURE 7.4. The developmental triad that emerges at the end of the first year in all normal children (from Tomasello, 2003, with permission).

the relationship is not locked into a dyadic struggle, both individuals will normally raise tentative topics of possible common concern and, as soon as we both acknowledge a joint topic, we are on the road toward cooperation. As Tomasello has emphasized, there is, of course, also an internal feeling that accompanies the experience of cognitive sharing with other human beings, and it is arguably that subjective feeling that will motivate us to engage with others and treat others as compatriots in a civilized society, but the crucial difference between dyadic competition and triadic cooperation has an explicable cognitive basis.

7.4. MORAL MINDS

A recent trend in neuroscience is the experimental testing of hypotheses concerned with morality and ethics by presenting people with "toy" ethical dilemmas and asking for their considered evaluation (often while measuring brain activity). For abundant examples of such research using situations familiar in contemporary society, Marc Hauser's *Moral Minds* (2006) and John Mikhail's *Elements of Moral Cognition* (2010) are good places to start, but in fact the core ideas are as old as philosophy and religion. The novel claim that has motivated this field is that a "universal moral grammar" is an *innate* feature of the human mind – on a par with and modeled after the linguist's "Universal Grammar" (Chomsky, 1985). Roger Scruton (1997) has argued for an innate "tacit knowledge" of musical grammar, and Ian Dutton (2009) and V. S. Ramachandran (2011) have presented the case for an "art instinct." And now, Mikhail (2007, 2010) and a new breed of cognitive scientists maintain that, details of the "parameter settings" of specific cultures and upbringings aside, human beings have an inborn sense of right and wrong with regard to behavior in a society of other, similarly encultured human beings.

In the present context, it is of interest to note that the core argument concerning the cognition underlying the "moral mind" is once again triadic. As stated most coherently by David Hume (1739/1978), moral judgments inevitably entail the interactions of three actors: an agent, a recipient and, crucially, a spectator. If the "spectator" is taken to be an omniscient god, then the rules of morality can be stated as religious laws of righteous behavior between any two actors – laws that have been "given" to human beings but not to other species. The simplest formulation of such a law is usually stated as the Golden Rule: Do unto others as you would have them do unto you (because God is watching). The Golden Rule itself is a foundational concept in most of the world's religions – explicitly stated in Buddhism, Confucianism, Taoism, Judaism, Christianity and Islam. In terms of recommended behavior, it is explicitly a triadic principle: How should you as an agent interact with a recipient in light of God's omniscience? If the tables were turned, can you expect the recipient to act as you have – again in light of an observing God? For the secular, evolutionary, cognitive scientist unwilling to postulate the existence of a metaphysical God, however, it is an evasion of the essential psychological questions to state the rules of morality as God-given laws. Quite aside from the perennial issue of metaphysics, the cognitive psychologist wants to know what the cognitive operations are that rumble through our minds when we make ethical/moral decisions.

As philosophers have long noted, the cognitive psychology of ethics turns out to be more than dyadic, and the Golden Rule can be reformulated such that it does not invoke metaphysical players, that is, stated as the interaction between an agent and a recipient as witnessed by a third-party human being. The moral judgment is then determined by the relations among comparable, but not necessarily omniscient human agents. Hume expressed the dynamics underlying an innate moral grammar in explicitly triadic terms. In the first place, there is an "event" that occurs between two actors. (Generalizing to events involving one-to-many or many-to-one does not change the fundamental dynamic.) But, no event occurs in a vacuum: There is inevitably a social context, to which both the agent and the recipient return. So, in the second place, there is an "objective," presumably unbiased, if not omniscient observer who is in a position to evaluate the right and wrong of the actions between the actors (Figure 7.5). (Again, generalizing the observer to a society of people does not change the dynamic. In the minds of both the agent and the recipient, there is an external, uninvolved observer who is, in principle, in a position to evaluate the fairness or unfairness of the event.)

FIGURE 7.5. The cognitive triad entailed in consideration of any moral dilemma: an agent, a recipient and a third-party observer of the event. Here, the agent has killed a recipient, and the entire event has been witnessed by an observer. In the animal kingdom, "might makes right" and there is effectively no thirdparty to evaluate the justice of the outcome (after Hauser, 2006, with permission).

The roles of reason and emotion and their causal influence on the ultimate judgment are central to many discussions of morality and ethics (Figure 7.6). Both Hauser (2006) and Mikhail (2007) contrast the roles of reason and emotion in Hume's understanding with those of Immanuel Kant (2001) and David Rawls (1971), and argue that Rawls more closely expresses the reality of the innateness of moral judgments.

The details of the mechanisms leading to judgments of what actions are permissible, obligatory or forbidden is where the argument for the "moral mind" becomes interesting... and controversial. Crimes of passion are generally evaluated differently from premeditated infractions, and notions of justice get complex as ever-changing cultural norms and differing versions of righteous behavior are brought into consideration. Those with presumed access to the mind of god may be willing to state categorically whether or not revenge killings, stoning for adultery, preemptive military strikes, meat-eating or pacifism are the "correct" course of action for all human beings for all time, but for most people there is no alternative but to continue to wrestle with the cognitive triads in our minds: "In the presence of a fair-minded observer, is the action of this agent in relation to that recipient permissible, obligatory or forbidden?" The "moral mind" argument is that the right course of action in the specific circumstances – taking into consideration not only one's own needs and desires but also those of one's immediate rival – and is intuitively known by all members of the society in which they are embedded.

It is relevant to note that, for all arguments of "innate" human talents, no serious scholar maintains that high-level moral cognition itself is coded in DNA or that the mechanisms, much less treatment of violations of the social code, should be pursued at the level of genetics. The core argument

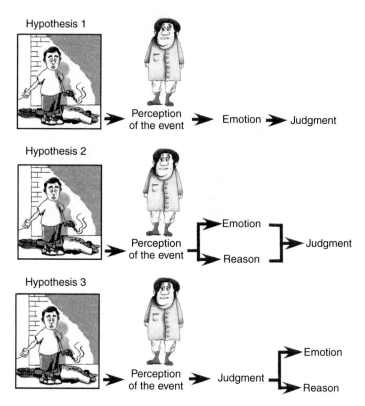

FIGURE 7.6. Contrasting hypotheses of the mechanisms of the moral mind. Hypothesis 1 elevates the role of emotion as the factor determining the moral judgment, as argued by Hume (1978). Hypothesis 2 emphasizes the inevitable conflicts between reason and emotion in reaching the judgment, as noted by Kant (2001). Hypothesis 3 is the "universal moral mind" argument that sees the judgment as "innate," and inevitably followed by *post hoc* rationalizations and emotional responses (after Hauser, 2006, with permission).

is more accurately characterized as an attempt to show that, even for such high-level capabilities as ethical judgments, there is a potential cognitive competence that all human beings have "at birth" and that appears to be absent in all other animal species. As is also true for language, music and art, normal people who have experienced the rudiments of a socialization process during the first few years of life learn what is fair and unfair (grammatical, euphonious or pleasing to the eye). In this regard, the argument for an "innate" morality (language competence, appreciation of music and art, etc.) is somewhat imprecise, for the inborn capacity needs to be nurtured into fruition. The emphasis on "innateness," however, is appropriate because learning the cognitive triad does not depend on explicit training or elaborate indoctrination. With repeated exposure to the social activities of compatriots engaged in triadic processes, all normal children pick up on the rules of "proper behavior" – the triads of social interactions that are the local cultural norms.

Whether or not we, as individuals, ponder the great moral dilemmas in the manner of Raskolnikov, human beings become moral creatures in ways that other creatures apparently do not. To say that there must be a genetic component to this competence is a truism, but the genetic endowment is arguably one for triadic cognition – the ability to hold different perspectives in mind and to weigh the importance of all factors. What has been emphasized by most scholars concerned with the "moral mind" is that "weighing the factors" is essentially automatic (Hypothesis 3 in Figure 7.6). Similarly, we do not need to ponder complex details about music, art or language to achieve some sense of musical, artistic or linguistic meaning. The normal development of the moral faculty in human beings inevitably leads to the ability to make seemingly automatic judgments of justice within a specific cultural context – judgments that are as innate, indubitable and "obvious" for human beings as are the instincts of power and sex for animal species.

7.5. INTELLIGENT NEURAL NETWORKS

In artificial neural networks, the classic example of an "intelligent" network is the so-called exclusive-OR (XOR) architecture. For anyone not already immersed in the world of artificial neural nets, the idea that a four-neuron network could be interesting, much less intelligent, may seem bizarre, but there is a simple triadic lesson to be drawn from such simple neuronal circuits.

The possible relevance of artificial neural networks started with nineteenth century research in the field of formal logic. That work remained academic until technological progress in the twentieth century meant that small-scale electronic circuits could be built and tested. It soon became apparent that the simple (Boolean) logic functions (AND, OR and NOT) could be implemented by allowing for appropriate levels of excitatory and inhibitory effects between simulated "neurons" in the input and output layers of electronic "brains." Such two-component circuits formed the basis for early "perceptron" neural networks and it was argued that a Turing machine which responds to an input by turning an output neuron "on" or "off" illustrated in principle the universality of mechanical logic: "artificial intelligence." The possibilities for artificial intelligence in robots

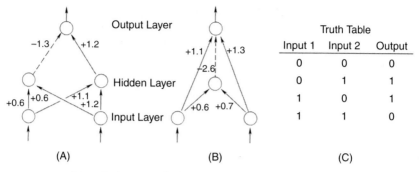

FIGURE 7.7. (A and B). Two of many possible instantiations of the XOR neural network. All neurons have activation thresholds of 1.0. Whenever the total input to any neuron exceeds that threshold, the neuron fires and sends an excitatory signal (solid arrows) or inhibitory signal (dashed arrows) of specified strength to other neurons. Although the input neurons are themselves binary on/off switches, the output of the network is determined by the balance of inhibitory and excitatory signals among all three layers. By definition, the XOR network must respond as shown in the Truth Table (C).

were then robustly explored in the science fiction literature, but further academic work revealed some genuine problems. Marvin Minsky and Seymour Papert (1969) formally showed that, using the simple, two-layer perceptron architectures, there are extreme limitations on the information-processing possibilities of machines. Specifically, the exclusive-OR (XOR) logic function could *not* be implemented in a Turing machine. Since XOR is the simplest version of a "conditional association," its absence in a robotic brain would imply a distinct lack of intelligence. For a brief moment, the AI dream was shattered, but it was soon rejuvenated with the slightly more complex neural network architectures of the 1980s (Anderson, 1995).

The logic of XOR is straightforward and can be stated as a circuit that responds when either of two input neurons is active, but not when both input neurons are active or inactive (Figures 7.7A and 7.7B). Trivial, to be sure, but no matter what levels of activation and inhibition are designed for a perceptron with direct linkages between the two input neurons and the one output neuron, it cannot produce the complete set of correct results for the four permutations of possible inputs (the Truth Table in Figure 7.7C). Fortunately, from theoretical work on toy neural networks it became apparent that the XOR problem could be solved by employing a so-called hidden layer wedged between the input and output layers (Rumelhart et al., 1986).

The essential role of so-called hidden layers (consisting of neurons that are neither input nor output devices) is now well-understood, and it is not an exaggeration to say that the nonsymbolic aspects of modern AI are built on (perceptron-like) association mechanisms *plus* the inhibitory effects of hidden layers. Stated conversely, neural nets that do not exploit hidden layers are limited to associational (correlational) effects, are unable to perform conditional associations and remain most definitely unintelligent. It is important to note that XOR logic can be trained in many animals and is not a uniquely human capability (e.g., Anderson et al., 2006). Indeed, at a microscopic (neuron) level, XOR is ubiquitous and such circuits are at work throughout the nervous system (e.g., in edge detection in the retina [Livingstone, 2002]). At a macroscopic (brain) level, however, the ability to utilize conditional associations appears to be a primate talent.

The simplicity and explicitness of neural networks have made the three-layer architecture (specifically, the importance of hidden-layers and the role of inhibition) well known. Surprisingly, however, red-herring neuronal networks that do *not* do what their creators intended are also widespread in the scientific literature (theoretical biology, theoretical cognitive science and even theoretical nuclear physics) and are a clear indication that the available computer technology that makes neural network simulations so easy is often used without any attempt at understanding the "cognitive" mechanisms they employ (Cook et al., 1995; Cook, 1995a, 1995b). So, despite the continuing allure of machines that seem to "understand" phenomena that even human beings do not, an important distinction needs to be made between dyadic and triadic neural network mechanisms. The former are capable of nothing more than associations, whereas the latter show sparks of "intelligence."

7.6. COLOR PERCEPTION

In the visual arts, the significance of the triadic insights of the Italian "geometers" of the early fifteenth century can hardly be exaggerated, insofar as the practical rules of linear perspective led inextricably to the illusory depth and spatial complexity of most subsequent art and gave convincing 3D depth to flat pictures. Such pictures, paintings and drawings gain a depth dimension that immediately *prevents* us from seeing them as 2D patterns – which, of course, they still are. But "real art" nearly always employs color (black-and-white photography being the main exception), and color too is a triadic talent. Specifically, even though most mammals have dichromatic (essentially grayscale) color perception, and many insects and fish have four or more distinct wavelength-sensitive color receptors, trichromatic color perception is something that *Homo sapiens* share with other primates. In this

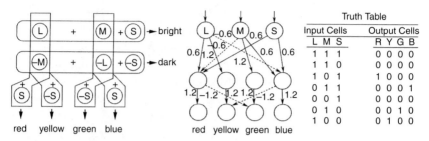

FIGURE 7.8. (A) A model of color perception (after Davidoff, 1995, with permission). There are three types of light-sensitive neurons in the retina – responding to short (S), medium (M) and long (L) wavelength photons. The greater the stimulation of any of these receptors, the "brighter" is the visual field, but different combinations of the different receptors lead to the perception of specific colors: L − M + S = red, L − M − S = yellow, M − L − S = green, and M − L + S = blue. (B) These relations can be expressed as a neural network that gives one of four color perceptions for any combination of three inputs. Each neuron in the network receives stimulation of varying magnitude. When the total stimulation exceeds a threshold value of 1.0, a neuron is activated and sends stimulation to downstream neurons, ultimately leading to perceptual outputs. Many alternative configurations and weightings of the connections in (B) can produce the empirical Truth Table (C), but the network is nonlinear, so that a hidden-layer with inhibitory effects is required.

respect, primate color perception is unusual for its simplicity and efficiency, while still allowing for the multidimensionality of color.

The basic retinal physiology of trichromatic color perception is understood at the molecular and cellular levels, but the psychology of color remains surprisingly mysterious. We do not in fact perceive simply the three colors corresponding to the three wavelengths of maximal photonic sensitivity (green, yellow and purplish). And we do not simply blend those pure colors to produce all color variations – as on an LCD monitor. Instead, the human visual system uses the three types of color receptor to calculate *relative* color strengths – thus allowing for "color constancy" over a wide range of bright and dark environments.

Moreover, with sensitivity to three distinct wavelengths of photons, it is a curious fact that we typically divide the color world into not three, but four main categories. Blue and yellow are "opponent" colors and do not mix, and there is similar opponency between red and green. That means that we cannot perceive a bluish yellow or a yellowish blue. And we don't even have a word for mixtures of red and green. The other combinations of long, medium and short wavelength detection are, however, possible (reddish blue, greenish yellow, etc.). The slight inconvenience of not having perceptions of opponent color combinations is more than compensated for

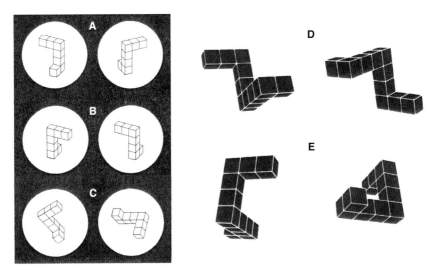

FIGURE 7.9. Examples of mental rotation tasks. Most people make a "same" or "different" judgment by rotating one object to the orientation of the other. Which one do you rotate? First try the easy examples of A, B and C. (Same, different, same, right?) Now try D and E while asking yourself which of the two objects (left or right) you use as the immobile reference and which undergoes rotation. A rather unusual "handedness" test, but most righthanders rotate the object on the right, most lefthanders rotate the object on the left.

by the fact that colors can be determined over a huge range of intensities. By *not* perceiving color as a literal reflection of the absolute levels of activation of specific hue-receptor neurons, we are able to "interpret" colors on the basis of relative activations. The theory of such interpretations is shown in Figure 7.8A, and a possible neural network implementation is shown in Figure 7.8B. Alternatives are possible, but the input–output relations of the Truth Table in Figure 7.8C are the perceptual reality and must be produced in any model of color perception.

7.7. MENTAL ROTATION

The ability to generate a mental image and then manipulate it in 3D space is a human talent that is often measured in IQ tests. Imagine the shape of Florida, and then rotate that image 180 degrees. Where is Miami? In fact, rotations and translations on a 2D plane are not that difficult – as is well known by several generations of laboratory pigeons, but rotations in 3D space can be taxing (Figure 7.9).

On the basis of many clinical studies over the past century, it is well known that the right hemisphere is involved in a variety of visuospatial tasks – particularly the completion or identification of geometrical shapes and part–whole relations. What has come as a surprise from modern brain-imaging studies is that the left hemisphere is heavily involved in mental rotation. In picture completion tasks, where active *manipulation/ transformation* of the visual information is *not* required, patients with left hemisphere damage (intact right hemispheres) perform better than those with right hemisphere damage (Mehta et al., 1987). But, when the visuospatial task is an active one, involving the generation of visual images (Kosslyn, 1994; D'Esposito et al., 1997) or the manipulation or rotation of images in depth (Alivisatos & Petrides, 1997; Gill et al., 1998; Wexler et al., 1998), the left cerebral cortex is more strongly activated than the right.

In fact, any task of this kind consists of several subtasks, making simple left/right dichotomies inaccurate. In mental rotation, there are at least three subtasks – image generation, rotation and comparison – each of which may be lateralized to the left or right (Gill et al., 1998). A simple manifestation of the uneven hemispheric involvement in mental rotation can be easily experienced. Figure 7.9 shows several pairs of geometrical objects. First try examples A, B and C by deciding if the paired geometrical shapes are the same or different. Most people can do this task without difficulty, but some practice certainly helps. Next try D and E, while asking yourself how you actually made your decision. Again, most people can indicate which of the two objects was used as a stable (immobile) reference, and which was actively rotated into the orientation of the reference ... and then the judgment of same or different becomes "obvious."

In a controlled experiment using a series of similar stimuli, we have found (Cook et al., 1994) that right handers have an overwhelming tendency to rotate the object on the right, and left handers to rotate the object on the left (viewing object pairs, as in Figure 7.9). Right handers (who are generally more practiced in manipulating real, material objects with the right hand) usually find it easier and "more natural" to mentally rotate the object on the right and to leave the object on the left alone as the reference to compare against (and vice versa for left handers). This bias is arguably an ingrained habit – where the dominant hemisphere functions as the executive – the active manipulator and sequentializer of information for both real and imagined behavior, and the nondominant hemisphere acts as the reference – maintaining an image in mind for comparison with the manipulated image in the left hemisphere. But is this talent triadic?

On the one hand, it is evident that only two visual images are involved. If the task were simply to judge whether or not the 2D images on the printed page are identical, then the cognition would entail the perception of just these two images and a judgment concerning their similarity. For the 3D mental rotation task, however, a direct comparison of the retinal images will not give the correct answer. To answer correctly, both of the images must be understood to be 3D objects and at least one rotated in 3D space to evaluate their structural similarity. As such there are arguably three processes involved in mental rotation: (i) perceiving the 2D images as 3D objects, (ii) mentally rotating at least one object to the orientation of the other object and (iii) judging their similarity.

Rotation of 2D images in the plane of the printed page and making same/different judgments can be performed by animals (Fagot, 2000), but the full 3D mental rotation talent appears to be uniquely human.

7.8. SUBITIZING

I had a friend in grammar school, Scot, who was always honored for being "the smartest kid" in class. From the mundane to the exceptional, from confusion to crisis, we led parallel lives with Scot at the intellectual forefront and the rest of us back in the pack. As it turned out, friend Corky was the smartest anyway, but one nagging doubt about my own mental apparatus was instilled in me by Scot at the billiards table at a young age. Why it was important to count the billiard balls still escapes me, but, in the course of amateurish rivalry at the table, it became apparent that Corky and I counted the balls in groups of three, while Scot managed in groups of four. On reporting this fact to friend Chuck, Chuck's salt-of-the-earth father dismissed the alleged ability for grouping-by-fours as typical adolescent bragging, but the nagging doubt had already been deeply implanted in my brain. Could the natural operation of my own mind be a simpler version – yes, a stupider version – of an operation that smarter people can do in a smarter way?

On the one hand, it seems entirely possible that different brains "subitize" – are aware of the precise numbers of objects in a visual (tactile or auditory) scene – in different ways. Glancing at the billiard table or the kitchen table with only a few objects on it, one immediately grasps their numerosity without ploddingly counting them one-by-one. And if there are a dozen or so objects on the table, we are likely to see pairs and small groups containing known numbers of objects. It is therefore conceivable

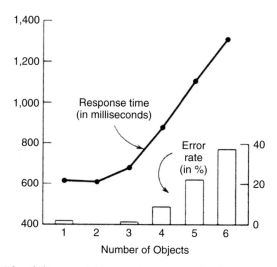

FIGURE 7.10. The ability to subitize can be measured either in terms of response times or error rates. Both measures indicate essentially perfect, automatic performance for one, two, or three items, but a significant slowdown and significant increase in errors with a fourth item (from Dehaene, 1997, p. 67, with permission).

that different people and different species use different basic strategies for counting. For normal people, two at-a-time seems easy, and three at-a-time is doable, but four at-a-time is, in my mind, inevitably three-plus-a-bit. As a three-plus-something-additional process, counting "in fours" is then possible, but, if three is certain, four is actually a bit of a guess (Figure 7.10). On the other hand, the addition of fours does have its advantages: instead of jumping back and forth between even and odd numbers when adding threes, adding fours will always give totals among the even numbers. Still, the initial grouping of the objects into fours is somehow a more difficult task than doing it by threes.

Counting billiard balls is not one of the important issues of life, and even the psychology of subitizing might not be a valid measure of intelligence, but the question of how many mental processes one can simultaneously comprehend is a question that can be answered empirically. Certainly, if there are grounds for concluding that the unusual features of human cognition relative to monkey cognition are due to an unprecedented ability to handle three processes simultaneously, then the future of mental evolution could well be in the direction of further numerosity. The issue is yet unsettled, but the cognitive psychology of mathematics, in general, and the issue of subitizing, in particular, are well described by Brian Butterworth (1999) and Stanistlas Dehaene (1997).

7.9. FOUR-BODY COGNITION?

Abstract arguments about the perception of numerosity may be inherently inconclusive, but the world of harmony perception provides some tangible evidence concerning the reality of four-body and even five-body perceptions. Specifically, there are several four-chords that have known musical uses and whose perceptual properties may help us to distinguish between the logical alternatives of "four-as-a-unit" and "three-as-a-unit plus one." The most obvious examples are the so-called seventh chords. The two seventh chords that are widely used in both classical and popular music are referred to as the minor seventh and the dominant seventh. The minor seventh is a minor triad plus the seventh tone of the minor scale, whereas the dominant seventh is a major triad plus the same seventh tone of the minor scale. (If the seventh tone of a major scale is used, the chord is a "major seventh" and has a very different sonority. Because the seventh tone in a major scale is only a semitone below the tonic, the major seventh chord contains rather strong dissonance and is less frequently used.)

Perceptually, both of the commonly used seventh chords retain their major or minor character, whereas the seventh provides a dab of unsettled dissonance (the whole-tone interval between the minor seventh and the first upper partial of the tonic). The obvious question is: Are the seventh chords essentially triads-plus-a-little-dissonance or is there a new acoustical feature that we hear in the four-chord as-a-whole?

One interesting possibility is that (in addition to their inherent tonality, dissonance, tension and modality characteristics) the seventh chords embody within them a directionality in which to proceed for harmonic resolution. It is well known that the major and minor seventh chords are most frequently followed by tonic chords. Unlike the augmented triad – that is a raw statement of unresolvedness with no implication of the direction in which to proceed for resolution – the seventh chords are unresolved, but contain within them an implied direction. In this view, perception of the seventh chords would provide us with essentially four levels of auditory phenomena. That is, there is (i) the tonality of the individual tones themselves, (ii) the consonance and dissonance of the intervals among the tones, (iii) the major/minor/tension modality of the three lower tones in the seventh chord and finally (iv) an overall impression of "directionality." The tendency to hear the seventh chords as calling out for resolution would therefore be a four-body harmonic effect that transcends even the modality issue.

Alternatively, perhaps four-body perception is not possible, and all that we actually hear are recognizable tones, consonant intervals and sonorous triads, "plus a little dissonance" thrown in at various locations in the various four-chords. In the latter case, the fundamental unit of harmonic perception remains the triad – where we readily acknowledge the major/minor/chromatic flavor of the underlying three-chord, and then perceive the spice of the seventh tone added onto the triad. In this view, the demand for resolution of even the seventh chords is a lower-level dissonance effect.

The theoretical argument against the idea that there is a four-tone "unit" in musical harmony is that there is, in the world of music, no widely accepted term for a four-tone feature. If there were a four-tone quality that is as salient and as universally recognized as major or minor modality, undoubtedly musicians would have invented a word for it. To be sure, to say that a chord is a seven-chord indicates that it is not a resolved, settled harmony but rather a harmony with major/minor modality but without resolution. For the seventh chords, however, there is nothing comparable to the generic label of major/minor mode and nothing comparable to the half-millennia debate on the meaning of major and minor affect. At the very least, we can conclude that any purported four-tone feature is more subtle than the modal feature that virtually everyone (including 4 year olds) can hear.

What nonetheless remains uncertain is whether or not there is a weaker, but perceptually real harmonic feature that can be described objectively in terms of four-tone acoustical structures. There clearly are distinct musical features described as two-tone consonance/dissonance and three-tone tension/modality. In the model described in Chapter 2, those two features were then added together to obtain an overall instability score using a weighting factor of 1/5, in order that the theoretical results correspond well with empirical data. In a similar manner, it may be that a four-tone feature exists and can be perceived by careful listeners, but this feature may also require a numerical weighting suitable for its subtlety. As such, it may yet be real and perceptible, but would be a truly subtle spice enjoyed only by those who are interested in savoring this musical morsel.

A slightly more optimistic view is that the overwhelming tendency to resolve the seventh chords to the corresponding major or minor tonic chord suggests that the well-known "directionality" of seventh chords is inherent to these chords, discussed under the rubric of "resolution" and is therefore most definitely not a topic for musical sophisticates only. In any case, some experimental work has already been done in this direction (Kuusi, 2002), and answers may yet be forthcoming.

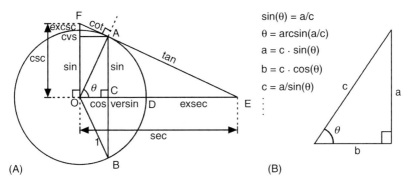

FIGURE 7.11. (A) All of the trigonometric functions of an angle, theta (θ), can be constructed geometrically in terms of a unit circle centered at O. (B) Each of these functions can be explained simply in terms of the relationships within one triangle containing a 90 degree, right angle. Taken in doses of one triad at a time, all normal children can learn trigonometry.

7.10. TRIGONOMETRY

The dilemma that teachers at all levels of education face is the need to impart knowledge of a sufficient complexity that students "learn how to learn" and can then pursue higher goals on their own, but not to torture unwilling/unprepared minds with facts and figures that have no meaning for them. At the earliest stages of mathematics, the basics of arithmetic (counting, addition, subtraction, multiplication and division) arguably do not provide students with the ability to learn; arithmetic must be memorized and is essential groundwork for numerical literacy, as is the alphabet essential for verbal literacy. But higher-level mathematics, specifically, calculus and the trials of differentiation and integration, are already specialist topics that many young minds, for reasons of personal development and environmental setting, cannot grasp. The dividing line comes at trigonometry: "triangular measures."

The importance of trigonometry is that it has real-world applications and leads into the full complexity of calculus and solid geometry, but its central ideas can be explained in terms of a triad of geometrical elements (Figure 7.11). The essential point is quite simply that, given two unambiguous facts about the edge lengths or angles of a right triangle, the other edge lengths and/or angles can be computed. By repeated application of this triangular insight, the geometry of extremely complex structures can be deciphered and understood. So, prior to addressing the real-world problems of architecture, civil engineering and moon shots, the triadic mental manipulations of trigonometry that lie at the heart of solid geometry need to be mastered.

For the educator, the most important distinction is between principles that are not easily explained at a more fundamental level – for example, the basics of number theory and the formulaic procedures that are required to perform arithmetic – and principles that can be explicated and trained as cognitive triads. "How to memorize" probably cannot be explained, so that all memorization tasks required for arithmetic must be trained at a young age when the teacher's praise and the act of mastery itself will be sufficient rewards. But, even young children are eager to exercise their triadic capabilities. As soon as basic arithmetic competence has been achieved, the basics of geometry can be absorbed, and eventually teaching of the triadic manipulations of trigonometry should follow. Recursive application of the triadic methods can then be applied indefinitely by students who find the motivation to understand more complex mathematics.

8

Conclusion

Many questions remain unanswered concerning the importance of triadic mechanisms for understanding specifically *human* perception and cognition, but there is the distinct possibility that the leap from plain-vanilla, dyadic associational mechanisms to the subtleties of the human mind has a triadic basis. Such "high-mindedness" is made possible, in the first place, by (i) the expansion of the human neocortex, (ii) the unprecedented emergence of trimodal association cortex and, ultimately, (iii) the development of three-element cognitive algorithms of various kinds. The "mental trick" that we, as a species, seem to have developed (if perhaps not yet completely mastered) is how to draw inferences from juggling three simultaneous streams of information.

Clearly, the summary of the already-known forms of triadic perception and triadic cognition in these chapters will not be the final statement on the human "brain code," but I would argue that it is the beginning – the gateway through which an understanding of the cognitive phenomena that occur beyond the reach of automatic, seemingly "mindless," associations can be obtained. Needless to say, the dyads involved in paired (simultaneous or sequenced) associations are relevant to all aspects of neuronal information processing and are the groundwork on which triadic mechanisms are built, but it appears that all of the interesting aspects of specifically *human* psychology entail three-way processes – triadic perception, trimodal perception, triadic cognition and "conditional associations" – processes that cannot be coherently stated as multiple or serial dyadic associations. It is specifically the triadic processes that appear to be our forte – and difficult or entirely absent in the mental lives of animal species.

There are also many unusual features of the human body that underlie our behavioral uniqueness. Those supporting somatic features (the bipedal stance, the oppositional thumb, the anatomy of the vocal tract and so on) and the underlying dyadic neuronal functions (Hebbian learning, short- and long-term memory, attentional mechanisms and so on) have, for good reason, received close scrutiny in ethology, comparative anatomy, artificial neural networks and human psychology for many decades. But, despite many scholarly attempts, the qualitative leap from "monkey cognition" to human cognition has not been explained either by somatic structures or by the quantitative increases in dyadic brain mechanisms that increases in neuron numbers bring. "Fancy bodies" are not enough, "big brains" are not enough and even "big associational cortex" is not enough.

What appears to have catapulted *Homo sapiens* to utter dominance on planet Earth are (i) the hard-wired substrate of polymodal association cortex where visual, auditory and touch information converges and (ii) a revolution in neuronal software that can be best summarized as well-placed inhibitory processes that prevent competitive Hebbian learning from becoming the one-and-only neuronal process at work in the brain. Note that the inhibition is not simply the suppression of competing processes or motor activity but rather the displacement or delay of ongoing, dyadic associational mechanisms that transform the cacophony of multiple, conflicting processes into an ordered hierarchy.

Our unusual cognitive abilities do not of course stop at the triadic level. To be sure, we can hear the affective modality of a three-tone chord, we can see the imaginary depth implied by three aligned visual cues printed on a flat sheet of paper, we can understand the causality of events described in a three-word sentence, we can strike a match to start a fire and we can follow a pointing finger to focus our attention on a topic chosen by someone else. Each of those core talents is explicable in a detailed, analytic way that is, on its own, fully understandable, but each talent is not "important" unless it occurs within a larger context. When a sufficient context is also provided, the triadic talents lead us into music, art, language, tool usage and social cooperation. And, at that level, the full-blown, embedded triadic phenomena have true psychological significance that gives our biological lives social meaning.

My conclusion is therefore that triadic cognitive mechanisms are the events that, to begin with, require the attention of researchers in *human cognitive neuroscience*. The study of cognitive triads falls within the realm of hard-headed, reductionist science at its best: It entails simplifying complex phenomena to their essential constituents in such a way that the higher-level phenomena can be described in terms of lower-level mechanisms. Importantly, such "reduction" remains firmly *within the realm of cognitive psychology* – and is not the absurd reductionist pseudoscience of trying to explain human cognition in terms of genes, molecules or quantum mechanics.

What is new within this greatly simplified, reductionist approach to cognitive psychology is the distinction between "two-body" and "three-body" mechanisms. Both are real and, in principle, can be delineated in detail down to the neuronal level, but it is essential to distinguish between types of behavior that are guided by two-body associations (or multiple and/or serial two-body associations), on the one hand, and three-body associations, on the other. Moreover, both the two-body and the three-body processes must be distinguished from other processes that are too complex to be described more precisely than to say that they are "context-dependent." As important as such complex processes may also be, they yet require deconstruction – reduction into their fundamental components before rigorous analysis will become possible. Two important but extremely complex topics that have not been discussed in these chapters come to mind: the cognitive psychology of mathematical equations and that of literary metaphor. What, in essence, is the nature of the mental process of considering the relationship (equality, inequality or similarity) between two entities?

The availability of various noninvasive techniques for studying brain activation in controlled situations means that a detailed understanding of the cognitive mechanisms underlying these and other kinds of human perception and cognition is today a real possibility. But it will require special efforts to decipher the brain code by designing experiments that are not "domain-specific." Of course, individually, each and every cognitive scientist enters the field with rather particular, nearly-always domain-specific questions in mind – questions about language, music, vision, social interactions or whatever. Answers that are stated solely in the framework of the relevant domain may be entirely valid and important in their own right, but significant contributions to a scientific understanding of "the brain code" must be general and stated in terms that have cross-domain and cross-modality meaning. Already, progress has been made at this rather abstract level – primarily in studies on the conditional associations in primate brains and those exhibited in preverbal human infants. Indeed, the issue of dyadic versus triadic mechanisms is already a major research theme (Deacon, 1997; Fuster, 2007; Gomez, 2004; Penn & Povinelli, 2007; Povinelli, 2000; Tomasello, 2003, 2008; Tomasello & Call, 1997) – although not always discussed in those terms.

At the very different levels of material organization studied in atomic physics and cell biology, delineation of precise causal mechanisms was in fact achieved during the twentieth century. In the 1920s and 1930s, the ideas of quantum physics and, in the 1950s and 1960s, the ideas of molecular genetics reached maturity. The core insights in those academic disciplines were eventually generalized to provide the foundations for rigorous sciences of physical matter and biological life – known today as the "quantum code" and the "genetic code," respectively. In those fields, as well, the actual motivations of individual researchers may well have been to isolate particular substances or cure specific diseases, but the answers came in the form of general, abstract codes – the utter simplicity of which no one had anticipated. Ultimately, the fundamental insights in atomic physics were concerned with nothing more complex than systems containing neutrons, protons and electrons – and described precisely by the Schrödinger wave-equation. At the cellular level, a similar simplicity in the four-letter genetic alphabet and the three-letters per "word" mechanism for translation into protein molecules resulted in conceptual clarity. To the surprise of all, the seemingly boundless diversity of both molecular structures and living forms found their respective unifications in quantum mechanics and informational macromolecules – leading to sophisticated technologies based on the now well-established codes. A similar "brain code" undoubtedly underlies human cognition – *not* simplification to the level of "dyadic associations only" (which might suffice for a behavioristic description of animal cognition) and certainly *not* simplification to the molecular level, but elucidation of a small number of explicable dyadic and triadic mechanisms underlying certain kinds of characteristically-human thought processes.

Where does the idea of triadic cognition fall within the last century of research in human psychology? Already by the early part of the twentieth century, the revolt against speculative introspectionist psychology had begun, and the behaviorist paradigm ultimately had the salutary effect of emphasizing the importance of empirical data and experimental testing. Some of the behaviorist tendencies were extreme, however, and it required several decades to reestablish many of the more subtle aspects of human psychology – consciousness, affect and "higher" cognition – as topics suitable for empirical research.

The higher-level topics that the behaviorists shunned were rejuvenated in the 1960s partly as a consequence of the demonstration of cognitive specializations of the left and right cerebral hemispheres. Laterality effects were of course already well known in clinical neuropsychology, but the surprising dominance of the right hemisphere for the "feminine" talents of art and music, face recognition, contextual processing, tone of voice and emotions in general made them suddenly accessible to any researcher with rudimentary electronic equipment. Ultimately, the laterality fad also went too far – with the abnormalities of a relative handful of patients having undergone surgical separation of the cerebral hemispheres somehow elevated to exemplars of the internal contradictions of the human soul! In the normal brain, of course, interhemispheric crosstalk over the corpus callosum produces behavioral unity, and the mechanisms of integration are far more important for understanding the mind than the demonstration of failures of integration produced by brain damage. In any case, the modern technology for noninvasive brain imaging has made it almost easy to ask questions about the localization of various forms of higher-cognition, but the challenge remains to ask the right questions.

If our ultimate goal is to decipher the *human* "brain code," it is important to appreciate the fact that essentially all of our characteristically-human, higher-level cognitive capacities have unambiguous laterality aspects: the left hemisphere is dominant for language and tool use (handedness), musical rhythm and various tasks requiring precise motor sequentialization; the right hemisphere is dominant for the perception and production of speech prosody, harmony perception, face recognition, pictorial depth perception and other tasks where sequentialization is less important than an understanding of the entire configuration. The cerebral lateralization underlying language and handedness has been known for many generations, while that for music and art has been noted only more recently: "Harmony in music is an analogue of perspective in painting. Each produces what is experienced as depth; each is right hemisphere-dependent" (McGilchrist, 2009, p. 419).

Since all of these high-level functions – music, language, tool use and art – demand the participation of the talents of both hemispheres, it of course makes no sense to speak of left-brain and right-brain as if they were independent of one another. While lower-level tasks may be less lateralized, the characteristically *human* aspects of human cognition have definable left and right hemisphere components, and it is specifically the coordination of sequential tasks within the configurational whole that is normally undertaken for behavior driven by high-level cognition. Attention to the mechanisms of integration of the lateralized pieces will be essential for establishing a "brain code" – comparable in scope and rigor to the already well-established genetic code at the cellular level and the quantum code at the atomic level (Figure 8.1).

RECOMMENDED READING

In the course of writing this book, I have been repeatedly reminded of the fact that many excellent texts have been published on related topics.

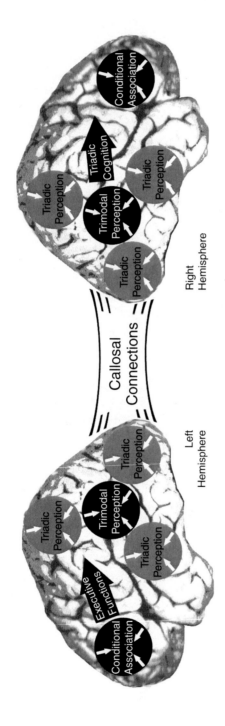

FIGURE 8.1. The full complexity of the brain code is apparent when we include the horizontal dimension of two comparable cerebral hemispheres undertaking slightly different perceptual and executive functions. Brain-imaging studies suggest a stronger involvement of the right than the left hemisphere in the processes of harmony and pictorial depth perception, but, for the executive control processes of speech and tool use, the left hemisphere is dominant. Callosal connections are thickest between regions of association cortex, suggesting the importance of interhemispheric exchanges in higher-level cognition.

Questions about the origins of human talents in the arts and sciences are as old as the written word, but the answers have changed decade by decade, as the methodologies of inquiry have changed. For the purposes of orientation and getting up to speed on these diverse topics, I recommend the books listed here. Without exception, they are indeed books, usually written by one author, and not necessarily available online. Technical articles, summaries in the popular press, hard-hitting blogs and visual images accessible on the Internet can also be useful, but it remains the case in the early twenty-first century that, if you are interested in understanding the well-thought out views of scholars whose normal workday includes reading the technical literature, the best sources are published books. Of course many unserious works are also published in book form; however, as a rule thorough and scholarly arguments that have been carefully constructed over a number of years can be found only as books. Even though electronic versions are rapidly becoming available, the bottom line is that valuable resources are rarely free.

In the field of music perception, there are many excellent advanced texts – notable among which are those by Meyer (1956), Cooke (1959), Narmour (1990), Scruton (1997), Temperley (2001) and Huron (2006). Connections between language and music are drawn by Wallin et al. (2000) and Patel (2008). General surveys of a high quality have been written by Storr (1992), Jourdain (1997) and Ball (2010). For basic musical acoustics, Sethares (2005) is the place to start, and, for psychoacoustics, Parncutt (1989) is enlightening. The recommended iconoclast on music is Pleasants (1955).

In visual psychology, the best survey of pictorial depth perception is that provided by Howard and Rogers (2002). Rather technical, but thoughtful texts on vision and art have been written by Gombrich (1960/2000), Arnheim (1969), Hoffman (1998), Zeki (1999) and Livingstone (2002). An edited book by Hecht et al. (2003) and Kubovy's (1986) discussion of perspective delve into some of the more difficult conceptual issues. Books specifically on shadows by Baxandall (1995) and Casati (2003) are exceptional. A textbook by Purves and Lotto (2003) on visual perception is essential reading, and the readable iconoclasts on art are Wolfe (1975) and Gablick (1977).

Outstanding books on the evolution of mind that are both rigorous and comprehensible to the nonspecialist have been written by Donald (1991, 2001), Mithen (1996), Tattersall (1998, 2008), Klein and Edgar (2002), Oppenheimer (2003) and Coolidge and Wynn (2009). More challenging discussions of brain structures and experimental techniques include books by Deacon (1997), Tomasello (2003, 2008), Fuster (2007), Gomez (2004) and Penn and Povinelli (2007). The triadic nature of human social interactions has been emphasized by Tomasello (1999). The most comprehensive, edited book is that from Lock and Peters (1999).

Laterality issues have been discussed in numerous single- and multiauthor books, among the best being those by Jaynes (1976), Corballis (1991), Hellige (1993), Ivry and Robertson (1998), Zaidel and Iacoboni (2003) and Hughdahl and Westerhausen (2010). McGilchrist (2009) has written a coherent, modern overview of human laterality with emphasis on the evolution of higher functions: highly recommended. Crow (2002) is the essential iconoclast.

Many excellent books on linguistics are available – the most readable of which are by Pinker (1994), Baker (2001) and Deutscher (2005). Intimations that head-phrase structure is the essence of Universal Grammar can be found there, but that point is not written in the underlined italics that are warranted. Jackendoff's career-long (2010) emphasis on the (tripartite) parallel architecture of language is remarkable. And books on the structure and evolution of language by Bickerton (1990, 2009) are essential reading.

Most essays on specifically human consciousness competently review the current lack of consensus on the problem of subjectivity, but the best critiques and clearest statements about red herrings and dead-ends are those by Searle (1997, 2004). Books by Baars (1997), Koch (2004), Dennett (1996), and Edelman and Tononi (2000) are enlightening on various psychological, brain science and philosophical issues, but only Damasio (2010) has hinted at a possible neuronal solution to the qualia conundrum. Finally, for the most coherent discussion of synchronization thus far, a book by Malsburg et al. (2010) is recommended. Koob (2009) is the glial iconoclast.

REFERENCES

Agnati, L. F., Zoli, M., Stromberg, I., & Fuxe, K. (1995) Intercellular communication in the brain: Wiring versus volume transmission. *Neuroscience* 69, 711–726.

Aleksander, I. (1996) *Impossible Minds: My Neurons, My Consciousness.* London: Imperial College Press.

Alivisatos, B., & Petrides, M. (1997) Functional activation of the human brain during mental rotation. *Neuropsychologia* 35, 111–118.

Allman, J. (2000) *Evolving Brains.* New York: W. H. Freeman.

Anderson, B., Peissig, J. J., Singer, J., & Sheinberg, D. L. (2006) XOR style tasks for testing visual object processing in monkeys. *Vision Research* 46, 1804–1815.

Anderson, J. A. (1995) *An Introduction to Neural Networks.* Cambridge MA: MIT Press.

Anderson, J. R. (1983) *The Architecture of Cognition.* Cambridge MA: Harvard University Press.

Arbib, M. A. (2002) The mirror system, imitation and the evolution of language. In *Imitation in Animals and Artifacts* (C. Nehaniv & K. Dautenhahn, eds.) Cambridge MA: MIT Press. pp. 248–299.

Arnheim, R. (1954/1974) *Art and Visual Perception.* Berkeley: University of California Press.

(1969) *Visual Thinking.* Berkeley: University of California Press.

Aron, A. R., Robbins, T. W., & Poldrack, R. A. (2004) Inhibition and the right inferior frontal cortex. *Trends in Cognitive Science* 8, 170–178.

Baars, B. (1988) *A Cognitive Theory of Consciousness.* Cambridge: Cambridge University Press.

(1997) *In the Theater of Consciousness: The Workspace of the Mind.* New York: Oxford University Press.

Baker, M. (2001) *The Atoms of Language: The Mind's Hidden Rules of Grammar.* New York: Basic Books.

Ball, P. (2008) Facing the music. *Nature* 453, 160–162.

(2010) *The Music Instinct.* London: The Bodley Head.

Banich, M. T. (1997) *Neuropsychology: The Neural Bases of Mental Function.* Boston: Houghton Mifflin.

Barlow, H., & Morgenstern, S. (1948) *A Dictionary of Musical Themes.* New York: Crown.

Baron-Cohen, S. (1995) *Mindblindness*. Cambridge MA: MIT Press.
Bartlett, T. (2010) Document sheds light on investigation at Harvard. *The Chronicle of Higher Education*. Vol. 56, Pp. 3–5.
Bates, E., & MacWhinney, B. (1989) *The Cross-Linguistic Study of Sentence Processing*. Cambridge: Cambridge University Press.
Baxandall, M. (1995) *Shadows and the Enlightenment*. New Haven CT: Yale University Press.
Bayne, T. (2010) *The Unity of Consciousness*. New York: Oxford University Press.
Bernstein, L. (1976) *The Unanswered Question: Six Talks at Harvard*. Cambridge MA: Harvard University Press.
Berthelet, A., & Chavaillon, J. (eds.) (1993) *The Use of Tools by Human and Non-human Primates*. Oxford: Clarendon Press.
Bianki, V. L. (1993) *The Mechanisms of Brain Lateralization*. Amsterdam: Gordon and Breach.
Bickerton, D. (1990) *Language and Species*. Chicago: University of Chicago Press.
    (2009) *Adam's Tongue: How Humans Made Language, How Language Made Humans*. New York: Farrar, Straus & Giroux.
    (2010) On two incompatible theories of language evolution. In *The Evolution of Language: Biolinguistic Perspectives*. (R. K. Larson, V. Deprez & H. Yamakido, eds.), New York: Cambridge University Press, pp. 199–210.
Bloom, J. S., & Hynd, G. W. (2005) The role of the corpus callosum in interhemispheric transfer of information: Excitation or inhibition? *Neuropsychology Review* 15, 59–71.
Boeckx, C. (2006) *Linguistic Minimalism*. Oxford: Oxford University Press.
Bolinger, D. L. (1978) Intonation across languages. In *Universals of Human Language: Phonology* (J. H. Greenberg, C. A. Ferguson & E. A. Moravcsik, eds.) Palo Alto: Stanford University Press, pp. 471–524.
Brenner, S. (1974) The genetics of *Caenorhabditis elegans*. *Genetics* 77, 71–94.
Brooks, D. I., & Cook, R. G. (2010) Chord discrimination in pigeons. *Music Perception* 27, 183–196.
Brown, S. (2000) The 'musilanguage' model of music evolution. In *The Origins of Music* (N. L. Wallin, B. Merker & S. Brown, eds.) pp. 271–300, Cambridge MA: MIT Press.
Butterworth, B. (1999) *The Mathematical Brain*. London: Macmillan.
Cairns-Smith, A. G. (1996) *Evolving the Mind*. Cambridge: Cambridge University Press.
Calvert, G., Spence, C., & Stein, B. E. (eds.) (2004) *The Handbook of Multisensory Processes*. Cambridge MA: MIT Press.
Casati, R. (2003) *Shadows*. New York: Random House.
Chalmers, D. (1996) *The Conscious Mind*. Oxford: Oxford University Press.
Chomsky, N. (1975) *Reflections on Language*. New York: Pantheon.
    (1985) *Knowledge of Language: Its Nature, Origin and Use*. New York: Praeger.
Chordia, P., & Rae, A. (2008) A large-scale survey of emotion in raag music. *Proceedings of ICMPC*, Sapporo, Japan.
Cook, N. D. (1984) Homotopic callosal inhibition. *Brain and Language* 23, 116–125.
    (1986) *The Brain Code: Mechanisms of Information Transfer and the Role of the Corpus Callosum*. London: Methuen.

(1995a) Artefact or network evolution? *Nature* 374, 313.
(1995b) Correlations between input and output units in neural networks. *Cognitive Science* 19, 563–574.
(1999) Simulating consciousness in a bilateral neural network. *Consciousness & Cognition* 8, 62–93.
(2001) Explaining harmony: The roles of interval dissonance and chordal tension. *Annals of the New York Academy of Science* 930, 382–385.
(2002) *Tone of Voice and Mind: The Connections Between Music, Language, Cognition and Consciousness*. Amsterdam: John Benjamins.
(2008) The neuron-level phenomena underlying cognition and consciousness: Synaptic activity and the action potential. *Neuroscience* 153, 556–570.
(2009) Harmony perception: Harmoniousness is more than the sum of interval consonance. *Music Perception* 27, 25–42.
Cook, N. D., & Fujisawa, T. X. (2006) The psychophysics of harmony perception: Harmony is a three-tone phenomenon. *Empirical Musicology Review* 1, 106–126.
Cook, N. D., Callan, D. & Callan, A. (2002) Frontal areas involved in the perception of harmony, *8th International Conference on Functional Mapping of the Human Brain*, June 2–6.
Cook, N. D., Frueh, H. R., Mehr, A., Regard, M., & Landis, T. (1994) Hemispheric cooperation in visuospatial rotations. *Brain and Cognition* 25, 240–249.
Cook, N. D., Frueh, H., & Landis, T. (1995) The cerebral hemispheres and neural network simulations. *Journal of Experimental Psychology: Human Perception and Performance* 21, 410–421.
Cook, N. D., Fujisawa, T. X., & Takami, K. (2006) Evaluation of the affective valence of speech using pitch substructure. *IEEE Transactions on Audio, Speech and Language Processing* 14, 142–151.
Cook, N. D., Fujisawa, T. X., & Konaka, H. (2007) Why not study polytonal psychophysics? *Empirical Musicology Review* 2, 38–43.
Cook, N. D., & Hayashi, T. (2008) The psychoacoustics of musical harmony. *American Scientist* 96(4), 311–319.
Cook, N. D., Hayashi, T., Amemiya, T., Suzuki, K., & Leuman, L. (2001) Effects of visual-field inversions on the reverse-perspective illusion. *Perception* 31, 1031–1037.
Cook, N. D., Yutsudo, A., Fujimoto, N., & Murata, M. (2008a) On the visual cues contributing to pictorial depth perception. *Empirical Studies of the Arts* 26, 67–90.
(2008b) Factors contributing to depth perception: Behavioral studies on the reverse perspective illusion. *Spatial Vision* 21, 397–405.
Cooke, D. (1959) *The Language of Music*. Oxford: Oxford University Press.
Coolidge, F. L., & Wynn, T. (2009) *The Rise of Homo Sapiens: The Evolution of Modern Thinking*. New York: Wiley-Blackwell.
Cooper, G., & Meyer, L. B. (1960) *The Rhythmic Structure of Music*. Chicago: Chicago University Press.
Corballis, M. C. (1991) *The Lopsided Ape*. New York: Oxford University Press.
(2009) The evolution of language. *Annals of the New York Academy of Science* 1156, 19–43.

Corballis, M.C. (2010) *The Recursive Mind*. Princeton: Princeton University Press.
Crick, F. C. (1994) *The Astonishing Hypothesis*. New York: Scribners.
Crick, F. C., & Koch, C. (2003) A framework for consciousness. *Nature Neuroscience* 6, 119–126.
Crow, T. J. (2000) Schizophrenia as the price that Homo sapiens pays for language: A resolution of the central paradox in the origin of species. *Brain Research Reviews* 31, 118–129.
 (2002) *The Speciation of Modern Homo sapiens*. London: British Academy.
Cruttendon, A. (1981) Falls and rises: Meanings and universals. *Journal of Linguistics*, 17, 77–91.
Cushing, J. T. (1994) *Quantum Mechanics: Historical Contingency and the Copenhagen Hegemony*. Chicago: University of Chicago Press.
Dahlhaus, C. (1978/1989) *The Idea of Absolute Music*. Chicago: The University of Chicago Press.
Damasio, A. (2010) *Self Comes to Mind: Constructing the Conscious Brain*. New York: Pantheon.
Damisch, H. (1987/1995) *The Origin of Perspective*. Cambridge MA: MIT Press.
Davidoff, J. (1995) Color perception. In *The Handbook of Brain Theory and Neural Networks* (M. A. Arbib, ed.). Cambridge MA: MIT Press, pp. 210–215.
Deacon, T. W. (1997) *The Symbolic Species*. New York: W. W. Norton.
Dehaene, S. (1997) *The Number Sense: How the Mind Creates Mathematics*. New York: Oxford University Press.
Dennell, R. (1983) *European Economic Prehistory*. London: Academic Press.
Dennett, D. C. (1991) *Consciousness Explained*. Boston: Little, Brown.
 (1996) *Kinds of Minds: Towards an Understanding of Consciousness*. London: Weidenfeld & Nicolson.
Desain, P. & Honing, H. (2003) The formation of rhythmic categories and metric priming. *Perception* 32(3), 341–365.
D'Esposito, M., et al., (1997) A functional MRI study of mental image generation. *Neuropsychologia* 35, 725–730.
Deutscher, G. (2005) *The Unfolding of Language: An Evolutionary Tour of Mankind's Greatest Invention*. New York: Henry Holt.
DeWaal, F. (2005) *Our Inner Ape*. New York: Penguin.
Dickens, P., et al. (2007) Clarinet acoustics. *Acoustics Australia* 35, 1–17.
Donald, M. (1991) *Origins of the Modern Mind*. Cambridge MA: Harvard University Press.
 (2001) *A Mind So Rare: The Evolution of Human Consciousness*. New York: Norton.
Dubreuil, J. (1642) *Practical Perspective*. London: Bowles & Carver. http://books.google.com.
Duffin, R.W. (2007) *How Equal Temperament Ruined Harmony (and Why You Should Care)*. New York: Norton.
Dunbar, R. (1996) *Grooming, Gossip and the Evolution of Human Language*. London: Faber & Faber.
Duncan, J. (2010) *How Intelligence Happens*. New Haven: Yale University Press.
Dutton, D. (2009) *The Art Instinct: Beauty, Pleasure and Human Evolution*. New York: Bloomsbury.

Eberlein, R. (1994) *Die Entstehung der tonalen Klangsyntax*. Frankfurt: Lang.
Eccles, J. C. (1994) *How the Mind Controls the Brain*. Berlin: Springer.
Edelman, G. M. (1989) *The Remembered Present*. New York: Basic Books.
(2004) *Wider Than the Sky*. New Haven CT: Yale University Press.
Edelman, G. M., & Tononi, G. (2000) *A Universe of Consciousness: How Matter Becomes Imagination*. New York: Basic Books.
Efron, R. (1990) *The Decline and Fall of Hemispheric Specialization*. Hillsdale, NJ: Lawrence Erlbaum.
Engel, A. K., & Singer, W. (2001) Temporal binding and the neural correlates of sensory awareness. *Trends in Cognitive Science* 5, 16–25.
Engel, A. K., Fries, P., Konig, P., Brecht, M., & Singer, W. (1999) Temporal binding, binocular rivalry and consciousness. *Consciousness and Cognition* 8, 128–151.
Escher, M. C. (1986) *Escher on Escher: Exploring the infinite*. New York: Abrams.
Fagot, J. (ed.) (2000) *Picture Perception in Animals*. Philadelphia: Taylor & Francis.
Fink, M. (1999) *Electroshock: Healing Mental Illness*. Oxford: Oxford University Press.
Fox, P. T., et al. (1996) A PET study of the neural systems of stuttering. *Nature* 382, 158–162.
Frosch, R. (2002) *Meantone Is Beautiful: Studies on Tunings of Musical Instruments*. Bern: Peter Lang.
Frost, G. T. (1980) Tool behavior and the origins of laterality. *Journal of Human Evolution* 9, 447–59.
Fujisawa, T. X. (2004) PhD Thesis, Kansai University (in Japanese).
Fujisawa, T. X., & Cook, N. D. (2011) The perception of harmonic triads: An fMRI study. *Brain Imaging and Behavior* 5, 109–125.
Fukushima, S. (2008) Master's Thesis, Kansai University (in Japanese).
Fuster, J. M. (2007) *The Prefrontal Cortex*. New York: Academic Press.
Gablick, S. (1977) *Progress in Art*. New York: Rizzoli.
Gabrielsson, A., & Juslin, P. N. (2003) Emotional expression in music. In *Handbook of Affective Sciences* (R. J. Richardson et al., eds.), Oxford: Oxford University Press, pp. 503–534.
Galaburda, A. M. (1995). Anatomic basis of cerebral dominance. In *Brain Asymmetry* (R. J. Davidson & K. Hugdahl, eds.). Cambridge MA: MIT Press, pp. 51–73.
Gazzaniga, M. S. (2000) Cerebral specialization and interhemispheric communication: Does the corpus callosum enable the human condition? *Brain* 123, 1293–1326.
Ghazanfar, A. A., & Schroeder, C. E. (2006) Is neocortex essentially multisensory? *Trends in Cognitive Sciences* 10, 278–285.
Gibson, K. R., & Ingold, T. (1993) *Tools, Language and Cognition in Human Evolution*. New York: Cambridge University Press.
Gill, H. S., O'Boyle, M. W., & Hathaway, J. (1998) Cortical distribution of EEG activity for component processes during mental rotation. *Cortex* 34, 707–718.
Glickstein, M., & Berlucchi, G. (2010) "Corpus callosum." In *The Cognitive Neuroscience of Mind* (Reuter-Lorenz, Baynes, K., Mangun, G.R., & Phelps, E.A., ed.), MIT Press, Cambridge MA, pp. 3–24.
Goldman, R. F. (1965) *Harmony in Western Music*. New York: Norton.

Gombrich, E. H. (1960/2000) *Art and Illusion*. Princeton NJ: Princeton University Press.

(1995) *Shadows: The Depiction of Cast Shadows in Western Art*. London: National Gallery Publications.

Gomez, J. C. (2004) *Apes, Monkeys, Children and the Growth of Mind*. Cambridge MA: Harvard University Press.

Green, H. S., & Triffet, T. (1997) *Sources of Consciousness: The Biophysical and Computational Basis of Thought*. Singapore: World Scientific.

Greenberg, J. H. (1968) Some universals of grammar. In, *Universals of Language* (J. H. Greenberg, ed.), Cambridge MA: MIT Press.

Gregory, R. L. (1966/1997) *Eye and Brain*. Princeton NJ: Princeton University Press.

Grodzinsky, Y., & Amunts, K. (eds.) (2006) *Broca's Region*. New York: Oxford University Press.

Gruesser, O.-J. & Landis, T. (1991) *Visual Agnosias and Other Disturbances of Visual Perception and Cognition*. London: Macmillan.

Guiard, Y. (1980) Cerebral hemispheres and selective attention, *Acta Psychologica* 46, 41–61.

Hameroff, S. R. (1994) Quantum consciousness in microtubules: An intra-neuronal substrate for emergent consciousness? *Journal of Consciousness Studies* 1, 91–118.

Hameroff, S. R., & Penrose, R. (1996) Conscious events as orchestrated space-time selections. *Journal of Consciousness Studies* 3, 36–53.

Hanslick, E. (1891/1986) *On the Musically Beautiful*. Indianapolis: Hackett.

Harrison, D. (1994) *Harmonic Function in Chromatic Music*. Chicago: The University of Chicago Press.

Hatton, G. I. & Parpura, V. (2004) *Glial-Neuronal Signaling*. Boston: Kluwer.

Hauser, M. D. (2006) *Moral Minds: How Nature Designed Our Universal Sense of Right and Wrong*. New York: HarperCollins.

Hayashi, T., Umeda, C., & Cook, N. D. (2007) An fMRI study of the pictorial cues (perspective lines, shading, shadows) contributing to depth perception in the reverse perspective illusion. *Brain Research* 1163, 72–78.

Hecht, H., Schwartz, R., & Atherton, M. (eds.) (2003) *Looking Into Pictures: An Interdisciplinary Approach to Pictorial Space*. Cambridge MA: MIT Press.

Heilbron, J. L. (2010) *Galileo*. New York: Oxford University Press.

Hellige, J. B. (1993) *Hemispheric Asymmetry: What's Right and What's Left*. Cambridge MA: Harvard University Press.

Helmholtz, H. L. F. (1877/1954) *On the Sensations of Tone as a Physiological Basis for the Theory of Music*. New York: Dover.

Hershenson, M. (1999) *Visual Space Perception: A Primer*. Cambridge MA: MIT Press.

Hevner, K. (1936) Experimental study of the elements of expression in music. *American Journal of Psychology* 48, 246–68.

Hilgard, E.R. (1980) The trilogy of mind: Cognition, affection, and conation. *Journal of the History of Behavioral Sciences* 16, 107–117.

Hill, H., & Johnston, A. (2007) The hollow-face illusion: Object-specific knowledge, general assumptions or properties of the stimulus? *Perception* 36, 199–223.

Hills, P. (1987) *The Light of Early Italian Painting*. New Haven CT: Yale University Press.

Hoffman, D. D. (1998) *Visual Intelligence*. New York: W. W. Norton.
Holloway, R. L. (1995) Toward a synthetic theory of human brain evolution. In *Origins of the Human Brain* (J. P. Changeux & J. Chavaillon, eds.). Oxford: Clarendon Press, pp. 42–55.
   (1999) Evolution of the human brain. In (A. Lock & C. R. Peters, eds.) *Handbook of Human Symbolic Evolution*. Oxford: Blackwell, pp. 74–125.
Howard, I., & Rogers, B. (2002) *Depth Perception*. Toronto: I Porteus.
Howe, C. Q., & Purves, D. (2005) *Perceiving Geometry*. Berlin: Springer.
Hughdahl, K., & Westerhausen, R. (eds.) (2010) *The Two Halves of the Brain*. Cambridge MA: MIT Press.
Hulse, S. H., Bernard, D. J., & Braaten, R. F. (1995) Auditory discrimination of chord-based spectral structures by European starlings (*Sturnus vulgaris*). *Journal of Experimental Psychology* 124, 409–423.
Hume, D. (1739/1978) *A Treatise on Human Nature*. Oxford: Oxford University Press.
Hurford, J. R. (2007) *The Origins of Meaning*. Oxford: Oxford University Press.
Huron, D. (2006) *Sweet Anticipation: Music and the Psychology of Anticipation*. Cambridge MA: MIT Press.
   (2008) Lost in music. *Nature* 453: 456–457.
Isacoff, S. (2001) *Temperament*. New York: Vintage.
Ivry, R. B., & Robertson, L. C. (1998) *The Two Sides of Perception*. Cambridge MA: MIT Press.
Jackendoff, R. (1997) *The Architecture of the Language Faculty*. Cambridge MA: MIT Press.
   (2002) *Foundations of Language*. Oxford: Oxford University Press.
   (2010) *Meaning and the Lexicon: The Parallel Architecture 1975–2010*. Oxford: Oxford University Press.
Jaynes, J. (1976) *The Birth of Consciousness in the Breakdown of the Bicameral Mind*. Boston: Houghton-Mifflin.
Jeeves, M. A. (1994) Callosal agenesis – A Natural Split-Brain. In *Callosal Agenesis: A Natural split brain?* (M. Lassonde & M. A. Jeeves, eds.). New York: Plenum, pp. 285–300.
Jelinek, A. J. (1977) The lower Paleolithic: Current evidence and interpretations. *Annual Review of Anthropology* 6, 11–32.
Johnson-Frey, S. H. (2004) The neural bases of complex tool use in humans. *Trends in Cognitive Sciences* 8, 71–78.
Jones, G. T. (1974) *Music Theory*. New York: Barnes and Noble.
Jones, E. G. & Powell, T. P. S. (1970) An anatomical study of converging sensory pathways within the cerebral cortex of the monkey. *Brain* 93, 793–820.
Jourdain, R. (1997) *Music, the Brain, and Ecstasy*. New York: William Morrow.
Juslin, P. N., & Laukka, P. (2003) Communication of emotions in vocal expression and music performance: Different channels, same code? *Psychological Bulletin* 129, 770–814.
Kaas, J. H., & Collins, C. E. (2003) The resurrection of multisensory cortex in primates. In *The Handbook of Multisensory Processes* (G. Calvert, C. Spence & B. E. Stein, eds.). Cambridge MA: MIT Press, pp. 285–293.
Kameoka, A., & Kuriyagawa, M. (1969) Consonance theory: Parts I and II. *Journal of the Acoustical Society of America* 45, 1451–1469.

Kandel, E. R., Frazier, W. T., & Coggeshall, R. E. (1967) Direct and common connections among identified neurons in Aplysia. *Journal of Neurophysiology* 30, 1352–1376.
Kant, I. (2001) *Lectures on Ethics*. New York: Cambridge University Press.
Kastner, M. P., & Crowder, R. G. (1990) Perception of major/minor: IV. Emotional connotations in young children. *Music Perception* 8, 189–202.
Kauppinen, R. A., & Williams, S. R. (1998) Use of NMR spectroscopy in monitoring cerebral pH. In *pH and Brain Function*. New York: Wiley-Liss, pp. 605–619.
Kemp, M. (1997) *The Science of Art*. New Haven CT: Yale University Press.
Kinsbourne, M. (1982) Hemispheric specialization and the growth of human understanding. *American Psychologist* 37, 411–420.
Klein, R. G., & Edgar, B. (2002) *The Dawn of Human Culture*. New York: Wiley.
Koch, C. (2004) *The Quest for Consciousness: A Neurobiological Approach*. Englewood CO: Roberts.
Koob, A. (2009) *The Root of Thought*. Upper Saddle River NJ: Pearson.
Kosslyn, S. M. (1994) *Image and Brain*. Cambridge MA: MIT Press.
Kubovy, M. (1986) *The Psychology of Perspective and Renaissance Art*. Cambridge: Cambridge University Press.
Kuusi, T. (2002) Theoretical resemblance versus perceived closeness: Explaining estimations of pentachords by abstract properties of pentad classes. *Journal of New Music Research*. http://www.informaworld.com/smpp/title~content=t713 817838~db=all~tab=issueslist~branches=31 − v3131, 377–388.
Ladd, D. R. (1996) *Intonational Phonology*. Cambridge: Cambridge University Press.
LeDoux, J. (2002) *Synaptic Self: How Our Brains Become Who We Are*. New York: Penguin.
Lerdahl, F. (2001) *Tonal Pitch Space*. New York: Oxford University Press.
Levelt, W. J. M. (1999) Producing spoken language: a blueprint of the speaker. In *The Neurocognition of Language* (C. M. Brown & P. Hagoort, eds.), (pp. 83–122). Oxford: Oxford University Press.
Levine, J. (2001) *Purple Haze: The Puzzle of Consciousness*. New York: Oxford. University Press.
Levitan, I. B., & Kaczmarek, L. K. (1997) *The Neuron: Cell and Molecular Biology*. New York: Oxford University Press.
Libet, B. (2004) *Mind Time: The Temporal Factor in Consciousness*. Cambridge MA: Harvard University Press.
Lichtheim, L. (1885) Über Aphasie. *Deutsches Archiv für Klinische Medicin* 36, 204–268.
Lieberman, P. (2006) *Toward an Evolutionary Biology of Language*. Cambridge MA: Harvard University Press.
Liepmann, H. (1908) Die linke Hemisphaere und das Handeln. In *Drei Aufsaetze aus dem Apraxiegebiet*. Berlin: Von Karger, pp. 17–50.
Livingstone, M. (2002) *Vision and Art: The Biology of Seeing*. New York: Abrams.
Llinas, R. R. (2001) *I of the Vortex: From Neurons to Self*. Cambridge MA: MIT Press.
Lock, A., & Peters, C. R. (1999) *Handbook of Human Symbolic Evolution*. Oxford: Blackwell.

Lynch, G., & Granger, R. (2008) *Big Brain: The Origins and Future of Human Intelligence*. New York: Macmillan.
MacLennan, B. (1996) Protophenomena: The elements of consciousness and their relation to the brain. *Journal of Consciousness Studies* 3, 409–424.
MacNeilage, P. F. (2008) *The Origin of Speech*. New York: Oxford University Press.
Macphail, E. M. (1987) The comparative psychology of intelligence. *Behavioral and Brain Sciences* 10, 645–656.
Malsburg, C. von der, Phillips, W. A., & Singer, W. (eds.) (2010) *Dynamic Coordination in the Brain: From Neurons to Mind* Cambridge MA: MIT Press.
Martin, B. (1986) Cellular effects of cannabinoids. *Pharmaceutical Review* 38, 45–74.
Mathieu, W. A. (1997) *Harmonic Experience*. Rochester, VT: Inner Traditions.
Maurer, D., et al. (2007) Neural correlates of processing facial identity based on features versus their spacing. *Neuropsychologia* 45, 1438–1451.
McDermott, J. (2008) The evolution of music. *Nature* 453, 287–288.
McFadden, J. (2002) Synchronous firing and its influence on the brain's electromagnetic field: Evidence for an electromagnetic theory of consciousness. *Journal of Consciousness Studies* 9, 23–50.
McGilchrist, I. (2009) *The Master and His Emisssary: The divided brain and the making of the Western world*. New Haven: Yale University Press.
McGinn, C. (2004) *Consciousness and Its Objects*. New York: Oxford University Press.
McManus, C. (2002) *Right Hand, Left Hand*. Cambridge MA: Harvard University Press.
Mehta, Z., Newcombe, F., & Damasio, H. (1987) A left hemisphere contribution to visuospatial processing. *Cortex* 23, 447–461.
Meyer, L. B. (1956) *Emotion and Meaning in Music*. Chicago: University of Chicago Press.
Mikhail, J. (2007) Universal moral grammar: Theory, evidence and the future. *Trends in Cognitive Sciences* 11, 143–152.
   (2010) *Elements of Moral Cognition*. New York: Cambridge University Press.
Minsky, M. & Papert, S. (1969) *Perceptrons*. Cambridge MA: MIT Press.
Mithen, S. (1996) *The Prehistory of the Mind*. London: Thames & Hudson.
Mitterauer, B., & Kopp, K. (2003) The self-composing brain: Towards a glial-neuronal brain theory. *Brain and Cognition* 51, 357–367.
Morton, E. W. (1977) On the occurrence and significance of motivation-structural roles in some bird and mammal sounds. *American Naturalist* 111, 855–869.
Nagel, T. (1986) *The View from Nowhere*. New York: Oxford University Press.
Narmour, E. (1990) *The Analysis and Cognition of Basic Melodic Structures*. Chicago: University of Chicago Press.
Neisser, U. (2001) Admirably adaptive, occasionally intelligent. In *Looking at Looking: An Introduction to the Intelligence of Vision* (T. E. Parks, ed.). Thousand Oaks CA: Sage Publications, pp. 75–82.
Ogura, A., & Machemer, H. J. (1980) Distribution of mechanoreceptor channels in the Paramecium surface membrane. *Journal of Comparative Physiology* A135, 233–242.

Ohala, J. J. (1983) Cross-language use of pitch: An ethological view. *Phonetica* 40, 1–18.

(1984) An ethological perspective on common cross-language utilization of Fo in voice. *Phonetica* 41, 1–16.

(1994) The frequency code underlies the sound-symbolic use of voice-pitch. In *Sound Symbolism* (L. Hinton, J. Nichols & J. J. Ohala, eds.), (pp. 325–347). New York: Cambridge University Press.

Oppenheimer, S. (2003) *Out of Eden: The Peopling of the World*. London: Constable.

Panofsky, E. (1927/1997) *Perspective as Symbolic Form*. New York: Zone Books.

Papathomas, T. V. (2007) Art pieces that 'move' in our minds –An explanation of illusory motion based on depth reversal. *Spatial Vision* 21, 97–117.

Parncutt, R. (1989) *Harmony: A Psychoacoustical Approach*. Berlin: Springer.

Passingham, R. (2008) *What Is Special about the Human Brain?* New York: Oxford University Press.

Patel, A. D. (2008) *Music, Language and the Brain*. Oxford: Oxford University Press.

Penn, D. C., & Povinelli, D. J. (2007) Causal cognition in human and nonhuman animals. *Annual Review of Psychology* 58, 97–118.

Penrose, R. (1989) *The Emperor's New Mind*. Oxford: Oxford University Press.

Perani, D., et al. (1995) Different neural systems for the recognition of animals and man-made tools. *NeuroReport* 6, 1637–1641.

Persichetti, V. (1961) *Twentieth Century Harmony*. New York: W. W. Norton.

Petersson, K. M., et al. (2004) Artificial syntactic violations activate Broca's area. *Cognitive Science* 28, 383–407.

Petrides, M. (2005) The rostral-causal axis of cognitive control within the lateral frontal cortex. In *Monkey Brain to Human Brain* (S. Dehaene et al., eds.). Cambridge MA: MIT Press.

Piattelli-Palmarini, M. (2010) What is language, that it may have evolved, and what is evolution, that it may apply to language? In *The Evolution of Language: Biolinguistic Perspectives* (R. K. Larson, V. Deprez & H. Yamakido, eds.). New York: Cambridge University Press, pp. 148–162.

Pinker, S. (1994) *The Language Instinct: How the Mind Creates Language*. New York: HarperCollins.

(1997) *How the Mind Works*. New York: W. W. Norton.

Piston, W. (1941/1987) *Harmony*. New York: Norton.

Pleasants, H. (1955) *The Agony of Modern Music*. New York: Simon & Schuster.

Plomp, R., & Levelt, W. J. M. (1965) Tonal consonances and critical bandwidth. *Journal of the Acoustical Society of America* 38, 548–560.

Pollard, C., & Sag, I. A. (1994) *Head-Driven Phrase Structure Grammar*. Chicago: Chicago University Press.

Povinelli, D. J. (2000) *Folk Physics for Apes*. New York: Oxford University Press.

Premack, D., & Premack, A. J. (2002) *Original Intelligence*. New York: McGraw-Hill.

Pulvermüller, F. (2002) *The Neuroscience of Language*. Cambridge: Cambridge University Press.

Purves, D., & Lotto, R. B. (2003) *Why We See What We Do*. Sunderland MA: Sinauer.

Ramachandran, V. S. (2011) *The Tell-Tale Brain*. New York: Norton.

Rameau, J.-P. (1722/1971) *Treatise on Harmony* (P. Gossett, trans.). New York: Dover.
Rawls, J. (1971) *A Theory of Justice*. Cambridge MA: Harvard University Press.
Reggia, J. A., Goodall, S. M., & Levitan, S. (2001) Cortical map asymmetries in the context of transcallosal excitatory influences. *NeuroReport* 12, 1609–1614.
Revonsuo, A. (2005) *Inner Presence: Consciousness as a Biological Phenomenon*. Cambridge MA: MIT Press.
Rizzolatti, G., & Sinigaglia, C. (2008) *Mirrors in the Brain: How Our Minds Share Actions, Emotions and Experience*. New York: Oxford University Press.
Roberts, L. (1986) Consonant judgments of musical chords by musicians and untrained listeners. *Acustica* 62, 163–171.
Rumelhart, D. E., & Hinton, G. E., & Williams, R. J. (1986) Learning internal representations by error propagation. In *Parallel Distributed Processing* (D. E. Rumelhart & J. L. McClelland, eds.) Cambridge MA: MIT Press, pp. 318–362.
Sacks, O. (1970) *The Man Who Mistook His Wife for a Hat and Other Clinical Tales*. New York: Touchstone.
 (1985) *The Man Who Mistook His Wife for a Hat*. New York: Knopf.
Satinover, J. (2001) *The Quantum Brain: The Search for Freedom and the Next Generation of Man*. New York: Wiley.
Savage-Rumbaugh, S., Murphy, J., Sevcik, R. A., Brakke, K. E., Williams, S. L., & Rumbaugh, D. M. (1993) *Language Comprehension in Ape and Child*. Monographs of the Society for Research in Child Development 233. Chicago: Chicago University Press.
Scherer, K. R. (1995) Expression of emotion in voice and music. *Journal of Voice* 9, 235–248.
Schick, K. D., & Toth, N. (1993) *Making Silent Stones Speak*. New York: Simon & Schuster.
Schoenberg, A. (1911/1983) *Theory of Harmony* (R. E. Carter, trans.). London: Faber & Faber.
Schulte, T., & Muller-Oehring, E. M. (2010) Contribution of callosal connections to the interhemispheric integration of visuomotor and cognitive processes. *Neuropsychology Review* 20, 174–190.
Scott, A. (1995) *Stairway to the Mind: The Controversial New Science of Consciousness*. New York: Copernicus.
Scruton, R. (1997) *The Aesthetics of Music*. Oxford: Clarendon.
Searle, J. R. (1992) *The Rediscovery of the Mind*. Cambridge MA: MIT Press.
 (1997) *The Mystery of Consciousness*. New York: NyRev.
 (1998) *Mind, Language and Society*. New York: Basic Books.
 (2004) *Mind: A Brief Introduction*. Oxford: Oxford University Press.
Seltzer, B., & Pandya, D. N. (1994) Parietal, temporal and occipital projections to cortex of the superior temporal sulcus in the rhesus monkey. *Journal of Comparative Neurology* 343, 445–463.
Sethares, W. A. (2005) *Tuning, Timbre, Spectrum, Scale*. Berlin: Springer.
Shlain, L. (1991) *Art and Physics: Parallel Visions in Space, Time and Light*. New York: Morrow.
Shkuro, Y., Glezer, M., & Reggia, J. A. (2000) Interhemispheric effects of simulated lesions in a neural net model of single-word reading. *Brain and Language* 72, 343–374.

Singer, W. (1998) Consciousness and the structure of neuronal representations, *Philosophical Transactions of the Royal Society (London), B: Biological Science* 353, 1829–1840.
Sloboda, J. (2005) *Exploring the Musical Mind*. Oxford: Oxford University Press.
Slyce, J. (1998) *Perverspective: Patrick Hughes*. London: Momentum.
Snyder, S. H. (1996) *Drugs and the Brain*. New York: Scientific American Library.
Solso, R. L. (1994) *Cognition and the Visual Arts*. Cambridge MA: MIT Press.
  (2003) *The Psychology of Art and the Evolution of the Conscious Brain*. Cambridge MA: MIT Press.
Sperber, D., Premack, D., & Premack, A. J. (eds.) (1995) *Causal Cognition*. Oxford: Clarendon Press.
Sperry, R. W. (1968) Hemisphere deconnection and unity in conscious awareness. *American Psychologist* 23, 723–733.
Stapp, H. P. (1993) *Mind, Matter and Quantum Mechanics*. New York: Springer.
Stein, B., & Meredith, M. A. (1993) *The Merging of the Senses*. Cambridge MA: MIT Press.
Stoichita, V. I. (1997) *A Short History of the Shadow*. London: Reaktion Books.
Storr, A. (1992) *Music and the Mind*. New York: Ballantine Books.
Stout, D., & Chaminade, T. (2007) The evolutionary neuroscience of tool making. *Neuropsychologia* 45, 1091–1100.
Striedter, G. F. (2005) *Principles of Brain Evolution*. Sunderland MA: Sinauer.
Suzuki, M., et al. (2008) Discrete cortical regions associated with the musical beauty of major and minor chords. *Cognitive, Affective & Behavioral Neuroscience* 8, 126–131.
Swanson, L. W. (2003) *Brain Architecture: Understanding the Basic Plan*. New York: Oxford University Press.
Tattersall, I. (1998) *Becoming Human*. New York: Harcourt Brace.
  (2008) *The World from Beginnings to 4000 BCE*. Oxford: Oxford University Press.
Taylor, J. G. (1999) *The Race for Consciousness*. Cambridge MA: MIT Press.
Temperley, D. (2001) *The Cognition of Basic Musical Structures*. Cambridge MA: MIT Press.
Terwogt, M. M., & van Grinsven, F. (1991) Musical expression of mood states. *Psychology of Music* 19, 99–109.
Tomasello, M. (1999) *The Cultural Origins of Human Cognition*. Cambridge MA: Harvard University Press.
  (2003) *Constructing a Language*. Cambridge MA: Harvard University Press.
  (2008) *Origins of Human Communication*. Cambridge MA: MIT Press.
Tomasello, M., & Call, J. (1997) *Primate Cognition*. Oxford: Oxford University Press.
Trainor, L. (2008) The neural roots of music. *Nature* 453, 598–599.
Trainor, L. J., & Trehub, S. (1992) The development of referential meaning in music. *Music Perception* 9, 455–470.
Ullman, S. (2000) *High-Level Vision*. Cambridge MA: MIT Press.
Velmans, M. (2000) *Understanding Consciousness*. London: Routledge.
Verkhratsky, A., & Butt, A. (2007) *Glial Neurobiology*. Chichester: Wiley.
Vollenweider, F. X., & Kometer, M. (2010) The neurobiology of psychedelic drugs: implications for the treatment of mood disorders. *Nature Reviews Neuroscience* 11, 642–651.

Volterra, A., Magistretti, P., & Haydon, P. (2002) *The Tripartite Synapse: Glia in Synaptic Transmission.* New York: Oxford University Press.

Walhout, M., Endoh, H., Thierry-Mieg, N., Wong, W., & Vidal, M. (1998) Insights from model systems: A model of elegance. *American Journal of Human Genetics* 63, 955–961.

Walker, E. H. (2000) *The Physics of Consciousness: Quantum Minds and the Meaning of Life.* Cambridge MA: Perseus Publishing.

Wallin, N. L., Merker, B., & S. Brown, S. (eds.) (2000) *The Origins of Music.* Cambridge MA: MIT Press.

Watson, J. B. (1913) Psychology as the behaviorist views it. *Psychological Review* 20, 158–177.

Wexler, M. (2005) Anticipating the three-dimensional consequences of eye movements. *Proceedings of the National Academy of Science (USA)* 102, 1246–1251.

Wexler, M., Kosslyn, S. M., & Berthoz, A. (1998) Motor processes in mental rotation. *Cognition* 68, 77–94.

Whiten, A. (1999) The evolution of deep social mind in humans. In *The Descent of Mind* (M. C. Corballis & S. E. G. Lea, eds.) New York: Oxford University Press, pp. 173–193.

Wolfe, T. (1975) *The Painted Word.* New York: Farrar, Straus & Giroux.

Wolpert, L. (2003) Causal belief and the origins of technology. *Philosophical Transactions of the Royal Society, London A* 361, 1709–1719.

Wood, W. B. (1988) *The Nematode Caenorhabditis Elegans.* Cold Spring Harbor NY: Cold Spring Harbor Laboratory.

Wrangham, R. (2009) *Catching Fire: How Cooking Made Us Human.* New York: Basic Books.

Wynn, T. G. (1996) The evolution of tools and symbolic behavior. In *Handbook of Human Symbolic Evolution* (A. Lock & C. R. Peters, eds.) Oxford: Blackwell, pp. 263–287.

Yarbus, A. L. (1959) *Eye Movement and Vision* (B. Haigh, transl., L.A. Riggs, ed.) New York: Plenum Press.

Zaidel, E., & Iacoboni M. (eds.) (2003) *The Parallel Brain: The Cognitive Neuroscience of the Corpus Callosum.* Cambridge MA: MIT Press.

Zatorre, R. J. (2001) Neural specializations for tonal processing. In *The Biological Foundations of Music.* (R. J. Zatorre & I. Peretz, eds.) *Annals of the New York Academy of Sciences*, Volume 930, pp. 193–210.

Zeki, S. (1999) *Inner Vision.* Oxford: Oxford University Press.

INDEX

Action potential, 257, 275–287, 293, 295, 300, 301
Aerial perspective. *See* Atmospheric perspective.
Agnati, L. F., 291, 299
Aleksander, I., 271, 298, 305
Alivisatos, B., 322
Allman, J., 191
Amunts, K., 223
Anderson, J. A., 318
Anderson, J. R., 24, 319
Anesthetics, 293
Apes, 1–3, 5, 6, 20–23, 121, 122, 177, 188, 190, 226, 239, 240, 252, 311, 312
Arbib, M. A., 252
Arnheim, R., 335
Aron, A. R., 211
Art, 1, 4, 10, 24, 120–174, 332
Association cortex, 25, 190–204, 330
Astronomy, 12, 116, 155–157
Atmospheric perspective, 128, 129, 158, 159
Augmented chords, 42–110

Baars, B., 262, 264, 272, 305, 336
Baker, M., 244, 336
Ball, P., 46, 48, 66, 81, 82, 118, 335
Banich, M. T., 224
Baron-Cohen, S., 312
Bartlett, T., 197
Bates, E., 239, 240
Baxandall, M., 150, 155, 335
Bayne, T., 274
Behaviorism, 212, 332
Bernstein, L., 36, 62
Berthelet, A., 213
Bianki, V. L., 185

Bickerton, D., 19, 225–235, 240–242, 247, 250, 336
Bipedal gait, 2, 227, 329
Birdsong, 15, 16, 67, 91, 95, 119
Bloom, J. S., 186
Boeckx, C., 243
Bolinger, D. L., 86
Brain size, 187–192
Brenner, S., 279
Brooks, D. L., 92, 118
Brown, S., 87
Butt, A., 301
Butterworth, B., 324

Cadences, 16, 27, 38, 102–107
Cairns-Smith, G., 282
Call, J., 213, 331
Calvert, G., 197
Casati, R., 148, 150, 151, 155, 156, 335
Causality, 211–213
Chalmers, D., 262–266, 271, 272, 304
Chaminade, T., 198
Chavaillon, J., 213
Chomsky, N., 20, 225, 226, 234, 241, 246, 313
Chordia, P., 67
Chromatic music, 15, 26, 62, 95
Circle of Fifths. *See* Cycle of Fifths.
Cognition, 255–257, 273–279, 301–303
Cognitive neuroscience, 4, 5, 8, 21, 126, 257, 276, 295, 330–333
Collins, C. E., 198
Color perception, 319–321
Coma, 293
Conditional associations, 9, 208–211, 317–319
Consciousness, 21, 250–306
Consonance, 26, 31–38, 43–119, 308

Context, 14, 23, 30, 38, 176, 177, 310, 331, 332
Cook, R. G., 92, 118
Cooke, D., 89, 335
Coolidge, F. L., 180, 197, 335
Cooper, G., 308
Cooperation, 5, 7, 23, 24, 311–313
Corballis, M., 183, 187, 217, 246, 336
Corpus callosum, 184–187, 333, 334
Crick, F., 264, 276, 278, 287, 288, 304
Crow, T. J., 182, 336
Crowder, R. G., 65
Cruttendon, A., 86
Cushing, J. T., 269
Cycle of Fifths, 101–105, 107
Cycle of Modes, 78–80, 83, 84, 104, 105, 109

Dahlhaus, C., 91
Damasio, A., 264, 303–306, 336
Damisch, H., 127, 131, 161
Davidoff, J., 320
Deacon, T. W., 190, 210, 228, 234, 331, 335
Dehaene, S., 324
Dennell, R., 205
Dennett, D., 262–265, 271, 272, 304, 336
Depth inversion. See Reverse perspective.
Depth perception, 7, 16, 24, 120–174
Desain, P., 308
D'Esposito, M., 322
Deutscher, G., 237, 239, 245, 336
Development, child, 22, 311, 312
DeWaal, F., 307
Diminished chords, 42–110
Dissonance, 26, 28–38, 43–119, 308
Donald, M., 24, 201, 223, 250, 287, 335
Dualism, 257, 265, 274
Dubreuil, J., 150
Duerer, A., 127
Duffin, R. W., 35
Dunbar, R., 226
Duncan, J., 260
Dutton, I., 313
Dyadic associations, 9, 12–14, 22–24, 212, 329–333

Eberlein, R., 76
Eccles, J., 268
Edelman, G., 264, 266, 304, 336
Edgar, B., 229, 335
Education, 23, 327–328
Efron, R., 184
Emotions, musical, 64–94, 113, 114, 332
Engel, A. K., 288, 289, 305

Evolution, 1–3, 6, 7, 24, 25, 90, 91, 175–215, 224–234
Excitable cells, 256–288, 295
Excitation, synaptic, 209, 210, 275–278, 282, 317–319
Explanatory gap, 267

Face perception, 309, 310, 332
Fagot, J., 122, 323
Feeling, 257, 258, 279–300
Fink, M., 293
Fire, 2, 201, 202
Four-body associations, 23, 325, 326
Fox, P. T., 183
Frequency code, 84–94
Frosch, R., 35
Frost, G. T., 181
Fujisawa, T. X., 47, 48, 66, 68, 69, 110–114
Fukushima, S., 101
Fuster, J. M., 208, 264, 331, 335

Gablick, S., 163, 335
Gabrielsson, A., 65
Galilei, V., 61, 117
Galilei, G., 61, 156
Gazzaniga, M. S., 185
Genetic code, 332
Gestalt psychology, 51, 308
Ghazanfar, A. A., 198
Gibson, K. R., 213
Gill, H. S., 332
Glia, 300, 301, 336
Goldman, R. F., 107
Gombrich, E. H., 149, 155, 335
Gomez, J. C., 213, 331, 335
Grammar. See Syntax.
Granger, R., 188
Green, H. S., 268
Greenberg, J. H., 238
Gregory, R. L., 133
Grodzinsky, Y., 223
Gruesser, O. J., 309, 310
Guiard, Y., 185

Hafted tools, 204–206
Hameroff, S. R., 261, 264, 268
Handedness, 177–215, 321–323, 333
Hanslick, E., 91
Harmonics, higher. See Partials, upper.
Harmony, 7, 10, 15, 26–119, 333
Harrison, D., 107
Hatton, M., 301

Hauser, M., 313–316
Hayashi, T., 39, 59, 93, 125, 171, 172
Head-driven phrase structure grammar.
    See Phrase structure.
Hebbian learning, 9, 330
Hecht, H., 161, 163, 335
Heilbron, J. L., 61, 117, 156
Heisenberg, W., 268, 269
Hellige, J. B., 185, 336
Helmholtz, H. L., 35, 37, 57, 61, 95, 115
Hershenson, M., 133
Hevner, K., 65
Hilgard, E. R., 12
Hill, H., 165
Hills, P., 149, 155
Hoffman, D. D., 127, 133, 335
Hogarth, W., 165
Holloway, R. L., 198, 214
Honing, H., 308
Howard, I., 127, 133, 335
Howe, C. Q., 136
Hughes, P., 164–166, 170, 171
Hulse, S. H., 92, 118
Hume, D., 314–316
Hurford, J. R., 222
Huron, D., 37, 48, 81, 103, 104, 118, 309, 335
Hynd, G. W., 186

Iacoboni, M., 184, 336
Illusions, visual, 51, 52, 164–173
Inhibition, synaptic, 181, 186, 187, 209, 210, 275–278, 282, 317–319, 330
Ingold, T., 213
Instability, harmonic, 38–119
Interrogatives, 87
Interposition. *See* Occlusion.
Intervallic equivalence, 68, 73, 75, 76, 78, 91, 94–96, 115
Intervals, pitch, 27–38
Intuition, 299–300
Isacoff, S., 35
Ivry, R. B., 185, 336

Jackendoff, R., 19, 217, 218, 222, 237, 336
Jaynes, J., 302, 304, 336
Jazz, 47, 68
Jeeves, M. A., 187
Jelinek, A. J., 200
Johnson, B., 130
Johnson-Frey, S. H., 207
Johnston, A., 165
Joint attention, 7, 22, 23, 311–313

Jones, E.G., 194, 195
Jones, G. T., 55
Jourdain, R., 335
Juslin, P. N., 65, 86

Kaas, J. H., 198
Kaczmarek, I. K., 285
Kameoka, A., 35, 36, 47
Kandel, E. R., 279
Kant, I., 315, 316
Kastner, M. P., 65
Kauppinen, R. A., 293
Kemp, M., 155
Kinsbourne, M., 186
Klein, R. G., 229, 335
Koch, C., 259, 264, 278, 288, 289, 304, 336
Kometer, M., 294
Koob, A., 300, 301, 336
Kopp, K., 300
Kosslyn, S. M., 322
Kubovy, M., 127, 131, 335
Kuriyagawa, M., 35, 36, 47
Kuusi, T., 326

Ladd, D. R., 86
Landis, T., 309, 310
Language, 3, 5–7, 11, 19–21, 23, 24, 216–254, 302, 333
Laukka, P., 86
LeDoux, J., 257
Lerdahl, F., 82
Levelt, W., 19, 34–36, 47, 86, 87
Levine, J., 265, 267, 304
Levitan, I. B., 285
Lichtheim, L., 223
Lieberman, P., 227, 229
Liepmann, H., 18, 206, 207
Linear perspective, 17, 127, 131–147, 151, 157–163, 166, 167, 319
Linguistics, 4, 10, 19
Livingstone, M., 319, 335
Llinas, R. R., 264, 272, 273, 290, 304
Lock, C. R., 336
Lotto, R. B., 123, 127, 136, 158, 335
LSD, 293, 294
Lynch, G., 188

Mach illusion, 154, 165
Machemer, H. J., 282
MacLennan, B., 264, 273
MacNeilage, P. F., 183
Macphail, E. M., 9

MacWhinney, B., 239, 240
Major mode, 15, 16, 26–28, 37, 39–119, 325, 326
Malsburg, C., 288, 336
Marijuana, 293, 294
Mathematics, 23, 299, 300, 327, 328
Mathieu, W. A., 107
Maurer, D., 309, 310
McFadden, J., 261, 285, 288, 289, 299
McGilchrist, I., 185, 186, 333, 336
McGinn, C., 264, 265
McManus, C., 182, 187
Mehta, Z., 322
Membrane permeability, 280–287
Mental rotation, 321–323
Meredith, M. A., 204
Meyer, L. B., 51–55, 61, 62, 83, 117, 308, 335
Mikhail, J., 313, 315
Minor mode, 15, 16, 26–28, 37, 39–119, 325, 326
Minsky, M., 318
Mirror neurons, 9
Mithen, S., 193, 202, 203, 335
Mitterauer, B., 300
Modality, 64–80
Moral mind, 7, 223, 313–317
Morton, E. W., 85, 86
Motion parallax, 16, 17, 121, 123–125, 166, 167
Muller-Oehring, E. M., 186
Murillo, B. E., 127
Music, 1, 4–7, 23–119, 332

Nagel, T., 265, 304
Narmour, E., 68, 309, 335
Necker cube, 51, 52, 172
Neisser, U., 122
Neural networks, 10, 317–321
Neurology, clinical, 11, 18, 19, 21, 206, 207, 222, 309
Neurons, 21, 22, 256–306, 317–319
Neurotransmitters, 257, 275, 293
Niceron, J. F., 152

Occlusion, 128, 133, 134, 158, 159
Ogura, A., 282
Ohala, J. J., 85–87
Oppenheimer, S., 188, 189, 231, 335
Orbitofrontal cortex, 113–115
Overlap. *See* Occlusion.
Overtones. *See* Partials, upper.

Pandya, D. N., 194
Panofsky, E., 160–163
Papathomas, T., 126
Papert, S., 318
Parncutt, R., 47, 335
Partials, upper, 28–38, 54–117
Passingham, R., 208
Patel, A., 82, 118, 335
Penn, D. C., 331, 335
Penrose, R., 261, 264, 269, 299
Perani, D., 184
Persichetti, V., 107
Perspective. *See* Linear perspective.
Peters, C. R., 336
Petersson, K. M., 225
Petrides, M., 211, 322
Phonetics, 19, 21, 220
Phrase structure, 217, 241–250
Piattelli-Palmarini, M., 247
Pictorial depth perception, 125–174
Pinker, S., 259, 260, 265, 304, 336
Piston, W., 105, 107
Pitch, 27–38, 120
Pleasants, H., 92
Plomp, R., 34–36, 47
Pointing, 23, 311–313, 330
Pollard, C., 242
Povinelli, D. J., 213, 331, 335
Powell, T. P. S., 194, 195
Premack, D., 213
Prosody, 221, 333
Psychotropic drugs, 293, 294
Purves, D., 123, 127, 136, 158, 335
Pythagoras, 16, 35, 61

Qualia, 21, 22, 255, 259, 260, 272, 273, 279, 287, 295–299, 302, 312, 336
Quantum mechanics, 11, 249, 264–271, 300, 303, 331–33

Rae, A., 67
Ragas, 62, 67, 94
Ramachandran, V. S., 200, 201, 211, 252, 260, 313
Rameau, J. P., 60, 61
Random mutation, 8, 25
Rawls, D., 315
Recursion, 20, 23, 215, 217, 246, 247, 265, 268
Reggia, J. A., 187
Religion, 2, 12, 297, 313, 314
Renaissance, European, 4, 15–17, 26, 48, 60, 61, 81, 95, 97, 115–119, 126–129, 150–157

Revensuo, A., 272
Reverse perspective, 164–173
Rhythm, 307–309, 333
Rizzolatti, G., 233, 252
Robertson, L. C., 185, 336
Robots, 20, 21, 261, 263, 271, 272, 318
Rogers, B., 127, 133, 335
Rumelhart, D. E., 193, 318

Sacks, O., 193
Sag, I. A., 242
Satinover, J., 268
Savage-Rumbaugh, S., 239
Scales, 35, 36, 40–42, 48, 49, 94
Scherer, K. R., 65, 86
Schick, K. D., 176
Schoenberg, A., 95–97
Schroeder, C. E., 198
Schulte, T., 186
Scott, A., 268
Scruton, R., 313, 335
Searle, J., 262, 265, 272, 304, 336
Self-consciousness, 258, 301–303
Seltzer, B., 194
Semantics, 19, 21, 219, 220
Sentience, 257, 258, 265, 279–291, 296, 298, 301, 303
Sethares, W. A., 47
Seventh chords, 45, 62–64, 71–74, 100
Shadows, 127, 128, 147–157, 159, 166, 167
Shkuro, Y., 187
Shlain, L., 156
Signaling, animal, 20, 175
Singer, W., 264, 273, 278, 288, 289, 305
Sinigaglia, C., 233, 252
Sloboda, J., 82, 118
Slyce, J., 165
Sonority, 38, 43, 44
Sound symbolism, 84–94, 110–113, 118, 221
Speciesism, 14
Sperber, D., 213
Sperry, R. W., 185
Stapp, H.P., 268
Stein, B., 204
Stereopsis, 16, 17, 121–123
Stoichita, V. I., 155
Storr, A., 335
Stout, D., 198
Striedter, G. F., 190, 191, 250
Subitizing, 323, 324
Subjectivity, 21, 22, 253–258, 262–275, 286, 291, 312, 336

Suspended chords, 42–110
Suzuki, M., 109
Symbolic thought, 6, 7, 24
Synapse, 257, 275–277, 286, 293, 295, 301–303
Synchronization, 22, 287–292
Syntax, 19–21, 219, 220, 235–241

Tattersall, I., 229, 335
Taylor, J. G., 271
Temperley, D., 309, 335
Tension, harmonic, 38–119
Terwogt, M. M., 65
Tetrads, pitch, 38, 62
Texture, 159
Thumb, 6, 329
Tomasello, M., 22, 24, 213, 230, 247, 250–254, 311–313, 331, 335, 336
Tonality, 28–38, 60
Tononi, G., 336
Tools (Tool use, Toolmaking), 1, 3, 5–7, 11, 17, 18, 23, 24, 175–215, 253, 333
Toth, N., 176
Trainor, L., 48, 65, 118
Transformational grammar. *See* Phrase structure.
Trehub, S., 65
Triadic associations, 8–14, 22, 24, 212–214, 329
Triadic grid, 39–119
Triads, pitch, 27, 38–119
Triffet, T., 268
Trigonometry, 327, 328
Trimodal cortex, 192–204
Tripartite architecture of language, 217–224
Tuning systems, 35, 41, 46, 48

Uncertainty principle, 268, 269
Universal grammar, 20, 217, 225, 241–250, 313, 336

vanGrinsven, A., 65
Velmans, M., 273
Verkhratsky, A., 301
Vollenweider, F. X., 294
Volterra, A., 301

Walhout, M., 279
Walker, E.H., 268
Wallin, N.L., 335
Watson, J.B., 225
Webern, A., 68, 95
Westerhausen, R., 336

Wexler, M., 322
Whiten, A., 229
Williams, S. R., 293
Wolfe, T., 130, 163, 335
Wolpert, L., 212, 252
Word-order, 19, 219, 235–241
Wrangham, R., 201
Wynn, T., 178, 180, 197, 335

XOR function, 193, 278, 317–319

Yarbus, A. L., 310

Zaidel, E., 184, 336
Zarlino, G., 60, 61
Zatorre, R. J., 113
Zeki, S., 335
Zombies, 297, 298